Evidence in Action between Science and Society

This volume is an interdisciplinary attempt to insert a broader, historically informed perspective into current political and academic debates on the issue of evidence and the reliability of scientific knowledge.

The tensions between competing paradigms, different bodies of knowledge and the relative hierarchies between them are a crucial element of the historical and contemporary dynamics of scientific knowledge production. The negotiation of evidence is at the heart of this process. Starting from the premise that evidence constitutes a central, but also essentially contested concept in contemporary knowledge-based societies, this volume focuses on how evidence is generated and applied in practice—in other words, on "evidence in action." The contributions analyze and compare different evidence practices within the field of science and technology, how they interlink with different forms of power, their interaction with and impact on the legal and political domain, and their relationship to other, more heterodox forms of evidence that challenge traditional notions of evidence. In doing so, this volume provides much-needed context and historical background to contemporary debates on the so-called "post-truth" society.

Evidence in Action is the perfect resource for all those interested in the relationship between science, technology, and the role of knowledge in society.

Sarah Ehlers is a postdoctoral researcher working on the global history of medicine, science and the environment at the Institute for the History of Science and Technology at the Deutsches Museum and an affiliated researcher at the Rachel Carson Center for Environment and Society at Ludwig Maximilian University of Munich.

Stefan Esselborn is a postdoctoral researcher at the Professur für Technikgeschichte at the Technical University of Munich. He is writing and teaching on topics in the fields of global and colonial history, the history of science and technology, the history of knowledge and expertise, and the history of risk and safety.

Routledge Studies in the History of Science, Technology and Medicine

Cold Science
Environmental Knowledge in the North American Arctic During the Cold War
Stephen Bocking and Daniel Heidt

Medical Memories and Experiences in Postwar East Germany
Treatments of the Past
Markus Wahl

Politics, Statistics and Weather Forecasting, 1840–1910
Taming the Weather
Aitor Anduaga

Social Class and Mental Illness in Northern Europe
Edited by Petteri Pietikäinen and Jesper Vaczy Kragh

Medicine and Justice
Medico-Legal Practice in England and Wales, 1700–1914
Katherine D. Watson

Lifestyle and Medicine in the Enlightenment
The Six Non-Naturals in the Long Eighteenth Century
Edited by James Kennaway and H.G. Knoeff

Germs in the English Workplace, c.1880–1945
Laura Newman

Pathogens Crossing Borders
Global Animal Diseases and International Responses, 1860–1947
Cornelia Knab

Evidence in Action between Science and Society
Constructing, Validating, and Contesting Knowledge
Edited by Sarah Ehlers and Stefan Esselborn

For more information about this series, please visit: https://www.routledge.com/Routledge-Studies-in-the-History-of-Science-Technology-and-Medicine/book-series/HISTSCI

Evidence in Action between Science and Society

Constructing, Validating, and Contesting Knowledge

Edited by Sarah Ehlers and Stefan Esselborn

Routledge
Taylor & Francis Group
NEW YORK AND LONDON

First published 2023
by Routledge
605 Third Avenue, New York, NY 10158

and by Routledge
4 Park Square, Milton Park, Abingdon, Oxon OX14 4RN

Routledge is an imprint of the Taylor & Francis Group, an informa business

Library of Congress Cataloguing-in-Publication Data
A catalog record has been requested for this book

ISBN: 978-1-032-03705-9 (hbk)
ISBN: 978-1-032-03706-6 (pbk)
ISBN: 978-1-003-18861-2 (ebk)

DOI: 10.4324/9781003188612

Typeset in Bembo
by MPS Limited, Dehradun

The Open Access version of chapter 6 was funded by DZHW.

Contents

PART II
Innovating Evidence: Contemporary Technoscientific Approaches

PART III
Governing Evidence: Evidence–Based Practice and Politics

PART IV
Contesting Evidence: The Politics of Heterodox Evidence

Figures

Contributors

Nadia Alaily-Mattar is a Research Associate at the Chair of Urban Development, Technische Universität München (TUM). Nadia is an architect and graduate of the American University of Beirut (AUB). She received her Master's degree in Housing and Urban Regeneration from the London School of Economics and Political Science, and her PhD in Planning Studies from the University College of London (UCL). Her research interests revolve around star architecture and its media effects, futures oriented urban planning, and the role of architecture in urban transformation processes.

Anna Apostolidou is a social anthropologist and works as an adjunct lecturer at Panteion University of Social and Political Sciences and at the Hellenic Open University. Anna holds a PhD in Social Anthropology (University College London, 2010) and a PhD in Distance Learning (Hellenic Open University, 2019). She has published on gender and sexuality, digital learning, refugee education, surrogate motherhood and fictional digital writing.

Christiane Arndt is an Associate Professor in the Department of Languages, Literatures and Cultures at Queen's University in Kingston, Canada. She focuses on nineteenth century German literature, the relationship of literature and photography, and medical representations in literature.

Diane Arvanitakis has developed her urban development expertise after graduating in the field of architecture, and specializing in governance of the public and development sectors. She draws her skills and experience from working in both developed and developing economies, through her contributions in academia, nongovernmental organizations and government sectors as a consultant, and private practice. Most recently she assists with teaching and research at the Chair of Urban development at Technische Universität München.

Clemens Blümel is a researcher at the German Centre for Higher Education Research and Science Studies (DZHW) and doctoral student at the Centre for Science and Technology Studies (CWTS). His research interests include: science and technology studies, innovation studies, network analysis, governance of biomedicine (with a specific focus on synthetic biology), as well as studies into digital scholarly practices.

Marcus B. Carrier is a research associate and PhD student in the subject area Historical Studies of Science at Bielefeld University, working in the DFG-funded research project "Forensic Toxicology in Germany and France during the 19th Century: Methods Development in Judicial Context".

Sascha Dickel is professor for Sociology of Media and Theory of Society at Johannes Gutenberg University Mainz. He is co-speaker of the DFG Research Group "Practicing Evidence – Evidencing Practice" and head of the BMBF funded collaborative project "Prototyping Futures." His research concerns socio-technical futures, citizen science, and digital communication. He was a member of the working group "Additive Manufacturing and 3D Printing" of the National Academy of Sciences Leopoldina as well as a member of the expert commission of the Federal Ministry for Family Affairs, Senior Citizens, Women and Youth on the topic "Young Civic Engagement in the Digital Age."

Sarah Ehlers is a postdoctoral fellow working on the controversies and negotiation processes of pest control in the Global South in the 1960s to 1980s at the Institute for the History of Science and Technology at the Deutsches Museum in Munich. She is an affiliated researcher at the Rachel Carson Center for Environment and Society at Ludwig Maximilian University Munich. She is a historian specializing in global history, the history of medicine, science and knowledge, and of global environments.

Stefan Esselborn is a Postdoctoral Researcher in History of Technology at the Technical University of Munich (TUM), and a member of the DFG Research Group "Practicing Evidence – Evidencing Practice." He has written, worked and taught on the history of colonial knowledge and expertise (particularly in the context of African Studies), the global history of science and technology, and the history of technical risk and safety. He is currently working on a history of risk research in Germany.

Salman Hussain is a Banting Postdoctoral Fellow in Anthropology Department, York University, Canada. His work focuses on human rights, political violence and protest movements, and gender and sexuality studies in South Asia. His current ethnographic and historical research examines how human body figures in the liberal projects of war-making, citizenship and human rights in South Asia and beyond.

Steffen Krämer is a postdoctoral researcher at the University of Konstanz in the Department of Literature, Art and Media Studies and at the Research Institute Social Cohesion. His doctoral research has focused on the history of diagrammatic media in epidemiology and medical geography. Further research areas include epistemological theories of affect and attention and socio-technical transformations of digital publics.

Kari Lancaster is Scientia Associate Professor at the University of New South Wales, and Honorary Associate Professor at the London School of Hygiene and Tropical Medicine. She leads a program of qualitative research focused on the development of critical approaches to the study of evidence-making practices, implementation science and intervention translations in health, especially in relation to drugs, hepatitis C, HIV and Covid-19.

John Lidwell-Durnin is a lecturer in the History Faculty of Exeter University (UK). His research focuses on agricultural improvement, scarcity, and famine within the British Empire in the eighteenth and nineteenth centuries.

Martina Löw is Professor of Sociology at the Technische Universität Berlin (TUB). Her areas of specialisation and research are sociological theory, urban sociology, space theory, and cultural sociology. She was visiting professor and held fellowships at universities in Gothenburg (Sweden), Salvador da Bahia (Brazil), St. Gallen (Switzerland), Paris (France), and Vienna (Austria). From 2011 until 2013 she was president of the German Sociological Association. Currently she is head of the Collaborative Research Center "Re-Figuration of Spaces."

Tim Rhodes is Professor of Public Health Sociology at the London School of Hygiene and Tropical Medicine (UK) and the University of New South Wales, Sydney (Australia). Drawing on critical social science methods, he leads a program of work investigating the social relations and evidence-making of health and care practices, especially linked to hepatitis C, HIV, drug use and addictions. He is currently using qualitative research methods to study the science and intervention of hepatitis C, COVID-19, and disease elimination, including in collaboration with global mathematical modelers and implementation scientists.

Julia E. Rodriguez is Associate Professor of History at the University of New Hampshire (USA). Her research focuses on the history of the human sciences in the late nineteenth and early twentieth centuries, including transatlantic scientific networks; the history of racial science and conceptions of humanity and personhood; the roots of cultural relativism; and the ethical dimensions of research on human subjects. She is currently completing a book about the nineteenth-century origins of Americanist anthropology.

Alexander Schniedermann is a research associate at the German Centre for Higher Education Research and Science Studies (DZHW). He investigates the characteristics of the review literature in science with a special focus on systematic reviews in biomedicine.

Arno Simons is a research associate at the German Centre for Higher Education Research and Science Studies (DZHW). In his academic work, Arno has explored various aspects of the relationship between science, technology, politics, and society. Currently, he is co-leading a project on the roles of review articles in different scientific fields and as boundary objects between science and politics.

Laura Stielike is a postdoctoral researcher in the research group "The Production of Knowledge on Migration" at the Institute for Migration Research and Intercultural Studies (IMIS) at the University of Osnabrueck. She works on knowledge production in the field of big-data-based migration research and governance, the relations between migration and development, postcolonialism and intersectionality, and discourse and apparatus analysis.

Alain Thierstein is a Professor for Urban Development at the Technische Universität München. He is also affiliated with EBP Schweiz AG, Zurich, as partner and senior consultant in the field of urban and regional economic development. His research interests are urban and metropolitan development; spatial impact of the knowledge economy, particularly the visualization of non-physical firm relationships, and spatial interaction of locational choices for residence, work, and mobility; and the role of star architecture for repositioning medium-sized cities.

Acknowledgments

This volume has grown out of the work of the interdisciplinary research group "Practicing Evidence – Evidencing Practice," supported by Deutsche Forschungsgemeinschaft (DFG). We would like to thank all members of the group who have contributed to this publication, either directly by discussing ideas and providing feedback, or indirectly through their general collaboration within the group.

1 Introduction: Evidence in Action

Sarah Ehlers and Stefan Esselborn

Follow the Evidence!

Evidence—long regarded as an important, but relatively mundane and uncontroversial ingredient in scientific knowledge production—has lately become a political rallying cry. When in 2017 and 2018, large crowds of citizens took to the streets in numerous places around the world to express their discontent with the new US administration led by Donald Trump, the political response to climate change, and the role of science in public discourse more generally, they were holding up signs with slogans such as "evidence, not arrogance," "trust the evidence," or "evidence trumps opinion." Probably the most memorable chant heard during these *Marches for Science* doubled down on the procedural detail: "What do we want? Evidence-based policy! When do we want it? After peer review!"[1] Similarly, during the early days of the COVID-19 pandemic, "follow the evidence" was not only the main advice offered by the American Medical Association (AMA) and other medical professionals, but became almost a sort of mantra on the part of those advocating for more stringent public health policies.[2] However, while (scientific) evidence seems ubiquitous as a slogan, it appears more unclear than ever what actually counts as such, and how and by whom this is determined. News reports, evening talk shows, and not least social media are awash with alleged "evidence" of wildly varying kinds and quality: on the effectiveness (or harmfulness) of face masks or vaccines, the reality of human-made climate change, the legitimacy of elections and the occurrence of voter fraud, or even the existence of various kinds of global (and sometimes extraterrestrial) conspiracies.

Fierce arguments about the conclusiveness of specific pieces of (scientific) evidence are of course not a new phenomenon. In this regard, historical controversies on issues such as the smallpox vaccination in the late nineteenth century, the carcinogenic effects of cigarette smoke and certain chemicals in the 1950s and 1960s, or the use of atomic energy in the 1970s and 1980s show many similarities to—and direct continuities with—the recent "climate wars" over the extent and causes of global warming, or the conflicts surrounding COVID-19 policies.[3] Given the ever-increasing

DOI: 10.4324/9781003188612-1

importance of technoscientific expertise in contemporary "knowledge so-cieties,"[4] it is not surprising that the validity of scientific knowledge has itself become a central battleground in a growing number of cases. In the words of the sociologist Peter Weingart, the advancing "scientification of society" came at the price of a "politicization of science." The intensified use of scientific knowledge as a legitimatory resource in political discourse tended to destabilize the impression of scientific neutrality and objectivity.[5]

This dynamic was one of the driving factors behind the perceived loss of public trust in scientific authorities, which numerous observers have deplored since the 1970s.[6] With regard to the position of evidence, this had a twofold effect. On the one hand, the reference to verifiable evidence became in-creasingly important not only for the validation of knowledge, but also for the self-legitimation of experts and institutions, who attempted to replace waning public confidence with increased transparency and better mechanisms of control. As the "'trust me' society" started to turn into a "'show me' society"—as one prominent chief executive claimed—that which could be shown acquired additional prominence.[7] On the other hand, the increased attention also served to highlight growing doubts about the concept of evi-dence itself. A variety of factors such as the growing complexity and hy-perspecialization of contemporary technoscience, insights into the (socially) constructed nature of scientific knowledge, as well as the rise of an anti-intellectualist and populist political discourse insisting on "alternative facts" have all made the idea of one discernible scientific "truth"—and the notion of evidence as a seemingly objective arbiter—increasingly problematic and contentious.[8]

Both parts of this dynamic are visible in the rise of the so-called "evidence-based practice" movement, arguably the main driver behind the sudden ubiquity of the term "evidence."[9] Its roots lay in the appearance of evidence-based medicine (EBM) in the early 1990s, a self-proclaimed "new paradigm" for clinical practice that aimed to improve the efficiency of therapeutic interventions by linking them directly and systematically to the newest and most reliable research findings.[10] Although it also inspired considerable resistance among clinical practitioners, the idea quickly spread to a large number of other professional fields. This resulted in the emer-gence of an astonishing number of new evidence-related subfields, such as evidence-based policy, evidence-based research, evidence-based education and social work, evidence-based management, evidence-based policing and prosecution, evidence-based design and architecture, evidence-based con-servation and philanthropy, and many more.[11] In spite of some considerable differences, they all share the goal of greater efficiency and transparency, a propensity for quantitative and statistical measurements, the widespread use of formal procedures and guidelines, and the introduction of a "pyramid of evidence" formally categorizing research findings by quality and reliability. On the other hand, critics have often pointed to what they see as a tendency toward inflexibility and formalism, a failure to take practical and personal experiences into account, as well as a scientistic adherence to outdated ideas of a pure and objective science clearly separable from the realm of politics and application.[12] Nevertheless, evidence-based practice has significantly transformed both professional practice and scientific research in a number of fields—thanks not least to its complementarity with wider societal trends toward (financial) efficiency, transparency, risk management, and data-driven rationality.[13]

There seems to be a need, therefore, for a broader historical, sociological, and cultural approach to the phenomenon of evidence. This volume is an attempt to provide one such, by means of drawing together new perspectives and original research on the topic. Growing out of an international conference held in Munich in February 2020 organized by the DFG research group "Practicing Evidence—Evidencing Practice,"[14] it reunites contributors from a wide variety of disciplines, working in very different fields of knowledge, on different historical periods, and in different geographical contexts. As they explore intersecting lines of inquiry, the chapters below are all concerned, in one way or another, with what counts (or does not count) as evidence in a given context. How do certain bodies of knowledge and certain practices of validating it become accepted in specific scientific fields or disciplines? In what ways are these practices influenced by their use as evidence—within the scientific community, but also in other contexts such as law, politics, or the media? How are scientific knowledge and scientific practices distinguished from other forms of knowledge production, and what connects them? What happens to practices of evidence when established knowledge is challenged?

While we cannot offer an easy solution to the problem of knowledge validation in the so-called "post-truth" society, nor an encompassing new epistemological model, we do believe that this book can not only generate new insights into the historical and contemporary dynamics of scientific knowledge production, but also insert some much-needed perspective into current political and academic debates. For this to work, however, we will first have to turn to a deceptively simple question: What do we actually mean when we talk about "evidence"?

What Is Evidence?

In spite of its undisputed centrality for almost all scientific and scholarly endeavors, evidence is a shifting and elusive concept, which cannot easily be reduced to one agreed-upon definition or theory. In legal theory and epistemological philosophy—the two fields that have arguably devoted the most systematic attention to defining the term—evidence is usually understood along the lines of "a good reason to believe."[15] In legal cases, presenting evidence to the court plays an important role in reaching a verdict, while in science, empirical evidence confirms or invalidates scientific hypotheses. Within this basic conception, one can find distinctions between a number of approaches. Philosophers disagree, for instance, whether evidence is best understood in an objectual manner (as a material thing), or a phenomenal manner (as an experience); whether it should be seen as a mental state (of belief), or whether it consists of facts and propositions.[16] Equally varied are the roles that evidence is thought to fulfill. Most authors emphasize that evidence not only acts as a "guide to truth" in the thought of a rational individual coming to a conclusion, but also holds an important intersubjective function as a "neutral arbiter," to which one can appeal when trying to persuade others and settle disputes over conflicting claims.[17]

However, the notion of evidence as proof is not without its own inherent contradictions. Indeed, the Latin meaning of *evidentia* originally referred to "clearness" and "obviousness," a quality possessed by *something that is known to be true without requiring further proof*.[18] Later, this idea became bound to scholarly authority: in early scholastic writings, as Ian Hacking has argued, "things could count as evidence only insofar as they resembled the witness of observers and the authority of book."[19] In the Early Modern era, scientific truths started to be seen less as derived from the authorities of their authors, but from things itself; more precisely, from impersonal scientific instrumentation and material devices.[20] Evidence therefore became "a matter of inferring one thing from another thing," defined as "one thing pointing beyond itself."[21] At least in its English-language form, this relational conception—in which evidence works as *proof for or against something else*—has come to dominate contemporary usage, while the notion of self-referential obviousness split off into the concept of "self-evidence." Nevertheless, the older association lingers on in the often normative use of the term, causing a certain amount of conceptual confusion—particularly since the vocabulary split does not necessarily exist in other languages.[22]

Given the newfound prominence of evidence as a sociopolitical issue today, in addition to its long-recognized centrality to (scientific) epistemology in general, one might assume that the phenomenon would also figure prominently in historical, sociological, and culturally informed studies of science. Curiously, this has so far not been the case. As late as the early 1990s, a prominent volume on the topic still found it "extraordinary how little direct attention" evidence had received in the history and theory of science.[23] This has only relatively recently started to change.[24] Most of the newer relevant work can be attributed to three major approaches and directions. First, as a highly specialized field of study, the study of evidence-based practice has been a growth area over the past decade. However, although topics such as random control trials in evidence-based medicine or mathematic models in evidence-based policy have attracted profound interest from scholars, they have rarely been examined from an interdisciplinary perspective.[25]

Secondly, scholars from science and technology studies (STS) and media studies have highlighted how evidence practices as well as scientific evidence itself are inextricably linked to the medium in which they are communicated. Focusing on the epistemological functions of models, charts, tables, and images in science, as well as more broadly on narratives and numbers, has added an important dimension to the study of scientific "facts" and their evidentiary modes.[26] Contrary to the representational and purely documentary modes that these media are often assumed to operate through, this strand of research has illuminated how specific aesthetic and material qualities bestow credibility and authority beyond that granted through their content. While, for example, different forms of visualization mobilize certain ideas about validity, narratives work to create coherence between a variety of different pieces of evidence from different sources that otherwise do not appear to fit together.[27]

Thirdly, the debate over trust in science and its importance for societal cohesion is having direct implications for the question of evidence. Political decisions not only have to be legitimized by scientific expertise; scientific research is also judged by its relevance to politics and society. Given the spread of scientific methods in the social, political, and economic fields, the related debate on scientification has shown the changing role of expert knowledge in science and society.[28] Moreover, as for example recent investigations into forensic cultures have shown, epistemic practices are deeply embedded in cultural contexts and political regimes. And—as encapsulated by the term "CSI effect"—they are often influenced by their own public image.[29] Several studies have explored the role of scientific expertise in politics and society with special attention given to experts, populism, and democracy. Many of these recently published works share an interest in how societies cope with decreasing trust in science and explore participatory formats of knowledge production.[30] A recent attempt to "democratize science" can be seen in citizen science, in which knowledge is coproduced. This again has important implications for the question of evidence.[31]

Evidence in Action

Notions of evidence are highly dependent on different chronological frames, cultural contexts, fields of application, and disciplinary cultures. Moreover, with its centrality to narratives of continuous scientific progress, the rise of evidence-based practice, and a polarized discourse of science denial and conspiracy theories, the term "evidence" has become normatively loaded and difficult to disentangle from political and cultural disputes. For these reasons, we propose to adopt a more flexible, analytical approach that seeks to explore not what evidence might be in principle, but how it works in practice. In *Science in Action*, one of the most canonical texts in the field of science and technology studies, Bruno Latour proposes to open up the "black boxes" of scientific and technological "facts" by investigating "science in the making," zooming in on specific places and times until the "clean distinction between context and content disappears."[32] Borrowing a leaf from his book, we will likewise start from the empirically specific, using a praxeological approach to gain insights into what one might call "evidence in action."

This approach shapes our perspective on the issue in two ways. On the one hand, it draws attention to the contingent and dynamic nature of evidence. What gets to count as evidence is not determined by some inherent and unchangeable quality of certain objects or bits of knowledge, but is subject to complex negotiation processes. Evidence is, in the words of one of the chapters in this volume, a "situated achievement," which must be presented, enacted, and often performed practically in a specific situation.[33] As part of an act of communication designed to convince others—whether scientific colleagues, political decision makers, or the general public—scientific evidence practices often comprise a certain amount of rhetorics, stagecraft, and *mise-en-scene*.[34] They not only involve a variety of human actors, but also other entities such as discourses and material objects. Even the seemingly simple and straightforward case of admitting a document as evidence in court, as one of the authors in this volume reminds us, presupposes not only the presence of the material object itself, but also the existence of the court as a forum in which to present it, and the services of an expert to make it "speak."[35] Furthermore, these socio-epistemic arrangements are themselves subject to change, as new stocks of knowledge or constellations of problems destabilize established certainties in an epistemic or social respect, and a new need for re-stabilization emerges in order to rehabilitate the established authorities or to provide new ones.

On the other hand, seen from a praxeological point of view, the site of production of evidence is not just the laboratory (or its functional equivalent in different disciplines), but also includes the "point of contact" with the envisioned (scientific and/or nonscientific) audience. Therefore, to borrow a concept from the philosopher Nancy Cartwright, we are not so much interested in evidence as an abstract phenomenon, but rather in "evidence for use."[36] More often than not, the intended use is already an important factor in determining the content and form of evidence. Therefore, production and use of evidence appear not so much as separate consecutive steps, but rather interact with each other in a constant feedback loop, blurring the distinction between (scientific) knowledge production and its application. This conception of evidence as something that oscillates between "science" and "society" not only highlights the function of evidence as a resource, but also the many ways in which scientific knowledge production interlinks with different forms of power in knowledge-based societies. In this regard, drawing on what Michel Foucault has described as a "regime of truth," evidence practices allow for reflection on the broader issue of knowledge, scientific understanding, and truth in modern society.[37]

To be sure, there are also particular challenges to a practice-based and context-bound definition of evidence. First, our approach faces the same problems that practice-oriented approaches to the study of science encountered: if our interest lies with historically specific dynamics, if we emphasize the local character of research and knowledge claims, do we have to give up on the (supposedly) universal, global, or translocal character of science and scientific knowledge?[38] As exemplified in recent debates about differences between science and knowledge, this distinction can be a productive one.[39] Drawing on these debates, we are interested in the ways in which scientific practices are distinguished from other forms of knowledge production.

Secondly, given that evidence in action concentrates on what counts (or does not count) as evidence in a specific context, our focus is not primarily on the macrostructural level. That said, the mix of historically and empirically rooted case studies gathered here nonetheless allows for reflection on larger-scale social processes and hierarchies in society. Accordingly, the question of how the larger context informs the construction and contestation of (scientific) facts is at the heart of our inquiry. For example, asking what happens when established knowledge is challenged is inextricably linked to the question of trust—in the context of modern knowledge-based societies specifically to trust in scientific institutions by the public. In this regard, studying practices of evidence sheds light on the relationship between science and the public.

Tensions of Evidence

Dealing with diverse empirical contexts, this book is structured by four basic tensions. All four are not so much to be understood in terms of polar opposites (although they sometimes appear as such), but rather as coexisting factors, the mixture of which is subject to ongoing negotiations. While they are clearly interrelated, each accentuates different aspects of evidence practices in action.

The first of these tensions concerns different kinds of evidence and their relative weight and validity. As the collected examples show, different evidence cultures exist(ed) in different scientific disciplines (say, anthropology and engineering) and at different historical moments.[40] Furthermore, very different things could be considered evidence within the same fields at the same time. Arguments arising from this heterogeneity are often framed in terms of an opposition between "hard"/tangible and "soft"/intangible evidence—with material objects, quantitative models, or the result of laboratory experiments usually counting on the side of the former, and narrative elements, aesthetic considerations, or personal judgment based on experience generally seen as examples of the latter. In practice, however, we find that in many instances both types coexist and complement each other. This volume seeks to compare the interplay of different types of evidence, and to draw out the performative component of enacting "hard" and "soft" evidence practices across various contexts.[41]

A second, related tension arises from the different contexts in which evidence is generated and deployed. Evidence practices, as we understand them, do not neatly follow the distinction between "science" and "application." Nevertheless, they are often regarded as being "translated" from one field (mostly "science") to others ("law," "policy," etc.). We would argue that this metaphor tends to underappreciate the co-constructive force of the "receiving" field.[42] Moreover, changes of context can also cause friction. More often than not, disputes over evidence practices arise out of (and therefore make explicit) the clashing logic between such different fields as chemistry and law, or epidemiology and policy. In this regard, the present volume highlights the mediating effect of evidence practices between different contexts.

Thirdly, as evidence practices do not occur in a social vacuum, they are inevitably interwoven with questions of power. By conferring the possibility to set standards and decide on the validity of evidence—but also simply by making resource-intensive ways of knowledge production possible in the first place—political power, financial strength, and social status can translate into influence on epistemic outcomes. In this regard, these different forms of capital could almost be considered as evidence practices in their own right.[43] This does not mean, however, that (scientific) evidence always necessarily has a stabilizing effect on existing power structures. As various examples from the volume show, evidence practices often proved highly portable and could be deployed as resources by marginalized groups as well as by those upholding the status quo. This takes on particular importance in the context of recent attempts to "decolonize" the hegemonic Western knowledge system by researching and incorporating epistemic perspectives, knowledge, and thinking from the Global South.[44]

All three of these tensions feed into a fourth, and perhaps most openly political one, which concerns the legitimacy of scientific actors and institutions. Since evidence does not speak by itself, it is intimately connected to the issue of trust in experts on the one hand, and (democratic) participation on the other. Since the 1960s, the inherent conflict between the hierarchical truth claims of scientific knowledge and its increasingly central political role has resulted in demands for greater transparency, democratic accountability, and a better representation of different societal groups and political positions in scientific knowledge production itself. While few today would disagree that these would in principle be desirable goals, whether and to what extent practical measures should be taken in their pursuit remains fiercely contested. Moreover, the growing politicization of doubts about the objectivity of science by populist movements, special interest groups, and conspiracy theorists has left many observers wondering about the potential negative consequences of the relativization of the special position of scientific knowledge for the democratic system.[45] These developments pose challenges for public understandings of science: given the complexity of modern science, who gets to decide what count as valid evidence practices, and on what grounds? How can we publicly embrace the inconclusiveness of science as well as the temporary and often even contradictory nature of scientific findings without undermining trust in the scientific method per se? How can one discern between healthy skepticism toward the epistemological establishment, and groundless conspiracy theories? Is formalized scientific evidence our best tool to ensure rational political action, or do we need to integrate new and different evidence practices if we want to make science and society more responsive and inclusive? These questions are of vital importance to our contemporary societies, and perhaps even the future of Western democracies themselves. However, as the assembled case studies also show, they are far from new.

Establishing, Innovating, Governing, and Contesting Evidence

Establishing evidence, the volume's first thematic section, deals with evidence practices in the formation of scientific disciplines and disciplinary cultures from a historical perspective. Combining case studies on the development of agricultural science, forensic toxicology, and anthropology, it focuses on social structures, practices, and processes that advanced and professionalized new fields of research. The chapters illuminate how practices of evidence in new scientific fields develop with some amount of historical contingency, yet are key to informing new disciplinary cultures. They not only produce and organize knowledge but control boundaries and make new disciplines and fields credible in society, as well as in the established scientific landscape.

John Lidwell-Durnin's chapter discusses different evidence cultures within the efforts to promote scientific approaches to agriculture in late-eighteenth-century Britain. He shows how the interplay of debates on evidence in natural philosophy, the role of direct observation and sensory perception in agriculture, and fears of war and famine gave rise to the Board of Agriculture in 1793. These early scientific cultivation practices were heavily dependent on the observations and expertise of practicing farmers. Yet, rather than producing knowledge, agriculturalists tried to filter out efficient methods and accurate observations from what they saw as superstition and false ideas in the practical knowledge of farmers. Evidence, Lidwell-Durnin argues, was not inviting consensus so much as fueling scientific debates, leading to an interplay of different evidence cultures that were developed, contested, and refined during the close of the eighteenth century.

Marcus Carrier examines forensic toxicology in nineteenth-century German states as a new evidence practice at the crossroads of scientific and legal contexts. As a branch of both legal medicine and analytical chemistry, forensic toxicology was developed as a field of practice to provide specific evidence for courts of law. Serving as an objective party, expert testimonies played a key role in establishing a crime. Carrier argues that testimonies of forensic toxicologists meant more than reproducing material evidence and explaining it in laypersons' terms. Rather, mirroring other types of judicial evidence, most notably eyewitness accounts, forensic toxicology shows the co-constructive power of the judiciary on scientific evidence practices.

Julia E. Rodriguez's chapter uses debates about the significance of land and place among Americanist anthropologists as a way of exploring scientific knowledge production and the development of evidentiary norms in nineteenth-century anthropology. Land, she argues, was seen as a vital source of evidence for scholars at this time, which could not only confer authenticity on artifacts, but also give scientific legitimacy to interpretations of relicts and objects. As a result, practices of knowledge production that relied on access to specific places changed what was considered to be scientific anthropology. Rodriguez's chapter thus highlights how anthropologists' engagement with land not only changed the ways in which scientific authority was authenticated but also fostered the professionalization of the field.

Turning to socio-epistemic constellations in the present, *innovating evidence*, the second cluster of chapters, focuses on attempts to adapt and reinvent forms of knowledge validation in the context of recent challenges such as digitalization, information overload, and interdisciplinarity in a time of hyperspecialization.

From a sociologist's perspective, Sascha Dickel investigates prototyping as an evidence practice that connects technoscientific designs with contexts of application. Interpreting prototypes both as epistemic objects and material artifacts that communicate the feasibility of technological projects, Dickel stresses the performative aspect of evidence practices. As such, prototyping performs an explicit or implicit opposition between the "hard"/tangible evidence of artifacts and the "soft"/intangible evidence of narrative scenarios. When prototypes are used as demonstrative evidence, they embody an epistemic authority of materiality over discourse. At the same time, Dickel argues, prototypes also need the "soft" evidence of narratives in order to make a convincing case for technoscientific possibilities.

Alexander Schniedermann, Clemens Blümel, and Arno Simons take reporting guidelines in biomedicine as a starting point to examine the relationship between scientific knowledge production and its utilization. Drawing on science studies, their chapter engages with what one might call a "hierarchy of evidence" that has been integral to the rise of so-called evidence-based medicine, including methodological innovations such as randomized controlled trials (RCTs) or systematic reviews (SRs). The rather recent phenomenon of reporting guidelines employs epistemic values such as reliability or transparency, and translates them into specific instructions. Addressing questions of science communication and epistemic change, Schniedermann, Blümel, and Simons present guidelines as tools of standardization that define what counts as evidence and mediate between knowledge producers and users. In doing so, they aim to highlight how the phenomenon of reporting guidelines is part of the ongoing renegotiation of the relation between the sciences and their contexts of application.

In their chapter analyzing the evidence base of the impact of star architecture projects, the interdisciplinary team of Nadia Alaily-Mattar, Diane Arvanitakis, Martina Löw, and Alain Thierstein discusses the challenges of multidisciplinary research. Addressing the incommensurability of evidence practices inherent in interdisciplinary cooperation, they present an impact model as a tool to integrate different disciplinary perspectives into a coherent body of knowledge. In doing so, they aim to point out the diverging yet complimentary nature of the evidence produced by the fields of architecture, cultural sociology, and economic geography, and to foster a scientifically grounded discussion of the failure or success of star architecture projects.

Governing evidence, the third thematic section, addresses evidence practices for and in the context of policy and governance. With contributions from the fields of public health, epidemiological surveillance, and migration governance, it deals with the move to evidence-based practice in different fields of public policy. Taken together, these case studies shed light on attempts to take ideology, opinions, and politics out of the policy process and thereby produce better policy outcomes. Yet, as the chapters show, the evidence base of evidence-based practice is far from neutral. These case studies illuminate how evidence—rather than being simply used for policy—is itself produced and performed in the process of policymaking.

The section begins with Kari Lancaster and Tim Rhodes explaining how evidence in policy is enacted and made legitimate through material-discursive practices. Drawing on examples from the management of the COVID-19 pandemic, the authors propose a framework for conceptualizing evidence and (political) interventions as entangled, and one in which evidence emerges through the situated relational dynamics involved in the performance of policy. Rather than conceiving of evidence and policy as distinct and separate domains to be bridged through translation, the authors shift our focus to how evidence is relationally made, thus working with the politics, contingencies, and uncertainties of implementation and policymaking.

From the perspective of media studies, Steffen Krämer's chapter discusses the role of charts, diagrams, and thresholds as "graphical evidence" in automatic outbreak detection in the context of epidemiological surveillance. Interestingly, the anticipatory surveillance practices he describes are at the same time evidence-making interventions. Making automatic outbreak suggestions before the confirmation of an epidemiological event, these practices determine when routine evidence procedures might be activated in the future, thereby negotiating evidentiary potential and creating a "virtual milieu" of evidence practices. In addition, Krämer emphasizes that the infrastructures of automatic early warning systems shape the evidentiary boundary work between data scientists, mathematicians, and epidemiologists.

In her chapter, political scientist Laura Stielike focuses on the increasing use of big data in international migration policy. For data scientists employing algorithms to analyze social media and mobile phone positioning data, migration has become a fascinating object of study. Conversely, states and international organizations have discovered big data as a source for evidence-based policymaking. Assuming the coproduction of migration knowledge instead of a simple translation from big data to evidence-based policy, Stielike asks us to look more closely at the interplay between evidence-making and policymaking, including the diversity of producers of knowledge and their unequal access to resources. Given that the conflictual relationship between migratory practices and the attempts to govern them is a highly political question, Stielike argues, the evidence base of migration policy cannot be seen as an objective source of expertise.

Contesting evidence, the volume's fourth thematic block, discusses the epistemology and politics of heterodox evidence. Dealing with "outsiders" and subversives, the chapters discuss the often-blurry boundary between scientific and nonscientific evidence practices, and their entanglement with questions of power and politics.

In the context of literature and culture, Christiane Arndt's chapter examines evidence production in the visual rhetoric of the historic anti-vaccine movement in nineteenth-century German print media. Despite medical successes and widespread hygiene education, major developments in the fields of photography and printing technology destabilized the popular scientific discourse by instigating the production and reception of "alternative facts" with respect to vaccine theories. Arguing that the images and texts used by the anti-vaccine league were presented like medical case studies, Arndt shows how anti-vaccinists made use of academically established evidential practices and, by doing this, were successful in promoting anti-vaccine material as evidence to large parts of the population.

Salman Hussain's chapter on missing person cases in Pakistan is an anthropological investigation into the use of documentary and visual artifacts as evidence against state violence and the secrecy surrounding it. Whereas forensic evidence performs scientific authority in forms legitimized by the state, Hussain proposes to engage with the materials collected and employed by the families of the "disappeared" as counterforensic evidence. Addressing how the question of evidence—its attainment, accuracy, and legitimacy—is tied up with the question of state and legal forms of power allows for a closer focus on subjugated voices in ethnographic research.

From a different ethnographic perspective, Anna Apostolidou poses the question as to whether practices of fabricating evidence can be employed as legitimate scholarly practices in the academic culture of contemporary anthropology. Based on fieldwork experience, empirical material, and digital storytelling practices from research on representations of surrogate motherhood, she explores what counts as trustworthy scientific expertise, and on what grounds, academically or otherwise. Since academic writing has often failed to do justice to research subjects or adequately engage with nonacademic audiences, Apostolidou proposes research-based fiction in the form of "Ethnographic Art(i)Facts" as a tentative resolution to the tension between a politics of truth and a politics of representation.

Evidence in the Knowledge Society

Reading these contributions together, the outline of a multifaceted picture starts to emerge, in which (although it is far from complete) we can already recognize some characteristic features, and find a number of open questions. Firstly, the chapters seem to confirm our initial observation that variance and context-specificity are in fact among the defining characteristics of evidence. What was considered and used as evidence changed considerably, not only over different scientific disciplines and areas of application, but also over time. Secondly, if the main function of evidence is to support knowledge claims, evidence *for something* is at the same time always evidence *for someone,* that is, for an imagined audience in a specific situation. Even in seemingly strictly "intra-scientific" cases—such as the Americanists's search for an evidence base to their discipline, or the introduction of reporting guidelines in biomedicine—the performance of evidence is inextricably linked to how science is socially embedded. Therefore, an investigation into evidence practices brings out the complex socio-epistemic structures and arrangements upon which they rely to be effective.

In spite of the volume's focus on the specific, as two historians we are of course tempted to wonder whether and how the developments described in the cases collected here could be related to wider time frames. Indeed, given the centrality of evidence to the validation of knowledge claims in the modern era, any broader history of evidence practices would at the same time probably constitute a history of the development of the knowledge society per se. While the present volume perhaps does not cover enough historical ground to support a general chronology, it does indicate certain historical conjunctions at which form and content of evidence changed substantially—such as the establishment of scientific method in the late Enlightenment, the "scientification of the social" during the nineteenth and early twentieth century, the partial destabilization of the positivist understanding of scientific authority in the 1970s, or the rise of big data and evidence-based practice since the 1990s. In this regard, the study of evidence offers a lens through which to view the changing relationship of science and society across time and, eventually, to cast a broader perspective on the emergence of today's post-truth debates.

If science is key to modern democracies, this is particularly true in times of crisis, when democratic leaders increasingly refer to scientific expertise to legitimize uncomfortable decisions. According to the sociologist Alexander Bogner, this results in an "epistemization of the political," as political disputes are negotiated as conflicts of knowledge, expertise, and qualification.[46] Is it a danger to democracy, as Bogner argues, if today's political or social divisions are outsourced to the realm of science, and ultimately treated as questions of evidence? One could argue that disputes about evidence—more so perhaps than political conflicts—tend to be fought with verbal arguments rather than outright violence. Yet, when political actors use scientific evidence to present their policies as inevitable and without alternative, they neither do justice to the complexities of science, nor to their role as representatives of the public. As a result, the credibility of both science and politics suffer. Moreover, turning political questions into questions of expertise does not make them less political. Scientific truths are not only complex and diverse, they also reflect epistemic hierarchies. This makes them difficult to challenge, particularly for marginalized communities.[47] Not just *what* counts as evidence, but also *whose* evidence counts will remain an open political question for the decades to come.

Notes

1 Tellingly, the protests were occasionally even referred to as "marches for evidence" by their organizers, see Rush D. Holt, "The March for Evidence," *Scientific American,* 13 April 2018, https://blogs.scientificamerican.com/observations/the-march-for-evidence/; Anne Q. Hoy and Andrea, Korte, "March for Science: Supporters of Science Stand Up for Evidence," 13 April 2018, https://www.aaas.org/news/march-science-supporters-science-stand-evidence.

2 Kevin B. O'Reilly, "Medical Consensus: Follow Evidence on Coronavirus Vaccine Review," https://www.ama-assn.org/delivering-care/public-health/medical-consensus-follow-evidence-coronavirus-vaccine-review; Nason Maani and Sandro Galea, "What Science Can and Cannot Do in a Time of Pandemic," *Scientific American,* 2 February 2021, https://www.scientificamerican.com/article/what-science-can-and-cannot-do-in-a-time-of-pandemic.

3 See, for example, Stanley Williamson, *The Vaccination Controversy: The Rise, Reign and Fall of Compulsory Vaccination for Smallpox* (Liverpool: Liverpool Univ. Press, 2007); Charles F. Wurster, *DDT Wars: Rescuing Our National Bird, Preventing Cancer, and Creating the Environmental Defense Fund* (Oxford: Oxford University Press, 2015); Naomi Oreskes and Erik M. Conway, *Merchants of Doubt: How a Handful of Scientists Obscured the Truth on Issues from Tobacco Smoke to Global Warming* (New York: Bloomsbury Press, 2010).

4 For the debate on the "knowledge society" see Daniel Bell, *The Coming of Post-industrial Society: A Venture in Social Forecasting* (New York: Basic Books, 1973); Gernot Böhme and Nico Stehr, *The Knowledge Society: The Growing Impact of Scientific Knowledge on Social Relations* (Dordrecht: Reidel, 1986).

5 Peter Weingart, "Verwissenschaftlichung der Gesellschaft—Politisierung der Wissenschaft," *Zeitschrift für Soziologie* 12, no. 3 (1983): 225–241.

6 The actual extent and contours of the much-discussed decline in public confidence in science are still surprisingly unclear; see, e.g., Georgine Pion and Mark Lipsey, "Public Attitudes toward Science and Technology: What Have the Surveys Told Us?," *The Public Opinion Quarterly* 45, no. 3 (1981): 303–316; Gordon Gauchat, "Politicization of Science in the Public Sphere," *American Sociological Review* 77, no. 2 (2012): 167–187.

7 The words of the zoologist and Monsanto CEO Hugh Grant, quoted by Steven Shapin, "The Way We Trust Now: The Authority of Science and the Character of the Scientist," in *Trust Me, I'm a Scientist*, ed. Pervez Hoodbhoy, Daniel Glaser, and Steven Shapin (London, 2004), 42.

8 See for example Lee McIntyre, *Post-truth* (Cambridge, Mass.: The MIT Press, 2018); Stuart Sim, *Post-truth, Scepticism & Power* (Cham, Switzerland: Palgrave Macmillan, 2019).

9 For an overview see Liz Trinder and Shirley Reynolds, eds., *Evidence-Based Practice: A Critical Appraisal* (Oxford: Blackwell Science, 2000), 42.

10 Evidence-Based Medicine Working Group, "Evidence-Based Medicine: A New Approach to Teaching the Practice of Medicine," *JAMA* 268, no. 17 (1992): 2420–2425; Trisha Greenhalgh, Jeremy Howick, and Neal Maskrey, "Evidence Based Medicine: A Movement in Crisis?," *BMJ* 348 (2014): g3725; Cornelius Bork, "Negotiating Epistemic Hierarchies in Biomedicine: The Rise of Evidence-Based Medicine," in *Weak Knowledge: Forms, Functions, and Dynamics*, eds. Moritz Epple, Annette Imhausen, and Falk Müller (Frankfurt: Campus Verlag, 2020); Catherine Pope, "Resisting Evidence: The Study of Evidence-Based Medicine as a Contemporary Social Movement," *Health* 7, no. 3 (2003): 267–282.

11 See for instance Justin O. Parkhurst, *The Politics of Evidence: From Evidence-Based Policy to the Good Governance of Evidence,* (London, New York: Routledge Taylor & Francis Group, 2017); Gary Thomas and Richard Pring, eds., *Evidence-Based Practice in Education* (Maidenhead: Open Univ. Press, 2010); Nathanael J. Okpych and James L.-H. Yu, "A Historical Analysis of Evidence-Based Practice in Social Work: The Unfinished Journey Toward an Empirically Grounded Profession," *Social Service Review* 88, no. 1 (2014) 3–58; Simon Moss and Ronald Francis, *The Science of Management: Fighting Fads and Fallacies with Evidence-Based Practices* (Bowen Hills: Australian Academic Press, 2007); Lawrence W. Sherman, "The Rise of Evidence-Based Policing: Targeting, Testing, and Tracking," *Crime and Justice* 42, no. 1 (2013): 377–451; D. Kirk Hamilton and David H. Watkins, *Evidence-Based Design for Multiple Building Types* (Hoboken: John Wiley & Sons Inc, 2009).

12 See for example Alexander Krauss, "Why All Randomised Controlled Trials Produce Biased Results," *Annals of Medicine* 50, no. 4 (2018): 312–322; Jon Glasby and Peter Beresford, "Commentary and Issues Who Knows Best? Evidence-Based Practice and the Service User Contribution," *Critical Social Policy* 26, no. 1 (2006): 268–284.

13 Liz Trinder, "Introduction: The Context of Evidence-Based Practice," in *Evidence-Based Practice: A Critical Appraisal*, ed. Liz Trinder and Shirley Reynolds (Oxford: Blackwell Science, 2000).

14 See also the conference report by Daniel Füger, "Practicing Evidence—Evidencing Practice. How Is (Scientific) Knowledge Validated, Valued and Contested?," H-Soz-Kult https://www.hsozkult.de/conferencereport/id/tagungsberichte-8741, updated 27.04.2020.

15 Thomas Kelly, "Evidence," in *The Stanford Encyclopedia of Philosophy*, ed. Edward N. Zalta (2016), https://plato.stanford.edu/cgi-bin/encyclopedia/archinfo.cgi?entry= evidence; Peter Achinstein, *The Book of Evidence* (Oxford: Oxford Univ. Press, 2003). For an overview of legal ideas see Hock L. Ho, "The Legal Concept of

Evidence," in *Stanford Encyclodedia of Philosophy*, https://plato.stanford.edu/entries/evidence-legal/, winter 2015; William L. Twining, *Rethinking Evidence: Exploratory Essays*, 2nd ed. (Cambridge: Cambridge University Press, 2006).

16 Logan Paul Gage, "Objectivity and Subjectivity in Epistemology: A Defense of the Phenomenal Conception of Evidence" (PhD dissertation, Baylor University, 2014), https://philpapers.org/rec/GAGOAS. Moreover, in scientific practice, explanatory power is often used as evidence. Michel Janssen, "COI Stories: Explanation and Evidence in the History of Science," *Perspectives on Science* 10, no. 4 (2002): 457–522.

17 Kelly, "Evidence."

18 Simon Schaffer, "Self Evidence," in *Questions of Evidence: Proof, Practice, and Persuasion Across the Disciplines*, ed. James Chandler, Arnold I. Davidson, and Harry D. Harootunian (Chicago: University of Chicago Press, 1994); Michel Foucault, "Orders of Discourse. Inaugural Lecture Delivered at the College De France," *Social Science Information* 10, no. 2 (1971): 201–208.

19 Ian Hacking, *The Emergence of Probability: A Philosophical Study of Early Ideas about Probability, Induction and Statistical Inference* (Cambridge: Cambridge Univ. Press, 1975), 33.

20 Simon Schaffer, "Self Evidence," *Critical Inquiry* 18, no. 2 (1992): 327–362; Lorraine Daston, "The History of Emergences," *Isis* 98, no. 4 (2007): 801–808. See also Lorraine Daston and Peter Galison, *Objectivity* (New York: Zone Books, 2009).

21 Hacking, *The Emergence of Probability*, 34.

22 Both the German *Evidenz* and French *évidence*, for instance, have retained the original notion of "obviousness" as their primary meaning. At least in the German case, recently this seems to be changing due to the adoption of the "English" meaning (referring to the proof itself) in the discussions around evidence-based practice ("evidenzbasierte Praxis"); see Matthias Kroß, "Klarheit statt Wahrheit," in *Die ungewisse Evidenz*, eds. Gary Smith and Matthias Kroß (Berlin: Akademie Verlag, 1998); Hans J. Sandkühler, "Kritik der Evidenz," in *Wissen, was wirkt: Kritik evidenzbasierter Pädagogik*, eds. Johannes Bellmann and Thomas Müller(Wiesbaden: VS Verl. für Sozialwiss, 2011).

23 James Chandler, Arnold I. Davidson, and Harry D. Harootunian, eds., *Questions of Evidence: Proof, Practice, and Persuasion Across the Disciplines* (Chicago: University of Chicago Press, 1994), 1.

24 One clear indicator of this new interest is the proliferation of research groups and initiatives dedicated to the concept, such as the ESRC project led by Mary S. Morgan on the "nature of evidence" (2004–2010, which later recentred on the notion of "facts"), the Berlin-based research group on "BildEvidenz" (2012–2021), or the DFG research group "Practicing Evidence—Evidencing Practice" (since 2017), which provided the impetus for the present publication. See also Karin Zachmann and Sarah Ehlers, eds., *Wissen und Begründen: Evidenz als umkämpfte Ressource in der Wissensgesellschaft* (Baden-Baden: Nomos, 2019).

25 See also the critique by Lancaster and Rhodes in this volume.

26 Bruno Latour, "Visualisation and Cognition: Drawing Things Together," *Knowledge and Society: Studies in the Sociology of Culture and Present* 6 (1986): 1–40; Edward R. Tufte, *Beautiful Evidence* (Cheshire, Connecticut: Graphics Press, 2006); Friedrich A. Kittler, *Gramophone, Film, Typewriter* Trans. Geoffrey Winthrop-Young and Michael Wutz (Stanford: Stanford Univ. Press, 1999); Johanna Drucker, *Graphesis: Visual Forms of Knowledge Production* (Cambridge, Mass.: Harvard Univ. Press, 2014); Angela N. H. Creager, Elizabeth Lunbeck, and M. N. Wise, eds., *Science without Laws: Model Systems, Cases, Exemplary Narratives* (Durham: Duke University Press, 2007).

27 Rüdiger Campe and Helmut Lethen, eds., *Auf die Wirklichkeit zeigen: Zum Problem der Evidenz in den Kulturwissenschaften* (Frankfurt am Main: Campus, 2015); Mary S.

Morgan and M. N. Wise, "Narrative Science and Narrative Knowing. Introduction to Special Issue on Narrative Science," *Studies in History and Philosophy of Science Part A* 62 (2017): 1–5.

28 Kerstin Brückweh et al., eds., *Engineering Society: The Role of the Human and Social Sciences in Modern Societies, 1880–1980* (Basingstoke: Palgrave Macmillan, 2012); Hans P. Peters, "Scientists as Public Experts," in *Handbook of Public Communication of Science and Technology*, eds. Massimiano Bucchi and Brian Trench (London: Routledge, 2008).

29 See, for example, Ian A. Burney and Christopher Hamlin, eds., *Global Forensic Cultures: Making Fact and Justice in the Modern Era* (Baltimore: Johns Hopkins University Press, 2019).

30 Naomi Oreskes, *Why Trust Science?* (Princeton: Princeton University Press, 2019); Gerard Delanty, ed., *Pandemics, Politics, and Society: Critical Perspectives on the Covid-19 Crisis* (Berlin: De Gruyter, 2021).

31 Bruno J. Strasser et al., "'Citizen Science'? Rethinking Science and Public Participation," *Science & Technology Studies* (2019): 52–76.

32 Bruno Latour, *Science in Action: How to Follow Scientists and Engineers through Society* (Cambridge: Harvard Univ. Press, 1987), 4.

33 Lancaster and Rhodes, "The Thing We Call Evidence: Towards a Situated Ontology of Evidence in Policy," this volume.

34 Stephen Hilgartner, *Science on Stage: Expert Advice as Public Drama* (Stanford: Stanford Univ. Press, 2000).

35 Hussain, "The Politics of Evidence: State Secrecy, Ambiguity and Counterforensic Practice in 'Missing Persons' Cases in Pakistan," this volume.

36 Nancy Cartwright, "Well-Ordered Science: Evidence for Use," *Philosophy of Science* 73, no. 5 (2006): 981–990.

37 Michel Foucault, "Truth and Power (1977)," in *Contemporary Sociological Theory*, ed. Craig J. Calhoun et al. (Oxford: Blackwell, 2007).

38 Olga Amsterdamska, "Practices, People, and Places," in *The Handbook of Science and Technology Studies*, 3rd ed., ed. Edward J. Hackett et al. (Cambridge: MIT Press, 2008).

39 Lorraine Daston, "The History of Science and the History of Knowledge," *KNOW: A Journal on the Formation of Knowledge* 1, no. 1 (2017); Christian Joas, Fabian Krämer, and Kärin Nickelsen, "Introduction: History of Science or History of Knowledge?" *Berichte zur Wissenschaftsgeschichte* 42 (2019), 2–3 .

40 Cf. Chandler, Davidson and Harootunian, *Questions of Evidence*; Karin Knorr-Cetina, *Epistemic Cultures: How the Sciences Make Knowledge* (Cambridge, Mass.: Harvard University Press, 1999); Karen Kastenhofer, "Converging Epistemic Cultures?" *Innovation: The European Journal of Social Science Research* 20, no. 4 (2007).

41 See, for example, Kari Lancaster, "Performing the Evidence-Based Drug Policy Paradigm," *Contemporary Drug Problems* 43, no. 2 (2016): 142–153.

42 On co-construction see Sheila Jasanoff, ed., *States of Knowledge: The Co-production of Science and Social Order* (London: Routledge, 2004); Nelly Oudshoorn, ed., *How Users Matter: The Co-construction of Users and Technologies* (Cambridge, Mass.: MIT Press, 2003).

43 See, for example, Oreskes and Conway, *Merchants of Doubt*; Stanley Aronowitz, *Science as Power: Discourse and Ideology in Modern Society* (Basingstoke: Macmillan, 1988).

44 See also Sarah Blacker, "Strategic Translation: Pollution, Data, and Indigenous Traditional Knowledge," *Journal of the Royal Anthropological Institute* 27, S1 (2021).

45 See, for example, Harry Collins et al., *Experts and the Will of the People: Society, Populism and Science* (Cham: Palgrave Pivot, 2020); Niels G. Mede and Mike S.

Schäfer, "Science-Related Populism: Conceptualizing Populist Demands Toward Science," *Public Understanding of Science* 29, no. 5 (2020): 473–491; Michael P. Lynch, *In Praise of Reason* (Cambridge, Mass.: MIT Press, 2012).

46 Alexander Bogner, *Die Epistemisierung des Politischen: Wie die Macht des Wissens die Demokratie gefährdet* (Ditzingen: Reclam, 2021), 16.

47 See for instance Rauna Kuokkanen, "Indigenous Epistemes," in *A Companion to Critical and Cultural Theory*, eds. Imre Szeman, Sarah Blacker, and Justin Sully (Chichester: John Wiley & Sons, Ltd, 2017).

Part I
Establishing Evidence: The Formation of Disciplinary Cultures

2 War, Wheat, and Crop Diseases of the Late Enlightenment: Contesting and Producing Evidence in Agriculture in Great Britain

John Lidwell-Durnin

Introduction

"The popular notion amongst farmers, that a barberry tree in the neighbourhood of a field of wheat often produces the mildew, deserves examination."[1]

In 1804, in a crowded lecture hall in London, the chemist Humphry Davy delivered a series of lectures on agricultural chemistry, during which he called for further investigation into the popular belief that common barberry could produce mildew in wheat fields.[2] The comment might seem speculative, academic, even trivial. But during Britain's involvement in the Napoleonic wars, the debate over barberry was a central scientific topic, and everyone from illiterate farmers to the president of the Royal Society weighed in on the question of whether or not the cultivation of barberry in Britain was threatening its food security. The idea that barberry shrubs, just by being "in the neighbourhood" of a field could affect it seemed, to many people, delusional. "I believe a similar prejudice formerly prevailed against beings equally harmless and innocent," wrote one farmer in defense of barberry, "many of whom were put to death by 'judicial authority', under the charge of *witchcraft*."[3] The agricultural press of the eighteenth century was voluminous: popular treatises on farming, encyclopedias, and periodicals flooded the print culture of London, although few practicing farmers looked to the advice emerging from agriculturalists as a means to increase their yields and income.[4] What agricultural print culture did provide was a space for political debate over land management and food production: the agriculturalist Arthur Young rose to fame not for his views on turnip cultivation, but for the tours of farmland he undertook in France on the eve of Revolution.[5] Disease, and its causes, were no less political than revolution. As agricultural writers, natural philosophers, botanists, and practical farmers took sides on the question over barberry, the fraught and complex means by which evidence was weighed, considered, or discounted in agricultural science, unraveled.

The food shortages and riots that affected Great Britain at the close of the eighteenth century (explored in their full significance by the historian

DOI: 10.4324/9781003188612-3

Adrian Randall's work on popular protest), threatened on numerous occasions to pitch the country into revolution (or at least, so it seemed to many within the political elite).[6] The causes of these shortages were complex, and historians disagree as much today as to their cause as did people at the time; population growth, crop failures, agricultural mismanagement, the engrossing of grain by landowners, disruptions in the food trade due to war.[7] But during bad seasons where excess rains, cold weather, or drought affected crops, the ravages of crop diseases on wheat fields were painfully visible to the public, and often positioned by contemporaries as playing key roles in driving shortages and risking famine. Famines were still regular occurrences both within Europe and within the wider British Empire in the eighteenth century, and while historians of famine are agreed that England did not suffer a "famine" again after the sixteenth century, the threat of famine remained immediate and politically potent.[8] The absence of a census or reliable data on food production before 1801 made these fears all the more acute.[9] This political interest in agriculture helped provide funding, state support, and widespread interest in experimental or scientific agriculture—derided as "book-farming" by sceptics and critics.[10] The "agriculturalists" or "agronomists" that wrote on agricultural matters for the scientific magazines, journals, treatises, and surveys that proliferated during this period all worked to enlist travel, circular letters, and surveys into this effort to develop a base of scientific agricultural knowledge.[11] This chapter explores the controversies and arguments that split these emerging scientific networks over the causes of diseases in wheat—particularly stem rust and its relationship to barberry. *Puccinia graminis* (stem rust) is a fungus that attacks wheat plants. Its life cycle incorporates *Berberis vulgaris,* or common barberry, as a host during winter in temperate climates. While many agricultural communities the world over understood that barberry had the power to affect and spread disease among wheat fields, agriculturalists remained deeply sceptical of these claims, rejecting them even when advanced by the president of the Royal Society, Joseph Banks.

This was not because agriculturalists were more inclined to rational explanations and practicing farmers more open to tradition and local wisdom. There were no experimental farms in Britain in the eighteenth century: the only people that possessed any knowledge on agricultural cultivation methods and practices were the farmers themselves. Moreover, this was an era where botanical and agricultural knowledge was often communicated and diffused in poetry, at least within the print culture active in Britain. Alongside the botanical poems of Erasmus Darwin, most of the agrarian and agricultural writers of the eighteenth century had a classical education and were familiar with Virgil's *Georgics,* and also the name of the Roman god Robigus, who protected wheat from rust and mildew in the Roman era.[12] In the eyes of many, the preventative measures and practices adopted by farmers in England to mitigate against the impact of the disease were no more effective or rational than the animal sacrifices practiced by

Romans at the ancient *robigalia*. In particular, the belief held by many farmers in Britain that *Berberis vulgaris*, or barberry shrubs, infected or transmitted mildew and rust to cereal crops, was viewed as symptomatic of superstition and ignorance.

When we look to larger historical studies of science in the eighteenth century, we see these contests over evidence and authority held significant political importance. Simon Schaffer has argued that science in the eighteenth century was founded upon (what he termed) "a practice of public display."[13] Spectacle and display provided evidence to fuel scientific debates, but were also arenas of political and social conflict. Evidence, far from inviting consensus, could threaten the status quo, inviting the powers of institutions—and sometimes the state—to work toward establishing consensus.[14] Jessica Riskin's work on the role of the senses and sensibility in scientific thought in the eighteenth century has shown how fraught questions of how to interpret evidence produced by the senses could be in this period.[15] Riskin and Schaffer were both concerned with the idea of spectacle as a small experiment being performed by a natural philosopher (or a charlatan) before the public: but outbreaks of wheat rust were also spectacles, witnessed by a wide public, where it was important for political powers to ensure some level of influence over how people understood the catastrophic operations of nature.[16]

A brief revision of agricultural production in Britain is important here. The eighteenth century saw most land in Britain in the hands of a select and wealthy few.[17] This is a period where landowners were in the process of petitioning parliament for the right to enclose land. Historically, tenants had use of commons pasture to keep cattle and livestock. In the fens, "common wastes" or "waste lands" (usually encompassing pasture, commons, fens, uncultivated farm land, and forest) provided land for peasants and tenants to hunt and grow crops—the enclosure of land by owners ushered in crop rotation, an emphasis on cereal cultivation, and was viewed by its critics as encouraging the depopulation of the countryside. By 1800, agriculture was still the largest employer in Britain and wheat was the largest component of national income.[18] Studies of the self-sufficiency of Europe have shown that despite the population explosion between 1750 and 1850 Europe was most certainly self-sufficient, although Britain had the largest deficits in grains, with imports averaging 1.4 million tons.[19]

If the island was struggling to produce sufficient grain, it was prolific in its production and diffusion of agricultural journals, pamphlets, and treatises. Joanna Innes, writing on social problems of the eighteenth century, regarded the success of agricultural printing as tied to a wave of concerns that brought agriculture to the focus of attention: rising food prices and the concern of advances in agricultural production on the continent.[20] Paul Warde has argued that in the eighteenth century, British agriculturists for the first time began to regard agronomy as "a generalised theory of the management of agrarian resources."[21] Agricultural writers often sought to identify principles and techniques that had universal merit, but even the least knowledgeable writers admitted that soil type and local circumstances would ultimately determine the suitability or unsuitability of cultivation methods: the main distinction between clay and calciferous soils of lime and chalk determined what farmers did and did not attempt to grow in their own regions.[22] Efforts to generalize and derive principles for agriculture were perpetually fractured and rendered uncertain by the question of local circumstances: practices that went against the grain of general opinion might prove somehow suited to some unknown or important aspect of locality. Other historians have pointed to the more immediate financial aims of agriculturalists. Successful writers like Arthur Young often began writing agricultural treatises in the first place in the hopes of finding wealthy sponsors to hire them as estate managers. Pamela Horn termed this generation of agricultural writers "propagandists," viewing the output as primarily political and self-serving, with little hope or actual intention of reaching the hands of farmers.[23]

Despite the proliferation of manuals and literature aimed at agricultural improvement addressed to "the public," it is obvious that most of the laborers and tenant farmers are not actually intended to be part of this public, despite the fact that they carry out most of the labor and planning in agricultural production, and possess most of the cultivation knowledge. Landowners, the clergy, estate managers, baileys, (those charged with overseeing practical management of larger farms), engineers (in charge of draining), gardeners and Yeomen constituted the public, although the question of how to talk to those that actually worked the land was a frequent one. The questions that plagued Britain in the eighteenth and nineteenth centuries concerning land management, particularly because all of the questions concerning improvement, progress, feeding the population, and ensuring the survival of the landed gentry against revolution and reform convene in the question of how agricultural land ought be cultivated.

If practical farmers were excluded from the agricultural sphere in principal, they still played a crucial role in producing knowledge and in implementing the kinds of practices and approaches that British agriculturalists hoped to see adopted throughout the kingdom. This led to an epistemological crisis: the validity of evidence was constituted by the education, social standing, and methods adopted by the farmer. This is unsurprising in itself. But the necessary role played by practical farmers in observing and informing agricultural writers on cultivation methods destabilized this culture. While practical farmers were often discounted in the print culture as illiterate, "ignorant," and stubborn, they still had some power in negotiating evidence practices. Agricultural writers such as William Marshall not only relied on agricultural laborers to conduct his experiments, but also designed and carried out cultivation experiments to test the veracity of practical knowledge according to scientific means. Scientific authorities like Joseph Banks invested in illustration and the use of microscopes to work to verify the claims of practical farmers along similar means. In this chapter, I show how the emergence of an agricultural print culture produced a discursive space in which evidence was produced and contested concerning the relationship between barberry, disease, and Britain's food production. Ultimately, this discursive space worked to conceal and obscure the relationship between barberry and crop disease, and it also worked to discredit and belittle the knowledge held by practical farmers, despite the fact that these persons were at the same time providing the vast majority of cultivation practices, observations, and techniques that agriculturalists drew upon in describing rational or experimental agriculture.

Crisis

It is important to provide some context for the arguments and debates over barberry during this period, as they were far from idle speculation happening within botanical, gardening, and agricultural circles. Britain had been dependent upon wheat imports since the 1750s, but war with France and reports of food shortages and famines on the war-torn continent provided extra motivation for the argument that Britain should convert its wastelands and pasture toward cereal cultivation, wherever possible.[24] The frequent crop losses that struck Britain in the latter half of the eighteenth century tended to encourage Britain toward increased wheat cultivation, rather than diversify cultivation—many agriculturalists urged the government and agricultural societies to nudge farmers toward other staple crops, like potatoes, oats, and barley—but the politics of bread ensured that the only viable political response was to work toward increasing wheat production.[25] After the failures of 1795, the Board of Agriculture established a committee on the improvement of these so-called "waste lands" with the direct aim of expanding cereal cultivation.[26] The belief that enclosure and cereal cultivation could provide Britain with military and economic domination over its competitors was nothing new—as Young argued in his first agricultural pamphlet in 1768, "it would be most political conduct to turn all the commons and sheep-walks in the kingdom into arable farms."[27]

Wheat prices rose and crashed during the eighteenth century, threatening food riots and shortages during bust periods and the financial viability of cereal cultivation when seasons were abundant. In 1786, wheat sold for an average of £575 pounds per tonne; by 1789 it was £772. The price fell to £603 in 1792 only to rise to £875 in 1795–1796.[28] The range of public debate over the cause of these price shifts demonstrates that there was no consensus as to whether or not the shortages and prices were primarily the result of bad weather and natural causes, agricultural incompetence and mismanagement, or human greed. As the harvest of 1793 looked to be disappointing, the *Evening Mail* reported in late September that "the immense quantity of horses in this country are, in a great measure, a principal cause of the high price of provisions."[29] In April 1796, members from Leicester and Worcester proposed a tax upon dogs, reasoning that uneaten food from the tables of the rich went first to their dogs, and secondly to the poor; if they felt pressed to keep fewer dogs, more scraps would be available for poor families.[30] As historians of rural riots have shown, many communities were convinced that the shortages were the direct result of hoarding by wealthy farmers and grain merchants, deliberately driving up prices in pursuit of wealth.[31] If newspapers and agricultural journals provided space for conjecture and speculation about contributing causes, the stamped press and institutions like the Board of Agriculture were largely agreed that (1) the shortages were very real, and (2) they were primarily the result of poor weather, but certainly aggravated by bad cultivation practices. The survival of the state and the protection of property both necessitated that cereal yields increase.

If agricultural writers pointed toward numerous and contradictory causes of the shortages, there was nonetheless widespread agreement that failed wheat crops were blighted, mildewed, and blasted. The significance of crop disease as a catalyst in the failures is evidenced by the speculative agricultural technology of the period. In 1789, Young reported that a citizen of Augsburg (north of Munich) had invented an engine that could clean smutty corn "in a few minutes." Excited by the news, Young speculated that the engine, if used across Britain, then even if smut couldn't be eliminated from agricultural practice, at least it would be possible to save more food from infected crops.[32]

An epistemological crisis accompanied these crop failures: there was no consensus as to what symptoms belonged to "mildew," "blight," "smut," and "rust," if these named one malady, several, or four separate diseases. Virgil's *Georgics* made reference to *robigo* attacking wheat, and was translated by Dryden as: "Soon was his Labour doubl'd to the Swain, and blasting Mildews blackened all his grain."[33] Grey, yellow, reddish, and black ailments, sometimes affecting the stem, sometimes the leaves, other times the kernel—the taxonomies we use to separate these fungi today were unavailable, although agriculturalists and poets alike were interested in what the different colors and areas of the plant affected might betoken. Since at least the time of Samuel Hartlib, agricultural writers and botanists had tried to define and untangle these diseases, but as reportage on outbreaks in Britain took up whatever terms locals used to name the malady affecting their crops, the terms remained puzzlingly interchangeable. A farmer named Mr Aiton wrote to *The Farmer's Magazine* to report failures to mildew in his area, but warned readers that: "… diseases termed blight, rust, mildew, &c. –Whether these ought to be ranked as one, or as so many different diseases in growing crops, is not for me to determine."[34] When the botanist and Anglican minister John Henslow tried to separate and distinguish the different fungi attacking cereal crops, he determined that "mildew" was the most common term used for the effects of *Puccinia graminis* (stem rust).[35]

Everywhere, wheat rust was felt to be increasing. Losses to crop disease were so thorough in Yorkshire in 1795 that the president of the Board of Agriculture, John Sinclair, said that farmers in the district should seek fresh seed from other counties "at any expense."[36] Within the public sphere, degeneration in seed stocks and appeals to miasmatic theories of infection were most popular, leading some newspapers to advise affected areas to import untainted seed. The *Telegraph* advised readers after the summer harvest failures of 1795 that:

> ... none but the best seed should be made use of ... In some parts of the kingdom, particularly in Yorkshire, the mill-dew has been much complained of. Any seed infected with that disorder ought to be avoided as much as possible, and untainted seed, at any expense, ought to be procured.[37]

Parliamentary records and discussion in the press affirm that 1796 saw farmers responding to what they understood to be government directions to grow wheat on their land, no matter if it interrupted their usual rotation schedule, with the result that cultivation of wheat (often on exhausted soil) certainly increased from 1796 onward. Few observers felt that the crops of 1796 provided evidence that wheat was thriving in Britain.

Echoing 1795, the crop failures in 1804 in Britain were observed by Joseph Banks to have been principally caused by "the mildew."[38] Arthur Young circulated a request for information on the English wheat crop that began with the observation that: "the mildew on wheat has rarely been so general or fatal as in the late crop of 1804."[39] In 1806, landowner and agriculturalist John Egremont proposed a series of measures to treating infected crops in the hopes that "that they would not be felt as a national calamity."[40] The unusual weather patterns in Britain from 1800 to 1817, including unseasonably cold and wet years, followed in 1816 by "the year without a summer," saw agricultural writers observing that "It is an unquestionable fact, that the Rust has made more considerable ravages within the last ten or twelve years."[41] The fact that war and shortages had pressed many farmers into growing wheat on exhausted land was observed at the same time, and lent renewed support to theories that the disease was produced by degeneration in the seed or depletion of the nutrients in the soil.

Cultures of Evidence

How, precisely, did agriculturalists propose scientific responses to the threat of wheat rust during this period? Recent historiography has shown how central agricultural production was not only to political discourse, but also to the philosophical discourse of the Scottish and European Enlightenment. Fredrik Jonsson has argued that the Scottish Highlands became the "practical laboratory" for many of the political and philosophical ideals of the Enlightenment.[42] Land management, agriculture, population studies and even efforts to reforest the mountainsides all provided a means where the ideals and politics of philosophers found a means of testing whether there were natural limits to improvement.[43] Many agriculturalists of the era prized experimentation as a means to test these limits, but all agreed that the state had failed to provide adequate financial support in this regard. While agricultural writers such as the intelligencer Samuel Hartlib had called for the establishment of schools for husbandry and agriculture since the 1650s, by the close of the eighteenth century Britain still lacked such an institution. Writing in 1799 and calling for the establishment of the same kind of husbandry school Hartlib had recommended 150 years prior, the agriculturalist William Marshall insisted that its political necessity was very clear: "England, at present, does not produce a supply of food for its own inhabitants."[44] Marshall, as Hartlib before him, viewed the potential school as a site of research and scientific education for farmers, estate managers, and even laborers—the possibilities of such an institution were pressed in utopian terms, "the whole kingdom will become systematized."[45] But it did not exist, and lurking behind descriptions of this idealized system was the complex web of print culture that constituted the exchange of agricultural knowledge among the educated classes.

With no experimental farms and very few farmers in a position to undertake the financial risks of experimenting with untested cultivation methods, agricultural print culture in the eighteenth century was composed of evidence provided by practicing farmers. Agriculturalists took it upon themselves to sift through testimony and observations, separating the useful from the spurious and the false. Many were eager to enlist philosophical concepts and methods in this effort to manage and limit the amount of influence practical farmers could have on the science taking shape in their publications. So-called "book farming," "experimental farming," or "the new husbandry," referred to a constellation of books and treatises, gentlemanly societies, fads, and communication networks which encouraged land owners to manage their estates according to rational principles and scientific knowledge, rather than to trust to the local practices that tenants and farm laborers were apt to follow.[46] As Jürgen Habermas has argued, this capitalist public that owned the land (and debated the virtues of experimental farming) were "a reading public."[47] Agricultural books and treatises abounded, but a few figures nonetheless sought to dominate the public debate over agricultural practice and to direct public opinion. Agricultural writers (such as William Marshall) worked to delineate and identify scientific agricultural practice by appeal to the methods and processes by which such literate farmers gained knowledge:

—the ILLITERATE FARMER either acts wholly by CUSTOM; or, if he observes advertently, trusts his observations to his MEMORY. The SCIENTIFIC FARMER, on the contrary, not only observes and records the useful information which occurs to him in the course of his practice, by INCIDENT; but discovers by EXPERIMENT, those valuable facts, which never did, nor ever might have come, incidentally, within his knowledge.[48]

Thus, experimentation—and proper consideration of evidence—were central to Marshall's vision of the habits and practices that distinguished the scientific farmer. These practices were all the more important given the absence of the kind of schools that could otherwise establish the methods and practices of agriculture. Whatever "evidence" in agricultural science might mean, agriculturalists recognized that the reading public were in the position of power to weigh and consider it.

Efforts to square and define "evidence" by philosophers in this period often served to reinforce the same means of consideration that agriculturalists were inclined to take concerning political economy, food, and the causes of rising prices. David Hume divided evidence into three kinds: evidence produced from knowledge, evidence that arises from the exercise of geometrical and logical proofs, and evidence that arrives "from probabilities."[49] The last form of evidence comprised all that evidence drawn from arguments of cause and effect, building on perceptions in the mind. Perhaps unsurprisingly, Hume's views on evidence and its role in the processes of reason led him to conclusions shared by most of his peers. Thus, in his historical account of agriculture in England, Hume viewed economic data of shifts in the price of grain as "sufficient proofs" that no cultivation practices had been developed to make any progress against weather fluctuations.[50] Likewise, Hume viewed high prices of wheat and periods of documented famine as "a certain proof of bad husbandry."[51] But if the increased attention to evidence and processes of reason made little difference on how Hume and his followers understood the operations of nature, the attention paid here to how evidence might sufficiently prove the causes of famine or the consequences of poor husbandry established ideals that agricultural writers like William Marshall sought to attain in their own work. Judicious attention to evidence and the process of argumentation could, Marshall believed, distinguish the products of imagination from genuine agricultural truths: a key ability to wield, in Marshall's eyes, when all the cultivation knowledge was wielded by those carrying out the work. But the limitations of such methods were hotly debated. Writing on individuals that argued from the basis of their own experience, the agriculturalist Arthur Young commented that: "I have nothing to say to gentlemen who are rather inclined to credit their own insulated experience upon some scrap of land, compared to whole counties."[52] Young certainly encouraged "superior minds" and those with the money to engage in it to undertake experiments. But since he viewed differences in soil and climate as hard to control, he doubted the ability of the best-intentioned farmers to be able to conduct trials that could be meaningfully repeated with hopes of success in other parts of the country where conditions would differ.[53]

The Debate over Barberry

Despite its occasional representation for witchcraft, Barberry was widely cultivated in Britain during the eighteenth century. The berries produced by the shrub were described in recipe books and gardening treatises all included advice, recipes, and uses for barberry during this period.[54] The eighteenth century also saw continued and rapid enclosure of commons and waste lands in Britain, a process that involved farmers searching for shrubs (like the barberry) to be enlisted in growing hedges that could both separate fields and provide a means for cultivating useful shrubs and trees.[55] As there were no other examples of a cultivated plant species infecting or attacking another from a distance, the theory promoted by some that barberry could attack wheat seemed counter to the logic of nature. But it also provided a means of distinguishing expertise from ignorance, so long as the theory remained unprovable.

In the confusion, many were quick to attribute to farmers and laborers whatever theory they hoped to argue against, thus the Anglican minister Robert Hoblyn asserted that "The practical farmer as invariably assigns it to the malignant effects of morning fogs, or, in other words, to atmospherical influence."[56] But such views were held by the wealthiest landowners in the country as well, including the Earl of Egremont, member of the Board of Agriculture, who observed that "the blight is generally immediately produced by excessive heat coming suddenly after much rain."[57] A farmer in the south of England wrote to Arthur Young that his crop was subject to rust and mildew, which Young attributed to the influence of sea fogs.[58] Every step of cultivation was scrutinized and held in suspicion. Crops that were drilled were suspected of being more prone to rust, while others argued that crops sewn without seed drills were more at risk.[59] Farmers and landowners alike speculated on varietal differences—a variety named "Creeping Wheat" from Yorkshire was celebrated as being less liable to rust, while varieties that were "woolly-eared" or "thick-chaffed" were singled out as being more at risk to the disease.[60] While some blamed sea airs, interviews with farmers in coastal areas were just as likely to support the belief that the salty airs preserved crops from rust.[61] An American correspondent suggested to Young that the rust might be attracted by clover, and that the increased cultivation of clover could explain the frequency of rust infections.[62]

The knowledge that wheat rust was transmitted from barberry shrubs to wheat crops was pervasive throughout farming communities in Britain, Europe, and North America. Agricultural writers and travelers, in compiling anecdotes and publishing letters from practicing farmers, frequently encountered records of how communities preserved knowledge of the dangers posed by barberry to wheat. Examples of such testimony abound in the agricultural press, but it is worth considering at least one example at length. A farmer in Leeds named John Baker recalled (in 1839):

> When I was a boy, I was taken by my father to a field of wheat, the middle of which, from side to side, was covered with mildew; a large barberry bush grew on one of the hedges near to a garden, and directly opposite to the portion of the field which was diseased, neither of the ends of the field being at all effected with the malady ... we attributed it to the plant in question ... Some years afterwards, I learnt from a botanical friend, to whom I had made some observations on the disease, that the mildew was a plant of the mushroom tribe, which fixed itself on the stem of the wheat, and that the same parasite found a harbour on the leaves of the barberry.[63]

Such testimonies frequently appealed to the geometrical aspects of the infection in relation to barberry shrubs—the nefarious influence of the plants could be directly observed in the patterns of disease in the fields. Baker learned from his father to see that this outbreak was downwind from the shrub, areas that were clear of the shrub also being clear of disease. Other practical farmers spoke of the relationship in similar ways. Young reported a landowner, upon purchasing a field, asking a farmer if it was prone to rust outbreaks. "Oh! Replied the man, "*it is in such, and such a line,* and in that direction you will find two barberry bushes, which always mildew some of the wheat whenever it is sown."[64] On the authority of the landowner (a Mr Sewell), the barberry bushes were removed and the field spared from future outbreaks. "Late ripening oats," observed a farmer, "equally near to barberry hedges, suffer most by the local mildew, which extends as far, in some years, as a quarter of a mile at least and sometimes more, from the diseased and infecting hedges."[65]

The farmer John Exter was particularly insistent that there was a geometric aspect to the relation of rust infections to wheat crops and the positions of barberry shrubs. In a detailed letter sent to Arthur Young, he explained that he had observed an outbreak of rust on his land that was most intense in the immediate circumference of the tree, growing fainter relative to distance. Most importantly, he explained that wheat growing behind barns and hay-stacks "where it could not be seen from the spot the bush grew" were unaffected. "It appears by this circumstance, that the influence of the Berbery is projected in a right line, as rays of light sent off from a luminous body, and that any refraction of its rays lessons, if not destroys, its effect."[66] When William Marshall experimented by planting a small crop of wheat around a barberry shrub, he likened the resulting rust infection to the tail of a comet, observing that the rust hit the field like a tail pointing downwind from the barberry.[67] These observations were debated in the scientific press for decades, and as late as the 1850s botanists still rejected them, arguing that these effects could easily be explained by appeal to the shadow cast by the barberry, or the wetness of the soil.[68]

Among the farming regions that had maintained strict communal policies against the cultivation of barberry in Britain was Norfolk. Visiting the county in the 1790s as an agricultural writer, William Marshall clearly had never previously heard of the idea and expected that his readers would be unfamiliar with it as well. He explained in his account of the travel that he laughed in the face of the farmer who first expressed the conviction.[69] Enquiring amongst other farmers in Norfolk, Marshall realized that they were "to a man, decided in their opinion."[70] Marshall subsequently undertook an experiment to test the influence of the shrub upon wheat, and though he observed (as described above) just the kind of geometric relation between shrub and infected crop that many had described, Marshall rejected the evidence from his own experiment. Having strongly rejected the knowledge accumulated by illiterate farmers in his earlier publications, it is reasonable to conclude that Marshall enlisted scepticism on this point because he feared that agricultural practice in Britain was conducted more according to error than it was to truth. There remained something irrational in the suggestion that barberry shrubs could blight a field of wheat. The operations of nature ought to be intelligible.

Contrasted with experimentation were efforts to compile evidence and observations of the putative link between barberry and mildew. In 1804, crops failed throughout Britain due to poor weather. Arthur Young returned to the circular letter as a means of trying to gather evidence and establish consensus. He had no fears that his readers would not engage with the questions on the causes of rust and mildew—the failures of 1800–1801 and the previous decade all ensured that correspondents in his agricultural network had sufficient direct experiences of crop failures to draw upon. Moreover, these particular crop failures were complicated by the approaching end of war with France, which spelled trouble for landowners and farmers that had profited from the soaring wheat prices of the past few years.[71]

Respondents could not be expected to answer all the questions; it was important to put the most important questions toward the top of the letter. Young's letter featured 12 questions. The first asked respondents to name the soil types most affected by mildew; the second question asked if early or late varieties were worse affected, the third asked after exposure to airs. It was only by the ninth question that Young demanded information on whether or not "you made any observations on the barberry, as locally affecting wheat?"[72] Sent to pastors in rural communities, land owners, farmers, and other members of the Board of Agriculture, the circulars provide testament to the variety of beliefs and suspicions related to the causes of the maladies affecting wheat. The responses showed that few within the agricultural writing community were persuaded by the snippets of testimony and observation supplied by practical farmers in support of the barberry theory. John Egremont reasserted his belief in response to Young that "the blight" was brought on by an excess of heat after rain. Thomas Estcourt also argued for climactic explanations, believing that chilling effects from cold rain led to the symptoms observed in affected fields. A farmer named George Sumner agreed with Egremont and Estcourt, saying that: "I saw the mildew on the chalk, on the wet loams and clays, and on the sand. I think, less on the wet loams and clays; but cannot say whether most on the sand or chalk."[73] Soil types all seemed equally affected, making it difficult to find the root cause in the earth. But the presence or absence of barberry shrubs caused tremendous confusion. Estcourt observed: "We have no barberry bushes in the hedges of this neighbourhood, but occasionally a great deal of mildew." A farmer in Leeds complicated the question further by observing that barberry seems to thrive on soils subject to mildew anyway; a farmer in Linton observed plenty of mildew and blight but noted that no barberry trees grew in the vicinity, an observation echoed by a farmer in Yorkshire. A farmer in Newmarket, inclined to the geometrical efforts to report outbreaks of rust, reported that he had observed; "in a line across the field also, issuing out of this semi-circle, the wheat was much injured, but least, furthest from the bush."[74] But in summarizing the reports, Young continued to reject the theory that barberry played a role in spreading crop disease.

Experimentation and Illustrating Evidence

Efforts to utilize print culture to develop a court of public opinion on the question provided opportunities for many people with direct experience of the impact of barberry on wheat cultivation to express their observations. In keeping with the period, as seen above, these observers appealed to geometrical metaphors and analogy to try and represent the veracity of their claims. It is worth noting that the unwillingness to promote the barberry theory was in keeping with the interests of the Board of Agriculture to promote better management and control over seed distribution in the country. John Sinclair believed that rust depended upon infected seed to be spread, and that preparation measures such as steeping could largely eliminate the disease.[75] Sinclair's authority led to his position being echoed decades later in authoritative encyclopedias and popular treatises, such as the widely read *Penny Magazine*, that steeping wheat in "certain solutions" virtually eliminates all fungus from crops.[76]

In 1792, Arthur Young included a translation of Felica Fontana's *Osservazioni sopra la Ruggine del Grano,* in the *Annals of Agriculture,* providing British readers with Fontana's argument that rust was, in fact, a parasitic plant.[77] For agricultural writers like Robert Forsyth (who peddled remedies for various crop and tree diseases to government boards), Fontana had demonstrated that rust was not composed of insect eggs, but rather "a great multitude of small plants."[78] Felice Fontana's experiments had a lasting influence on British efforts to understand rust. Joseph Banks, Allen

Thompson, and Humphry Davy were all authoritative names that had promoted versions of the plant theory by 1815.[79] The immediate difficulty with such theories was how this parasitical plant—wheat rust—could convey itself to plants the succeeding year, surviving a winter to infect next year's crop. Did the seeds of wheat rust lay dormant in the soil? Did they travel on the wind from warmer climates? Joseph Banks read the responses to Young's circular and corresponded with him on the question of rust, but his own aims were to argue the case of the practical farmer by enlisting microscopes and scientific illustration. He would put the full weight of British science behind the barberry theory.

Botanist on Captain Cook's *Endeavour* voyage and president of the Royal Society, Banks was very interested in developing scientific and experimental agriculture in Britain. He had attended Board of Agriculture meetings throughout the 1790s, and when the Royal Institution was established, Banks worked to position Humphry Davy as its resident chemist and encouraged him to write a series of lectures on agricultural chemistry.[80] As Banks used his status to promote patriotic agriculture, this work had immediate consequences in an era when the cost of bread had soared due to shortages and the continuation of war. And Banks had also accrued experimental success and the taste of political power in questions over Britain's wheat production. In 1789, Banks had been called on by the government to advise on the threat posed by "Hessian Fly blight" (barley midge), and was instrumental in urging the government to redouble its efforts to catch wheat imports that had evaded quarantine measures.[81] In the course of preparing his work on wheat rust, Banks asked Davy to analyze the nutritional qualities of spring and winter wheat, concluding that winter wheat was more nutritious. After the losses of wheat crops in the early 1800s to disease, Banks corresponded with Arthur Young, Thomas Knight, John Sinclair, and numerous other agriculturalists, as he sought to compile present views and to use his authority as president of the Royal Society to advance a position.[82]

Convinced from his reading of Felice Fontana that these diseases were the product of parasitic fungi (considered plants at the time), Banks discounted the arguments that sought to characterize the attack of insects and fungus as a symptom of a constitutional weakening or degeneration in the plant. Through an acquaintance with the classicist Richard Payne Knight, Banks had established contact with Richard's brother, Thomas Andrew Knight, a farmer in Hertfordshire that Banks would increasingly rely upon for support in furthering his influence in Britain's agricultural and horticultural societies.[83] Knight supported Banks' theory that rust was composed of microscopic, parasitic plants, although he rejected Banks' suggestion that it could be communicated from one species of plant (the barberry) to another.[84] Knight instead suggested that the seeds for rust dwelt in the ground, like those of other fungi and mushrooms. Having examined mushroom spores, he also proposed that infected fields produced the seeds of ruin on an apocalyptic scale:

I am therefore much inclined to believe that the parasitical fungus, which occasions every disease of this kind, enters the plant, in the first instance, by its roots ... a single acre of mildewed wheat would probably afford seeds sufficient to communicate disease to every acre of wheat in the British empire.[85]

But like others, Banks believed that the task of agricultural science in Britain was to put popular ideas to experimental test, and barberry presented for him an ideal case where traditional knowledge and understanding could be vindicated by expertise and intervention. Banks made an analogy to mistletoe to try and explain how the parasitic fungus might be capable of living on two different species.[86] How to explain the frequent observations of blight where no barberry shrubs were present? Banks knew from Young's circulars that many farmers and agriculturalists questioned how the theory could be valuable to understanding the causes behind affected fields when there were numerous cases of infection far from any observed plants. Banks speculated that the air could be "charged with seed for miles together," but he must have seen that this would provide little motivation for sceptics to admit the possibility that barberry played a role in the injury.[87] He also argued against the degeneration thesis, and perhaps more persuasively than the wind hypothesis, he claimed that 90% of wheat plants grown in a hothouse taken from infected plants grew disease free. If that was a reliable ratio, heredity and infected seed could never be the driving cause behind the malady.[88] We know from the postscript that Banks received considerable criticism for the suggestion that farmers could utilize seed from infected fields—a practice that many would engage in out of necessity anyway, but which nonetheless evidences the political urgency to providing effective advice.[89]

Knight was unconvinced; his own trials in growing wheat near a barberry shrub had resulted in infected wheat, but other wheat fields further away from the barberry were also infected.[90] Banks enlisted Franz Andreas Bauer (1758–1840) to produce illustrations of the parasite, and hoped that the illustrations would provide further evidence of how blight propagated and grew upon its host (see Figure 2.1).[91] Sharing the illustrations Bauer produced with Knight, Banks discovered that the effort to illustrate the theory met with the same resistance as the anecdotal testimony from practicing farmers.[92] The effort to involve the barberry in the explanation of the progress of the disease not only involved a reliance on "practical farmers," but it also worked against Knight's own beliefs that the causes of most ailments affecting crops were rooted in degeneration, and that the cultivation of new varieties was the most effective means of combatting disease.[93]

Banks was widely criticized for his belief that *Berberis vulgaris* could cause rust outbreaks. A reviewer in the Edinburgh Journal noted that "a good deal of liberal attack has been excited by these most important suggestions," not the least because Banks' suggestion that seed from infected fields could be utilized, which the reviewed duly noted would risk "a considerable

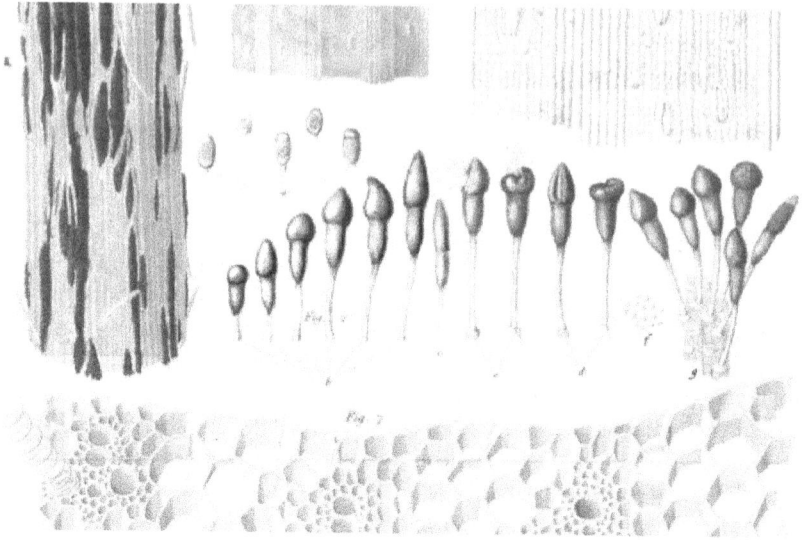

Figure 2.1 Illustration of wheat rust by Franz Bauer in Joseph Banks, *A Short Account of the Diseases in Corn* (London: Nicoll, 1806).

portion of the crop" on a lone author's views.[94] "He has evidently trusted to the commonly received notion ... without due inquiry," complained Aiton, likening the bias against barberry to the charge of witchcraft.[95] In a subsequent survey of the agriculture in Norfolk sponsored by the Board of Agriculture, Arthur Young undertook to interview farmers and gain knowledge of local practice. Young was well aware of Banks's publication, but it clearly had no weight in how he assessed the opinions of the farmers that he met and interviewed. Farmers like "Mr Margateson" of Norfolk that believed barberry could affect wheat fields were represented as being under the sway of "observations that could not deceive."[96] In other words, the farmer was deluded by his own experience. The surveyor visited the parish of Elsing where he learned that local farmers had joined together to extirpate barberry in their area—including on "the lands of those who are careless in this business."[97] Agricultural writers persisted in ridiculing Norfolk. The author of a land management guide published shortly after Banks' essay commented that he had "never perceived the berberry bush produce blights," but that he understood it was "a faculty for which the bush of Norfolk may perhaps have obtained an exclusive patent."[98] In refuting Banks, later writers appealed to division amongst practical farmers to make the case. After all, if practical farmers were considered to form a constitutive body, their disagreement on this point was evidence enough to discredit it. John Henslow took just this approach in 1841. "Even practical men," he wrote, "are by no means unanimous in denouncing the berberry."[99]

Conclusion

The debates over barberry revealed a deeper epistemological crisis in agricultural science at the close of the eighteenth century. Agriculturalists were not engaged in producing knowledge so much as they were sorting and filtering the good from the bad in the vulgar and practical knowledge of farmers. At the heart of this process was the political claim that Britain's food security was endangered in no small part by the fact that its farmers were unlettered and given to ignorant and contrary practices. There was a strong need to discover points at which practical farmers were not merely using inefficient methods, but where practical farmers were also revealed to be absorbed by superstition and false ideas: barberry presented just such an opportunity.

The history of debates over wheat rust invites comparisons with ideas of citizen science and public involvement in the production of knowledge—yet it is important to introduce here some important caveats. The close of the eighteenth century drew questions of citizenship, civic duty, and social order to the fore in Britain.[100] While it is tempting to view the farmers and laborers that held expertise on wheat rust as "citizens" contributing to science, part of what made this conflict of authority so important was that the agricultural elite were not so eager to view these laborers as citizens, or as contributors to a scientific practice that transcended the boundaries of property and political power.[101] In 1795, the president of the Board of Agriculture, John Sinclair, wrote a satirical story about a kingdom where law, order, and prosperity were all lost when the rural laboring class declared: "the land is ours, for we till it."[102] The very right to ownership of the land and its management was tied to the argument that agriculturalists *knew better*—regarding laborers and small farmers as "citizens" contributing to a larger scientific project would have run against the logic of the arguments made by the Board of Agriculture to support enclosure and increase the powers of wealthy farmers to finance land improvement projects. Sociologists and philosophers interested in citizen science can find in this period that by possessing command of evidence related to pressing scientific questions (like the causes of wheat rust), subjects gained leverage and importance that aided the creation of the very idea of "citizen."

Notes

1 Humphry Davy, *Elements of Agricultural Chemistry in a Course of Lectures* (London: Longman, 1813), 195.
2 Frank James, "'Agricultural Chymistry Is at Present in Its Infancy': The Board of Agriculture, The Royal Institution and Humphry Davy," *Ambix* 62, no. 4 (2016): 363–385.
3 Mr Aiton, "On Local Mildew," *Farmer's Magazine* (1815): 298.
4 Peter M. Jones, *Agricultural Enlightenment: Knowledge, Technology, and Nature, 1750–1840* (Oxford University Press: Oxford, 2016), 62–64.
5 Laurent Brassart. "Les enfants d'Arthur Young? Voyageurs agronomes en France au temps du Consulat et de l'Empire," *Annales historiques de la Révolution française* 3, no. 385 (2016): 109–131.

6 Adrian Randall, *Riotous Assemblies: Popular Protest in Hanoverian England* (CUP: Cambridge, 2008). See also Carl Griffin, *The Politics of Hunger: Protest, Poverty, and Policy in England, C. 1750–1840* (Manchester University Press: Manchester, 2020).

7 David Meredith and Deborah Oxley, "Food and Fodder: Feeding England, 1700–1900," *Past & Present* 222, no. 1 (February 2014): 163–214.

8 Patrick Hoyle, "Britain," in *Famine and European History*, ed. Cormac O'Grada and Guido Alfani (Cambridge: CUP, 2017), 156.

9 Patricia James, *Population Malthus: His Life and Times* (London: Routledge, 1979).

10 James Fisher, "The Master Should Know More: Book-Farming and the Conflict over Agricultural Knowledge," *Cultural and Social History* 15 (2018): 315–331.

11 Adam Fox, "Printed Questionnaires, Research Networks, and the Discovery of the British Isles, 1650–1800," *The Historical Journal* 53, no. 3 (2010): 593–621.

12 Mary Beard, J.A. North, and S.R.F. Price. *Religions of Rome: A History*, Volume 1 (Cambridge: Cambridge University Press, 1998), 45.

13 Simon Schaffer, "Natural Philosophy and Public Spectacle in the Eighteenth Century," *History of Science* 21 (1983): 2.

14 Classic studies on this include Mario Biagioli, "*Galileo's Instruments of Credit: Telescopes, Images, Secrecy,*" (Chicago: University of Chicago Press, 2007); see also Simon Schaffer and Steven Shapin, *Leviathan and the Air-Pump* (Princeton: Princeton University Press, 1985).

15 Schaffer (1983); see also Jessica Riskin, *Science in the Age of Sensibility* (Chicago: University of Chicago Press, 2002), 2.

16 For a discussion of enlightenment-era attitudes toward natural disaster, see Daniel Gordon, "Confrontations with the Plague in Eighteenth Century France," in *Dreadful Visitations: Confronting Natural Catastrophe in the Age of the Enlightenment*, ed. Alessa Johns (London: Routledge, 1999), 4.

17 See Joan Thirsk, *Agricultural Regions and Agrarian History in England, 1500–1750* (London: MacMillan, 1987).

18 Nick Crafts, *British Economic Growth during the Industrial Revolution* (Oxford: OUP, 1985), 144–159; Liam Brunt, "Nature or Nurture? Explaining English Wheat Yields in the Industrial Revolution, c.1770," *Journal of Economic History* 64, no. 1 (2004): 193–225.

19 George Grantham, "Agricultural Supply During the Industrial Revolution: French Evidence and European Implications," *The Journal of Economic History* 4, no. 1 (1989): 43–72.

20 Joanna Innes, *Inferior Politics: Social Problems and Social Policies in Eighteenth-Century Britain* (Oxford: OUP, 2009), 146.

21 Paul Warde, "The Invention of Sustainability," *Modern Intellectual History* 8, no. 1 (2011): 153–170.

22 Thirsk, *Agricultural Regions*, 5.

23 Pamela Horn, "Eighteenth Century Agricultural Development," *The Historical Journal* 30, no. 1 (1982): 313–329.

24 Hoyle, "Britain."

25 Thirsk, *Agricultural Regions.*

26 House of Commons, *Select Committee Appointed to Take into Consideration the Means of Promoting the Cultivation Improvement of the Waste, Uninclosed, Unproductive Lands of the Kingdom. First Report from the Select Committee Appointed to Take into Consideration the Means of Promoting the Cultivation and Improvement of the Waste, Uninclosed, and Unproductive Lands of the Kingdom. Ordered to Be Printed 23d December 1795* (London: J. Stockdale, 1796).

27 Arthur Young, *The Farmer's Letters to the People of England* (London: W. Nicoll, 1768), 16.

28 For data, see https://ourworldindata.org/grapher/wheat-prices-in-england?time=1755.1815 (accessed 5 July 2021).

29 *Evening Mail*, no. 716, 23 September 1793; *Morning Post*, no. 6381, 3 October 1793.

30 Parliamentary Papers, House of Commons, 5 April 1796, 364.

31 See Randall, *Riotous Assemblies*; see also Griffin, *Hunger*.

32 "Memorial," *Annals of Agriculture*, no. 11 (1789): 548–552.

33 John Dryden, *The Works of Virgil: Containing His Pastorals, Georgics, and AEneis* (London: Jacob Tonson, 1722), 220.

34 Mr Aiton, "On Local Mildew," *Farmer's Magazine* (1815): 292.

35 John Henslow, "On the Diseases of Wheat," *Royal Agricultural Society* 1 (1841): 9.

36 *Telegraph*, 231, 24 September 1795.

37 Ibid.

38 Joseph Banks, *A Short Account of the Diseases in Corn* (London: Nicoll, 1806), 22.

39 Arthur Young, "Mildew (Circular)," *Annals of Agriculture* 43, no. 252 (1804): 321.

40 Ibid., 36

41 Rev. Robert Hoblyn, "On the Diseases of the Plant of Wheat," *Letters and Papers on Agriculture, Planting, &c. Selected from the Correspondence of the Bath and West of England Society* XIV (1816): 85.

42 Fredrik Jonsson, *Enlightenment's Frontier: The Scottish Highlands and the Origins of Environmentalism* (Yale: Yale University Press, 2013), 8.

43 Ibid., 6.

44 William Marshall, *Proposals for a Rural Institute or College of Agriculture and the Other Branches of Rural Economy* (London: G. and W. Nicol, 1799), 4, 5.

45 Ibid., 16.

46 Fisher "Knowledge," 315.

47 Jürgen Habermas, *The Structural Transformation of the Public Sphere,* trans. Thomas Burger (Cambridge, Mass.: Polity Press, 1962), 23.

48 William Marshall, *Experiments and Observations Concerning Agriculture and the Weather* (London: J. Dodsley, 1779), 264.

49 Hume, *Treatise* 1.3.11.2, SBN 124.

50 Hume, *History of England*, App 4.56.

51 Ibid., App 1.44.

52 Arthur Young, *The Question of Scarcity Plainly Stated, and Remedies Considered* (London: B. McMillan, 1800), 16.

53 Arthur Young, *Rural Oeconomy, or Practical Essays on Husbandry* (London: Nicoll, 1770), 313.

54 Gerry Barnes, Diane Saunders, and Tom Williamson, "Banishing Barberry: The History of *Berberis vulgaris* and Wheat Stem Rust Incidence across Britain," *Plant Pathology* 69 (2020): 1193–1202.

55 Gerry Barnes and Tim Williamson, *Hedgerow History in England* (London: MacMillan, 2008).

56 Robert Hoblyn, "On the Diseases of the Plant of Wheat," *Letters and Papers on Agriculture, Planting, &c. Selected from the Correspondence of the Bath and West of England Society* XIV (1816): 59.

57 Arthur Young, "Mildew (Circular)," *Annals of Agriculture* 43 (1804): 323.

58 Arthur Young, "A Farming Tour in the South and West of England, 1796," *Annals of Agriculture* 29 (1797): 312.

59 Lewis Majendie, "On the Drill Husbandry," *Annals of Agriculture* 17 (1792): 422–432.

60 John Egremont, *Observations on the Mildew Suggested by Arthur Young* (London: J. Hatchard, 1806), 32, 37–38.

61 Ibid., 47.

62 Richard Peters, "Agricultural Enquiries on Plaister of Paris, in America," *Annals of Agriculture* 29 (1797): 458, 556.

63 John Baker, "Barberry Mildew," *Quarterly Journal of Agriculture* 9 (1839): 596.

64 John Egremont, *Observations on the* Mildew, 32; Arthur Young, *General View of the Agriculture of the County of Essex*, vol. 1 (London: W. Nicol, 1813), 301.

65 "Review of Agricultural Publications," *Farmer's Magazine* 17 (1816): 341–355.

66 John Exter, "Damage to Wheat Occasioned by the Berbery Plant," *Annals of Agriculture* 27 (1796): 543.

67 William Marshall, *The Rural Economy of Norfolk* (London: G. Nicol, 1787), 359–360.

68 John Wilson, *The Rural Cyclopedia, or a General Dictionary of Agriculture* (Edinburgh: Fullarton, 1847), 327.

69 Marshall, *Rural Economy*, 19.

70 Ibid., 22.

71 John Gascoine, *Joseph Banks and the English Enlightenment: Useful Knowledge and Polite Culture* (Cambridge: CUP, 1995), 84–85.

72 Arthur Young, "Mildew (Circular)," *Annals of Agriculture* 43, no. 252 (1804): 321.

73 Ibid., 329–330.

74 Ibid., 321–336.

75 John Sinclair, *Hints Regarding the Agricultural State of the Netherlands, Compared with That of Great Britain* (London: B. McMillan, 1815).

76 Bauer, "Fungus," *Penny Magazine* 2 (1833): 126–128.

77 Arthur Young, "Observations on the Mildew Affecting Corn," *Annals of Agriculture* 17 (1792): 232–280.

78 Robert Forsyth, *the Principles and Practice of Agriculture, Systematically Explained,* (Edinburgh: Arch. Constable, 1804), 158.

79 Robert Hoblyn, "On the Diseases of the Plant of Wheat," *Letters and Papers on Agriculture, Planting, &c. Selected from the Correspondence of the Bath and West of England Society* XIV (1816): 58–96

80 James, "Agricultural Chemistry"; Gascoigne, *Joseph Banks.*

81 Gascoigne, *Joseph Banks*, 114–115.

82 Guy Meynell, *Archives of Natural History* 11, no. 2 (1983): 209–221; see also H. W. Lack, *Annalen des Naturhistorischen Museums in Wien. Serie B für Botanik und Zoologie*, 112. Bd. (2010): 253–264.

83 John Lidwell-Durnin, "Inevitable Decay: Debates over Climate, Food Security, and Plant Heredity in Nineteenth-Century Britain," *Journal for the History of Biology* 52 (2019): 271–292.

84 Thomas Andrew Knight, "On the Prevention of Mildew in Particular cases," *Horticultural Society* (1813). In *A Selection of the Physiological and Horticultural Papers of the Late Thomas Andrew Knight* (London: Longman, 1841), 204–209.

85 Ibid., 205.

86 Banks, *A Short Account*, 12.

87 Ibid., 14–15.

88 Ibid., 16.

89 Ibid.

90 Thomas Andrew Knight to Joseph Banks, 20 March, 1806. The *Scientific Correspondence of Sir Joseph Banks, 1765–1820*, Vol. 5 (Ed. N. Chambers) (London: Pickering and Chatto, 2007): 465–468.

91 Banks to Bauer, April 1805, Natural History Museum Library and Archives, BL.A.MS 32439, 168; see also Franz Bauer, "memoranda and papers rel to diseases in Corn," Held at the Royal Botanic Gardens, Kew, NRA 25004.

92 Thomas Knight to Joseph Banks 23 August 1805. In *Scientific Correspondence of Sir Joseph Banks*, 447–450.

93 John Lidwell-Durnin, "Cultivating Famine: Data, Experimentation, and Food Security," 1795–1848, *The British Journal for the History of Science* 53(*2020*) no. 2, 159–181.

94 [Review] "Blight in Corn…," *Edinburgh Journal* October (1806): 148–151.

95 William Aiton, "On Local Mildew," *Farmer's Magazine* (1815): 292.

96 Arthur Young, *Agricultural Survey of Norfolk* (London: Nicoll, 1813), 298.

97 Ibid., 298.

98 John Lawrence, *The Modern Land Steward* (London: C. Whittingham, 1806), 379.

99 John Henslow, "Diseases of Wheat," *Royal Agricultural Society Journal* 1 (1841): 13–14.

100 See, for example, Joanna Innes, *Inferior Politics: Social Problems and Social Policies in Eighteenth-Century Britain* (Oxford: OUP, 2009).

101 See, for example, Bruno Strasser, *Collecting Experiments: Making Big Data Biology,* (Chicago: Chicago University Press, 2019) for a discussion of the contributions of citizens to big science in the twentieth century.

102 John Sinclair, "Nineveh: A Fragment," *Annals of Agriculture* 23 (1795): 3.

3 Presenting Chemical Practice in Court: Forensic Toxicology in Nineteenth-Century German States

Marcus B. Carrier

Practices and Their Practitioners

In May 1848, the local public medical officer Krauß in Tübingen in the south-west of today's Germany performed a postmortem on an infant. The child had been only 12 days old when it died. The mother had been found with the dying child, and with a small bottle of oil of vitriol in her pocket, also known as sulfuric acid. She confessed that she had killed her child using the acid because she could not have taken care of it, nor would any of her relatives have taken her in with the child. The case seems to have been as clear-cut as it gets. The culprit was found with the victim, immediately after the deed, in possession of sulfuric acid, and she confessed to the crime. In the postmortem, Krauß found no possible cause of death other than the corroded, in parts nearly dissolved intestines, which were according to him "irrefutably the characteristic signs of the effect of a concentrated mineral acid (sulfuric, hydrochloric and nitric acid)."[1] Although the medical evidence and the confession seemed clear and there were no doubts left, the local pharmacist Winter was commissioned by the court to chemically analyze the remaining contents of the small bottle found with the mother of the victim, as well as a sample of the dissolved intestines. He was asked to confirm, first, that the bottle had indeed contained sulfuric acid, and second, tmahat the intestines of the child also contained the same sulfuric acid as the bottle. Winter was able to confirm that the bottle did indeed contain sulfuric acid. However, he was unable to find any sulfuric acid in the intestines of the victim. The autopsy had been conclusive, so this lack of chemical evidence had to be explained, and Krauß did that in his remarks on the chemical analysis:

> Sulfuric acid is, after all, not an indestructible substance in an organism, as are arsenic, mercury, and other metals; rather, in the process of destroying organic substances, it itself breaks down chemically and ceases to be detectable.[2]

This example illustrates the two themes of this chapter. First, that chemical analysis held a special place in poisoning trials. It alone was seen as being

DOI: 10.4324/9781003188612-4

able to produce certain proof that a poisoning had taken place, and to distinguish poison from illness or other natural causes. This was the view of forensic toxicology pushed by forensic toxicologists themselves, and in the first section of this chapter the self-descriptions of forensic toxicologists will be the focus. It is in these self-descriptions where forensic toxicologists argue for their place in the courtroom.

Second, in their expert testimonies, forensic toxicologists focused on their own practice rather than any theoretical discussions. Chemical theory remained in the background and was not explained, normally not even mentioned in the expert testimonies. The emphasis of their written testimonies lay in actions and observations. The expert witness would interpret his actions and observations, but not theoretically explain them. Chemical knowledge in this sense was thus presented as a form of practical knowledge: through their analytical actions, experts produced chemical evidence. This encompasses both that they quite literally created material evidence in the form of the isolated poisonous substance as the product of their practices, and that they strongly relied on the descriptions of their practices as evidence for their own expertise. I will illustrate this focus on practice via two case studies. The first of these will take a close look at the language of a successful analysis, and highlight the focus on practice. The second one will focus on an expert testimony challenged in court, to show that the criticism of forensic expertise also focused on practice rather than chemical theory. For sources, I rely on court records of poisoning cases and especially on expert testimonies as well as on handbooks on forensic toxicology.

To clarify, when I use the term "forensic toxicologists" in the context of the nineteenth century, I do not refer to any kind of institutionalized discipline. Rather, I refer to a loosely defined expert community that consisted of roughly speaking two distinct groups. The first were the authors of textbooks; either on judicial medicine in general, which also covered the chemical analysis necessary in poisoning cases, or special textbooks for judicial chemistry. The second group was of those who would actually conduct chemical analyses in these cases and act as expert witnesses in court. For the most part, this group consisted of pharmacists—in many cases the local pharmacist at the place of trial or at the place of the crime—as well as the local medical officer (the *Physikus* or *Amtsarzt*). The former would usually do the analysis alone or—as was the case in the example cited above—in conjunction with the latter. Only very rarely would the medical officer do the chemical analysis alone and without help from a pharmacist. As was the case with the whole specialty of legal medicine, forensic toxicology was built on the demands of the judicial system. The context of application of the practitioners' knowledge and practices was clearly defined by judicial demands, as was, by extension, their research. However, in contrast to legal medicine, forensic toxicology was not institutionalized as a special discipline.[3] Rather, forensic toxicology stayed institutionally connected to either legal medicine or to pharmacy and analytical chemistry.

At first glance, a modern reader might be surprised that pharmacists were the ones conducting the analysis and not professional chemists. However, it is important to remember that nineteenth-century chemistry was strongly connected to pharmacy. In fact, as Christoph Meinel has shown, the strong connection to pharmacy, among other things, played an important role in the institutionalization of chemistry between the late eighteenth and the mid-nineteenth century.[4] During the nineteenth century most German states made at least some university courses more or less obligatory for aspiring pharmacists.[5] A lot of professorships in chemistry in the early nineteenth century were not solely professorships for chemistry, but rather professorships for chemistry *and* pharmacy. So pharmacists were not at all an odd choice for analytical practitioners in the courtroom. The idea was that the requirements of becoming a pharmacist were high enough to also ensure the quality of the chemical analysis.

Ian Burney has argued that chemical evidence was seen as a special case of evidence in criminal trials because it could make poisonings visible; it could produce and demonstrate the murder weapon:

> By enabling experts to present poison in its tangible, material form, chemical demonstration held out the promise of disrupting the poisoner's insidious designs. Through his reproduction of the equivalent of the bloodied dagger in his tubes and retorts, the toxicologist promised to translate this most ephemeral of crimes into a more conventional form of violence.[6]

I have argued elsewhere that this promise was important for the way methods were chosen in the judicial context. Systematically, methods were preferred which produced the poison in its material form, easily identifiable by nonchemists.[7] In this chapter, however, I will focus not on the choice of methods, but rather on the presentation of methods and chemical practice in written expert testimonies. In contrast to the outcome of the analysis, which should in the best case produce a result that was easily understandable, the described practices were never explained and must have been completely incomprehensible to laypeople. Perhaps somewhat counterintuitively, I will argue that this way of describing analytical practices also served the purpose of producing evidence that was easily manageable for the judicial system. As the product of the analysis was meant to reproduce "the equivalent of the bloodied dagger,"[8] the written testimony aimed to mirror eyewitness accounts. Rather than giving a comprehensive explanation of what they were doing, the experts matter-of-factly described their practices and observations. Furthermore, without boring laypersons in the laboratory with experiments lasting for hours or even days, this was a possibility to bring parts of the performance of the analysis into the courtroom.

Self-Descriptions of Forensic Toxicologists

Throughout the nineteenth century, textbook authors maintained—with very few exceptions—that the positive outcome of chemical analysis was the only certain proof that a poisoning had taken place. For example, in his *Lehrbuch der gerichtlichen Medicin* from 1812, the physician Adolph Henke (1775–1843) writes:

> Although it is necessary to consider observations of phenomena which are noticed on the living and dead body to enlighten the problematic question of an occurred poisoning and the death caused by it, the main focus of the investigation always concerns the retrieval and the nature of a toxic substance suitable for poisoning. For the only unalterable proof of an occurred poisoning is the truly found poison inside the body.[9]

Neither symptoms before the death of presumed victims nor postmortem observations were truly deemed sufficient by experts to conclude that someone had been poisoned. The symptoms of poisoning were not distinguishable enough from some diseases. The same was true for postmortem observations, such as inflamed intestines. As Henke explained, "It is possible for the appearances occurring before death as well as for alterations of the corpse to be caused by fierce and quickly killing diseases."[10]

Although the importance of chemical analysis in poisoning cases was emphasized throughout the nineteenth century, the argument shifted slightly during the 1840s. Whereas before—as exemplified by Henke—the argument was that other medical examinations were not nearly as useful as chemical analysis, textbook authors after the 1840s qualified this claim by emphasizing that this should not mean that other medical examinations were useless, or that a positive chemical analysis was necessary to establish poisoning. As Johann Ludwig Caspar (1769–1864) argued in 1860 in the second volume of his *Practisches Handbuch der Gerichtlichen Medicin*:

> It is, however, a non-justifiable leap of logic if one claimed that *only* the detection of the poison would give certainty in the diagnosis [of poisoning], since one would disregard all other evidence and supporting proofs ... and thus, teach a procedure concerning poisoning which is rightfully rejected in all the rest of medical diagnosis.[11]

Now, rather than underlining the usefulness of positive outcomes of chemical analyses, the authors were concerned with not giving the impression that a negative result of chemical analysis was proof of the absence of poison. False negatives should be considered. They could be easily explained, for example, by the very nature of some poisons, which were destroyed during the process of poisoning, as was the case with the sulfuric acid in the introduction. In other cases, there might have been no suitable method to distinguish certain poisons from substances occurring naturally in the body, as was the case for organic poisons during a large part of the nineteenth century.[12] Even if methods for detecting poisons existed, they might not have been sensitive enough to find poisons in the low concentrations necessary to cause harm in some poisoning cases.[13] Essentially, there were many different explanations why a chemical analysis might yield a negative result, even though a poison had been used. False negatives were of much wider concern for forensic toxicologists than false positives. Thus, forensic toxicology was rendered nearly useless for the defense, since only positive results—the detected poison—could really count as certain proof of anything. In practice, forensic toxicological analysis was normally used by the prosecution, not by the defense, although the defense could use counterexpertise in some cases as will be exemplified below.

This shift in argument during the 1840s can be seen as the result of the success of forensic toxicologists in establishing their importance in poisoning trials. As Hans Langbein has shown, after the sixteenth century, the rules of judicial evidence in the trial systems of continental Europe changed. Before that, the focus of judicial evidence had been direct evidence. This is the reason why, in Roman-canon law, torture played such an important role. In general, a defendant could only be convicted if either there were two reliable (male) eyewitnesses who had seen the crime itself, or the defendant confessed. Circumstantial evidence (or *indicia* in the judicial jargon of the time) could never be sufficient to find the defendant guilty of a crime. However, in cases where there was no confession, the circumstantial evidence could amount to a so called "half proof," which was seen as equivalent to one eyewitness report. In the presence of such a "half proof," the defendant could then be tortured following strict rules to secure a reliable (in the understanding of the time) confession.[14] By the seventeenth century, however, judicial practice had invented a different system of proof in the form of the so called *poena extraordinaria* (extraordinary or exceptional punishment), which enabled the Roman-canon judicial system to punish a defendant even if the torture did not lead to a confession, or if circumstantial evidence fell short of amounting to a "half proof." Technically, the defendant was then not punished for the crime itself but "on account of the suspicion amassed against him."[15] This meant that circumstantial evidence, which included most importantly for this chapter every kind of medical and chemical evidence, gained much more importance in practice. As a result, by the eighteenth century, circumstantial evidence could justify every sentence except capital punishment, which was still connected by law to sufficient direct evidence.[16] By the mid-nineteenth century even capital punishment in severe cases, such as murder using poison, could be passed based on circumstantial evidence alone.

In this context, the role of expert witnesses (traditionally physicians and other medical professionals) was strengthened. Here, the overemphasis on the importance of chemical analysis and the relative uncertainty of other possible medical evidence can be understood as the struggle to raise awareness of such importance in the minds of the judicial actors—judges, as well as the not necessarily chemically well-trained medical officers. After the 1840s, however, the standing of forensic toxicology was well established, and it seemingly became more important—from the vantage point of the textbook authors—to fight the overreliance on chemical analysis. It is not clear why the textbook authors stress this point rather continuously. Perhaps they were simply very concerned that a guilty person might go free because a judge did not understand the limits of chemical analysis. However, this also coincided with a general strategy to limit situations of expert disagreements in court. If only positive results from an analysis were regarded as proof of whether a poisoning had taken place or not, the defense could not hope to gain much by bringing in its own experts. Furthermore, if a negative result would not endanger the general conclusions of the investigation, chemical experts could be called upon even in cases where a positive result was not likely, such as the Häfele case from the beginning. In this way, toxicologists could try to ensure to be part of every investigation concerning poison.

Performing Evidence—Presenting Performance

The events of the first case study considered here took place in Ludwigsburg near Stuttgart in 1844.[17] Christiane Ruthardt was indicted for allegedly having killed her husband with arsenic. Her husband, Eduard Ruthardt, had died on 11 May 1844 after having been administered arsenic at least three times. At the time of the trial, Christiane Ruthardt had already confessed to the murder. She explained that her husband had spent a lot of money on expensive books, and had it in his mind to build a perpetual motion machine. At the same time he had, according to his wife, ruined his own health by reckless and unhealthy behavior. His doctor confirmed that Eduard Ruthardt suffered from gout. In fact, the doctor had also attributed Ruthardt's poor health in April and May 1844 to gout, causing him not to suspect poisoning and not to treat him accordingly.[18] Eduard Ruthardt's poor health and his large expenditure on his project were, in his wife's view, the cause of the family's low economic standing. She thought of leaving him, but feared that she would not have enough evidence in court to justify divorce. Instead, she confessed, she planned to kill herself and bought arsenic for this reason. However, her plans changed because she feared that Eduard would not take care of their child while working on his machine. So instead of killing herself she ended up killing her husband.

From the dossier, it is not clear what exactly led to the autopsy of the deceased after his death, but it took place on 11 May 1844, the day of Ruthardt's death. The results of the autopsy are not particularly interesting in detail, except that the examining doctors found grains of a white powder in the stomach. This was an important clue that arsenic had been used. Arsenic trioxide, or white arsenic powder, was the most commonly used poison in the nineteenth century. Katherine Watson has shown that in England and Wales between 1750 and 1914, among the 540 poisoning cases she analyzed, 237 poisonings—approximately 44%—were committed with arsenic compounds (arsenic trioxide, arsenic trisulfide or orpiment, and arsenic tetrasulfide or realgar). The next most common poisons were narcotics (opium and laudanum), which accounted for 52 cases, less than 10% of the cases analyzed.[19] Although there are no similarly thorough statistical numbers collated for the German states in the nineteenth century, the way arsenic was treated in German textbooks suggests that their authors also believed arsenic was the most used poison. As the physician Ignaz Schürmayer (1802–1881) argued in his *Lehrbuch der gerichtlichen Medicin* (1852), "Perhaps most poisonings are done with arsenic, which is possibly caused by the fact that its deadly effect, even in small doses, is common knowledge and that it can be secretly administered more easily than other toxic substances."[20]

The white powder in the Ruthard case was collected and given to two pharmacists named Franken and Schmidt to analyze, in the presence of two physicians, the judge, and three other judicial officials as witnesses.[21] The question of whether judicial officials should be present during the chemical analysis in poisoning trials was discussed excessively in textbooks. Usually, the textbook authors thought that the presence of judicial officials would be unnecessary in the best case, and detrimental to the analysis in the worst. As Schürmayer argued, "What is, for example, the judge supposed to do during a complicated and long-lasting chemical analysis besides being bored if he is not by chance knowledgeable about chemistry and, thus, has a personal interest in the matter!"[22]

Whether the presence of judicial personnel at the chemical analysis was necessary, however, was not a matter of opinion of the experts but a matter of the law—which is why the authors used their textbooks to try to in-fluence these laws. While exact provisions on this point varied, in most cases, the presence of the judge and/or judicial officials as witnesses for the observations was required. Their role was not to interpret these observa-tions; this was the duty of the expert.[23] Even so, this requirement could implicitly be read by forensic toxicologists as a sign that their testimony was still deemed less reliable than eyewitness accounts. Eyewitnesses were, obviously, not supervised during their observation and still trusted; experts who had learnt their trade, however, should be monitored by judicial of-ficials who admittedly were not able to interpret what the expert did and what the result of the analyses was. In this sense, the underlying conflict was not only about whether the judge had authority in judging scientific practices and thus monitoring chemical experts, but also about how to integrate expert testimony into a new trial system with shifting standards of evaluating judicial proof.

In the following, I will take a closer look at the practical section of the expert testimony of this case; that is, the part of the testimony after all persons present were listed and the integrity of the seal on the flask holding the white powder was confirmed. The flask contained "a number of small white, partly yellow colored sandy grains ... the look of which raised the suspicion of white arsenic."[24] Accordingly, only two tests, both looking for arsenic, were conducted during the analysis. The pharmacists started with "the test for arsenic following Hugo Reinsch."[25] This test, known as the Reinsch test, was developed by the chemist Hugo Reinsch (1809–1884) and first published in 1841.[26] In this case, the pharmacists even cited a textbook written by Reinsch in 1843,[27] which was uncommon in expert testimonies. Since there was no citation for the second test described below, it seems that the novelty of the Reinsch test was the reason for it requiring some further citation. The experts could not assume in any way that a judge would even be familiar with the name.

The procedure of the Reinsch test is described by the expert witnesses in the following way:

> Some grains were put into a small cylindrical glass to be tested for arsenic following Hugo Reinsch. .... To this end, some hydrochloric acid and distilled water were added into the same glass. After a bare small copper rod was placed into the glass which was then heated with an ethanol flame [*Weingeistflamme*] until boiling, one could see an iron gray coating on the rod where it was submerged in the liquid.[28]

This is close to Reinsch's own description of his test. The specimen should be boiled in hydrochloric acid and a piece of copper submerged in this solution, which should exhibit a coating. Depending on the color of this coating it can be determined if the solution contained arsenic, antimony, tin, lead, bismuth, mercury, or silver.[29] The iron gray coating described in the testimony also conforms to Reinsch's description of what he calls the *metallische Haut* ("metallic film")[30] on his copper rod, which looks like "being converted to an iron rod."[31] To further confirm the result, the experts in the Ruthardt case prepared an arsenic solution themselves and repeated the same test "whereupon [the copper] was covered again by a very similar iron gray coating."[32] This kind of comparison probe could serve as a visual aide and bolster persuasiveness of the tests.[33] For the purposes of this chapter, however, it is more interesting that there was no kind of chemical explanation of this test. Nowhere in the testimony did the experts explain what was happening, nor how the changing color of the copper rod could indicate arsenic. Only right at the end of the testimony is there a short remark that the white arsenic has been "transformed" (*verwandelt*) into metallic arsenic.[34]

I have argued elsewhere that normally the Reinsch test was not used exclusively in poisoning cases in the German states due to another test—the

Marsh test—which much better conformed to the practical needs or values expected from "good" forensic tests. I have argued that the Marsh test was preferred due to its higher sensitivity, selectivity, the possibility of using less of the sample, and because the outcome was a mirror of metallic arsenic rather than Reinsch's metallic film.[35] This was also true in the Ruthardt case, where the second test used by the experts was, in fact, the Marsh test. This test was developed by the British chemist James Marsh (1794–1846) and first published in 1836.[36] In its simplest version, a specimen is put into a glass vessel containing sulfuric acid and zinc, forming a gas which can be collected and burned. If there is arsenic present in the specimen, the burning gas is arsine, and a porcelain dish which is held over the flame will be coated with a metallic, blackish film. If no arsenic is present, the burned gas is hydrogen and no such film is visible. The experts in the Ruthardt case used a variation normally attributed to Jöns Jacob Berzelius (1779–1848).[37] Here, the gas was led through a glass tube before being burned. This tube could also be heated from the outside, and any arsenic present would form a blackish brown metallic coating inside the tube.

The experts began their experiment without any specimen. Having put only sulfuric acid and zinc inside the apparatus, the gas produced "burnt with the typical [*eigenthümlichen*] color of hydrogen gas"[38] and the heated tube also showed no change. This altered when the specimen—which the experts had dissolved in hydrochloric acid before—was added. This addition of "the arsenical acid liquid"[39] changed the color of the flame from yellowish to bluish. A porcelain dish held over the flame "was repeatedly covered by a brown-blackish, shiny film."[40] They continued their experiment by heating up the glass tube of the apparatus. Here the heat produced a blackish-gray film. The experts then channeled hydrogen sulfide through the same tube, "whereby the blackish-gray color changed instantly into a lovely yellow."[41] This last step was normally not used in poisoning cases. At least, this is the only case I have found. The chemical reaction used was a variant of a procedure introduced—to my knowledge—by Samuel Hahnemann (1755–1843) in 1786,[42] whereby an arsenical solution would form a yellow precipitate[43] with hydrogen sulfide. This test was commonly used in the nineteenth century, not normally after the Marsh test but rather as a pilot test before the actual analysis started. Here, though, it was used to further emphasize the arsenical nature of the sample.

Again, there was no actual explanation of why any of this should demonstrate the presence of arsenic. The observations are described in detail and the different colors in particular are mentioned multiple times. But the interpretation at the end simply states "that the solid grains in question found in the corpse … are arsenic."[44]

By focusing on practice and observations while nearly completely avoiding theoretical discussions and explanation, the experts adhered very well to the recommendations of the textbook authors—some of whom even openly admitted avoiding theoretical explanations completely in their textbooks, which they then defended as a virtue. For example, Franz Schneider (1812–1897) explained in the preface of *Die Gerichtliche Chemie* (1852):

> It is in the nature of the subject matter [forensic chemistry] that in a book which is exclusively concerned with judicial purposes every theoretical viewpoint is forbidden. ... If a representative of the law demands information from an expert, he addresses the question not to the person himself but rather to the science the latter represents. One should, therefore, not express one's subjective belief but only objective knowledge [*Erkenntnis*].[45]

One reason to avoid theory and focus largely on practice during legal cases was that theory was perceived as subjective and (potentially) contested, as Schneider suggests in the quotation above, whereas observations and, remarkably, also their interpretations apparently were not. This only makes sense with the state of chemical theory and analytical chemistry in the nineteenth century in mind. Analytical chemistry as it was taught in textbooks throughout the nineteenth century was essentially a collection of reactions empirically found to be specific for certain substances. Some reactions might include a theoretical discussion,[46] but in general it was only shown that certain substances reacted in a certain way under certain conditions and that other substances either react differently or not at all under the same circumstances. This was in a way the tradition of analytical chemistry when used for practical purposes. Furthermore, chemical theory in the mid-nineteenth century was in fact contested, beginning with questions of the weight of the elements, which was also connected to questions of how compounds were put together and how reactions were understood.[47] Here, the focus on empirical catalogues of observations was a way of not having to commit to a theoretical viewpoint or school, but concentrating on the facts everyone agreed upon. Of course, I do not want to argue that observations and analytical practices were not theory-laden. The identification and classification of "elements" and "compounds," which clearly are theoretical concepts, was an important use of analytical chemistry within scientific practice. Rather, I want to argue that the way experts presented their knowledge suggested objectivity, free from preconceived notions of theoretical ideas and that this move was deliberate. In short, theoretical discussions could potentially lead to an expert disagreement that was difficult to resolve, which the forensic toxicologists wanted to avoid.

The other reason for avoiding theoretical discussions was that observations visible to everyone, including laypersons, played an important role in the use of toxicological evidence in court which toxicologist/textbook authors wanted to highlight even more. As Ian Burney has argued, this constituted a shift "from a chemical tradition privileging bodily experience as the basis for reliable knowledge."[48] Whereas up until the early nineteenth century arsenic was primarily detected by its smell and taste, new tests—such as the liquid tests, the Marsh test and the Reinsch test—focused on the visual sense. Whereas the senses of smell and taste were thought to be in need of special training—to be "incorporated knowledge" as Burney puts it[49]—the new reliance on the visual sense promised a tangibility—"intuitive comprehensibility" or "*Anschaulichkeit*"[50]—which was in principle open to everybody and did not rely on special training.[51] To be sure, this shift in chemical knowledge was far from completed by the early nineteenth century. At least in the forensic context, smell continued to play an important role; be it as a test for arsenic in court cases[52] or in other forensic practices.[53] However, the general trend to remove "incorporated knowledge" in chemistry, and thus also in forensic toxicology during the nineteenth century seems to hold.

As exemplified in the Ruthardt case, in forensic toxicology, the practical knowledge—the know-how to perform the right procedures, as well as the interpretations of the observations—relied on the training and the knowledge of the experts; the observations themselves did (or should) not. The way the expert testimonies for the court were written and the evidence presented as detailed accounts of actions and observations resembled accounts of eyewitnesses of the procedures as a form of demonstration, in which everybody present would have seen the same. Although the toxicologists seemingly did not want to be supervised by laypersons in their laboratory, they still relied on the minute descriptions of their performance to present their expertise, thus bringing part of their performance into the courtroom. In the Ruthardt case, the court was satisfied by the evidence. Christiane Ruthard was sentenced to death[54] and decapitated on 27 June 1845.[55] In the second case study, I will focus on the importance of actions in the criticism of rival experts in court.

Challenging Evidence by Criticizing Practice

The defendant in the second case was Elisabeth Regine von Flandern, who was accused of the attempted murder of her husband using plant extractions in Stuttgart in 1852.[56] According to the bill of indictment, their ten-year-long marriage was full of problems. Her husband, Johann Rudolph von Flandern, accused her of squandering their money, and he complained that she would run away repeatedly, probably being unfaithful to him. Allegedly she began an affair with Ferdinand Kleinknecht, a former business associate of her husband. In 1852, Elisabeth and Johann Flandern agreed to a divorce. The prosecution alleged that Elisabeth Flandern tried to speed up the end of the marriage by trying to kill her husband. Apparently, Kleinknecht was not willing to wait for her, and was already proposing marriage to other women, reinforcing Elisabeth's sense of urgency. After having had another argument with her husband over some clothes, which he did not want to release to her until she paid a debt she owed him, she anonymously sent him food. Something in the appearance of the food made him suspicious, and he reported his suspicions to the authorities, handing over the food as evidence of his allegations.

The vegetables in the food were examined and found to contain Verbascum ("*Wollkraut*") which, however, is harmless. The prosecution alleged that the accused only mistakenly used Verbascum, and instead wanted to use henbane ("*Bilsenkraut*"), which would have been harmful. The prosecution argued—in accordance with the law—that since the defendant believed she used poison, and since the intended outcome of this use was her husband's death, this should count as attempted murder.

Up to this point, chemistry and forensic toxicology was of no use. The Verbascum had been identified by means of botanical identification, and since it is not poisonous, there were no poisonous substances to be found by chemical analysis. Nonetheless, the experts—a physician whose name unfortunately remained unreadable, and a pharmacist called Franken,[57] both from Stuttgart—conducted analyses for another possibly added substance, which might have been poison after all. The problem with organic poisons in general was that an elemental analysis could not yield a satisfactory result of which substance was used. Whereas with inorganic poisons, especially metallic poisons, the chemical analysis could find chemical elements where they do not occur naturally—such as arsenic in the digestive track—this was not the case for organic substances. Meaning the organic molecule could not be destroyed and its elemental parts analyzed, but rather the whole molecule had to be separated and isolated in order to be identified.[58] This was what the experts in the Flandern case claimed to have done to find poison. They asserted that they also had found coniine, the poisonous substance contained, for example, in hemlock.[59]

The defense attorney of Elisabeth Flandern asked the pharmacist Dr Haidlen to examine the first expert testimony of the prosecution, and to give his opinion on its reliability. It is this second expert testimony—which quotes extensively from the first one commissioned by the prosecution, and comments on the actions of the first experts—that is of particular interest here. As Haidlen states, the first experts started by separating the water-soluble parts of the food from the rest, which were then mixed with ethanol and "bitter almond" (i.e. probably benzaldehyde) and were boiled down multiple times. The material left after this procedure had "a disgusting, salty, pungent taste,"[60] and smelled similar to the urine of mice ("*Mäuseurin*"). This procedure was criticized harshly by Haidlen:

> The preceding course of the examination points to the experts' original aim to find some *non-volatile* poisonous substance, particularly a *non-volatile* alkaloid, in the vegetables. If they had thought of the presence of a *volatile* poison [such as coniine] at the beginning of the examination, they would not have boiled down the substance three times.[61]

This sets the tone for the rest of Haidlen's expert testimony. According to him, the actions of the first experts were not appropriate to claim the presence of coniine or some similar substance. All that substantiated their claim was the smell of mouse urine which was usually used, as Haidlen admitted, to describe the smell of coniine in chemical textbooks. However, he continued:

> This conjecture [that coniine was present] comes too late since if a volatile alkaloid and supposedly coniine was present in the vegetables, nothing would have been more suitable than the experiments of the experts to evaporate the largest part of it, and thus, to make a complete analysis impossible ....[62]

Notably, Haidlen did not attack the observations of the experts, not even their later experiments which seemingly corroborated the presence of coniine. His critique was based completely on his assessment of the first actions of the experts. These actions made all the other experiments useless. Yet contrary to the request of the attorney and for reasons still unknown, Haidlen was not summoned to court to publicly discuss his testimony. However, Christian Gmelin (1792–1860), professor of chemistry and pharmacy in Tübingen, was summoned instead, and the notes of the session summed up his opinion in the following way: "He was not able to decide what the experts might have found in the vegetables. It was no coniine."[63] He continued, citing the same reasons as Haidlen, that coniine is a volatile substance and the process of boiling down the sample would have been detrimental to finding coniine.

In the defense's rebuttle of the accusation against poisoning, all that was criticized were the actions of the first expert group, which were depicted as not suitable to support their claims. Any mistakes highlighted in the science were attributed to the individual experts and their actions, leaving the general credibility of forensic toxicology and chemistry untouched.[64] Such practical disagreements, I argue, were easily solvable either by the assessment of other chemists, or by repeating the experiment—which was however not necessary in this case. By focusing on analytical catalogues and empirical observations, the core science itself was not questioned in the courtroom. What can also be seen in this case is that for solving such problems authority and hierarchy among the experts themselves was important. It was not the pharmacist Haidlen who was summoned to criticize the first expert testimony in court, but Gmelin, a university professor for chemistry and pharmacy. To some extent, this approach also mirrors judicial practice, where in case of appeals or cassations higher judicial instances reexamine the facts or reinterpret the underlying law.

Presenting Chemical Practice

The special status forensic toxicologists claimed for themselves in poisoning cases was tightly connected to their own focus on practice. As Ian Burney[65] and I[66] have argued, the success of toxicological methods relied—on the one hand—on their ability to show in a visual sense the murder weapon, making the invisible crime of poisoning visible. On the other hand, however, it wasn't only the results that were observable, but also the specific actions of the toxicologists that led to the interpretable facts. As exemplified in the Ruthardt case, the expert testimonies were very detailed in describing all the actions which happened in the laboratory, while explaining none of them. In this sense, the expert testimonies resembled accounts of eyewitnesses, limited to the things an observer could see. The crucial difference between experts and eyewitnesses was that the former were able and allowed to interpret their observations in court, and their actions were needed to make the observations possible in the first place. The judges and other nontrained persons could see the changing color of a copper rod, but only experts were able to perform the Reinsch test. Chemical evidence could be seen by everyone, but only after chemical practice had made it visible.

The focus on practice also made it possible to confine expert disagreements to individual cases, as opposed to toxicological evidence practices as a whole. The Flandern case showed that, in case of a second (and differing) opinion, specific actions could and would be the target of criticism. Here, the actions of a forensic toxicologist had to be explained to a certain degree in order to show that they would not fulfill the intended purpose. However, there was no public disagreement on the theory, just on the ability of individual experts, thus not allowing any expert disagreements to publicly undermine the credibility of forensic toxicology as a whole. Individual mistakes were seen as a problem and there were calls in the textbooks for an educational reform of forensic toxicologists and, most importantly, for the state to employ professional legal chemists.[67] However, there were legal options to deal with questionable expert testimonies and practical disagreement seems to not have been as big a concern for the textbook authors as theoretical disagreement, which would not have been easily solvable. In the process of refuting another expert's testimony, the specific actions and practical knowledge were judged, not the theoretical beliefs, which were seen as "subjective"[68] and thus unimportant for the analysis. In this way, chemical analysis did not only produce material evidence, but was evidence in itself, which then could be reinterpreted or criticized in the same way that eyewitness accounts could be challenged by contradicting testimonies.

Acknowledgments

Financial support for this research was provided between 2017 and 2019 by a dissertation grant of the *Evangelisches Studienwerk Villigst*, and since 2019 by the DFG-project "Forensic Toxicology in Germany and France during the 19th Century: Methods Development in Judicial Context" (391910812).

Notes

1 "[Die Erscheinungen sind] unwiderlegbar die charakteristischen Vorzeichen der Einwirkung einer concentrirten Mineralsäure (der Schwefel-, Salz und Salpetersäure)." Krauß and Winter, "Expert Testimony in the Case of Margarethe Häfele, Tübingen, 26 May 1848," Archives of the Land Baden-Württemberg, Ludwigsburg (in the following abbreviated as LABW, StA LB), E 331 Bü 100, Qu. 26, 2; for the description of the case see also Prosecution, "Bill of Indictment of Margarethe Häfele, Tübingen, 9 August 1848," LABW, StA LB, E 331 Bü 100, Qu. 46. Unfortunately, other than their profession and last name, no additional information on the two experts in question is available from the sources.

2 "Die Schwefelsäure ist nämlich nicht wie Arsenik, Quecksilber und andere Metalle ein im Organismus untilgbarer Stoff; vielmehr wird sie, indem sie organische Substanzen zerstört, chemisch selbst zersetzt und hört dann auf, nachweisbar zu sein." Krauß and Winter, "Expert Testimony, Häfele Case," 3–4.

3 Although legal medicine too had its problems. See Hans-Heinz Eulner, *Die Entwicklung der medizinischen Spezialfächer an den Universitäten des deutschen Sprachgebietes* (Stuttgart: Ferdinand Enke Verlag, 1970), 159–179.

4 Christoph Meinel, "*Artibus Academicis Inserenda*: Chemistry's Place in Eighteenth and Early Nineteenth Century Universities," *History of Universities* 8 (1988): 89–115.

5 Axel Helmstädter, Jutta Hermann, and Evemarie Wolf, *Leitfaden der Pharmaziegeschichte* (Eschborn: Govi-Verlag, 2011), 108–114.

6 Ian Burney, *Poison, Detection, and the Victorian Imagination* (Manchester: Manchester University Press, 2012), 80.

7 Marcus B. Carrier, "The Value(s) of Methods in the Courtroom: Values for Method Selection in Forensic Toxicology in Germany in the Second Half of the Nineteenth Century," in *Global Forensic Cultures. Making Fact and Justice in the Modern Era*, ed. Ian Burney and Christopher Hamlin (Baltimore: Johns Hopkins University Press, 2019); Marcus B. Carrier, "The Making of Evident Expertise: Transforming Chemical Analytical Methods into Judicial Evidence," *NTM Journal of the History of Science, Technology and Medicine* (forthcoming).

8 Burney, *Poison*, 80.

9 "Zwar bedarf es zur Aufhellung der zweifelhaften Frage über geschehene Vergiftungen und den dadurch bewirkten Tod, auch der Betrachtung solcher Erscheinungen, welche am lebenden und todten menschlichen Körper wahrgenommen wurden, aber das Hauptmoment der Untersuchung betrifft immer die Auffindung und Beschaffenheit einer zur Vergiftung anzuwendenden giftigen Substanz. Denn der einzige unumstößliche Beweis einer geschehenen Vergiftung, ist das in dem Körper wirklich gefundene Gift." Adolph Henke, *Lehrbuch der gerichtlichen Medicin* (Berlin: Julius Eduard Hitzig, 1812), 329.

10 "Es können sowohl die vor dem Tode eintretenden Erscheinungen, als auch die Veränderungen der Leiche, möglicher Weise auch durch heftige und schnell tödtende Krankheiten hervorgebracht werden." Henke, *Lehrbuch*, 342.

11 "Indess ist es jedenfalls ein nicht zu rechtfertigender logischer Sprung, wenn man hiernach behauptet, *nur* die Auffindung des Giftes gäbe die Sicherheit der Diagnose, indem man hiernach die Zwischenmomente und unterstützendenden Beweise ausser Erwägung lässt, … und so hinsichtlich der Vergiftung ein Verfahren lehrt, wie es in der ganzen übrigen medicinischen Diagnostik mit Recht verworfen ist." Johann Ludwig Casper, *Practisches Handbuch der gerichtlichen Medicin*, vol. 2 (Berlin: Verlag von August Hirschwald, 1860), 408 (emphasis in the original).

12 For example, Bernhard Brach, *Lehrbuch der gerichtlichen Medicin* (Cologne: Franz Carl Eisen, 1850), 379–380; see also Sacha Tomic, "Alkaloids and Crime in Early Nineteenth-Century France," in *Chemistry, Medicine and Crime. Mateu J.B. Orfila*

(1787–1853) and His Times, ed. José Ramón Bertomeu-Sánchez and Agustí Nieto Galan (Sagamora Beach: Science History Publications, 2006).

13 See, for example, Eduard Caspar Jakob von Siebold, *Lehrbuch der gerichtlichen Medicin* (Berlin: Theodor Christian Friedrich Enslin, 1847), 452.

14 John H. Langbein, *Torture and the Law of Proof: Europe and England in the Ancien Régime* (Chicago: University of Chicago Press, 2006), 4 f.

15 Langbein, *Torture*, 47.

16 Langbein, *Torture*, 50–55.

17 The following discussion of the course of events is based on the bill of indictment Prosecution, "Bill of Indictment in the Ruthardt Case, Esslingen, 30 July 1844," LABW, StA LB, E319 Bü 159, Qu. 47. See also Eva-Kristin Waldhelm, "Anklage Mord – Vergiftungsfälle im Königreich Württemberg: Forensisch-toxikologische Nachweisverfahren in Giftmordprozessen unter Berücksichtigung strafrechtlicher Bestimmungen und sozialer Aspekte," (PhD diss., Technische Universität Braunschweig, 2013), 81–86. https://publikationsserver.tu-braunschweig.de/receive/dbbs_mods_00054615.

18 Ruthhardt's physician (name unreadable), "Testimony, Stuttgart, 14 March 1845," LABW, StA LB, Bü 159 E 319, Qu. 16.

19 Katherine Watson, *Poisoned Lives. English Poisoners and Their Victims* (London: Hambledon and London, 2004), 33.

20 "Vielleicht werden die meisten Giftmorde durch Arsenik ausgeführt, was wohl daher rührt, dass die tödliche Wirksamkeit desselben, schon in ganz kleinen Gaben, gemeinkundig, und er auch leichter, als andere giftige Stoffe, unbemerkt beizubringen ist." Ignaz Heinrich Schürmayer, *Lehrbuch der Gerichtlichen Medicin. Mit Berücksichtigung der neueren Gesetzgebung des In- und Auslandes, insbesondere des Verfahrens bei Schwurgerichten* (Erlangen: Ferdinand Enke Verlag, 1854), 242.

21 Franken and Schmidt, "Expert Testimony in the Ruthardt Case, Stuttgart 13 May 1844," LABW, StA LB, E 319 Bü 159, Qu. 7.

22 "Was soll z. B. der Richter bei einer complicirten und langedauernden chemischen Untersuchung thun, ausser sich langweilen, wenn er nicht etwa zufällig durch Besitz chemischer Kenntnisse persönlich Interesse daran nimmt!" Schürmayer, *Lehrbuch*, 20.

23 Legislation on this matter differed from state to state and changed over time. See Enno Poppen, *Die Geschichte des Sachverständigenbeweises im Strafoprozeß des deutschsprachigen Raumes* (Göttingen: Musterschmidt Verlag, 1984), 225–235, 275. Furthermore, these legal prescriptions were not always followed in practice. In the case discussed here, however, there were two pharmacists and a total of eight persons present to witness the analysis.

24 "Eine Anzahl kleiner weißer, theils gelblich gefärbter, sandartiger Körner …, deren Ansehen den Verdacht von weißem Arsenik erregten." Franken and Schmidt, "Expert Testimony Ruthardt."

25 "Die Probe auf Arsen nach Hugo Reinsch." Franken and Schmidt, "Expert Testimony Ruthardt."

26 Hugo Reinsch, "Ueber das Verhalten des metallischen Kupfers zu einigen Metalllösungen," *Journal für Praktische Chemie* 24, no. 1 (1841).

27 Hugo Reinsch, *Das Arsenik. Sein Vorkommen, die hauptsächlichsten Verbindungen, Anwendungen und Wirkungen, seine Gefahren für das Leben und deren Verhütung, seine Erkennung durch Reagentien, und die verschiedenen Methoden zu dessen Ausmittelung, nebst einer neuen von Jedermann leicht ausführbaren zu dessen Auffindung* (Nürnberg: Johann Leonhard Schrag, 1843).

28 "Einige Körnchen wurden nun in ein kleines Cylinder-Gläschen gebrachte, und hiermit die Probe auf Arsenik nach Hugo Reinsch … angestellt. Zu diesem Zwecke wurde in dasselbe Gläschen etwas Salzsäure mit destillirtem Wasser gegossen und nachdem ein blankes Kupferstäbchen in dasselbe hineingestellt war, jenes über der

Weingeistflamme bis zum Kochen erhitzt, worauf man einen eisengrauen Ueberzug über dasselbe, so weit es in die Flüßigkeit eingetaucht war, erkannte." Franken and Schmidt, "Expert Testimony Ruthardt."

29 Reinsch, "Verhalten," 247–250.

30 Reinsch, "Verhalten," 245.

31 "Wobei das Kupfer in ein Eisenstäbchen umgewandelt zu werden scheint." Reinsch, "Verhalten," 245.

32 "Woraufhin sich [das Kupfer] wieder mit einem ganz ähnlichen Beschlag belegte." Franken and Schmidt, "Expert Testimony Ruthardt."

33 For the use of practices of comparison in toxicological expert testimonies see also Carrier, "Evident Expertise."

34 Franken and Schmidt, "Expert Testimony Ruthardt."

35 See Carrier, "Value(s)," 46.

36 James Marsh, "Account of a Method of Separating Small Quantities of Arsenic from Substances with Which It May be Mixed," *The Edinburgh New Philosophical Journal* 21 (1836); on the Marsh test see also Katherine Watson, "Criminal Poisoning in England and the Origins of the Marsh Test for Arsenic," as well as José Ramón Bertomeu-Sánchez, "Sense and Sensitivity. Mateu Orfila, the Marsh Test and the Lafarge affair," both in *Chemistry, Medicine and Crime. Mateu J.B. Orfila (1787–1853) and His Times,* ed. José Ramón Bertomeu-Sánchez and Agustí Nieto Galan (Sagamora Beach: Science History Publications, 2006).

37 Jöns Jakob Berzelius, *Jahres-Bericht über die Fortschritte der physischen Wissenschaften. Eingereicht an die schwedische Akademie der Wissenschaften am 31. März 1837,* trans. Friedrich Wöhler (Tübingen: Heinrich Laupp, 1837), 192.

38 "[Das Gas] verbrannte mit der eigenthümlichen Farbe des Wasserstoffgases." Franken and Schmidt, "Expert Testimony Ruthardt."

39 "Die arsenhaltige saure Flüßigkeit." Franken and Schmidt, "Expert Testimony Ruthardt."

40 "Bedekte sich wiederholt mit einem braunschwaren, glänzenden Anflug." Franken and Schmidt, "Expert Testimony Ruthardt." Note that, strictly speaking, the experts already had anticipated the result before describing the procedure by describing the sample as "arsenical." However, this is only an interesting technicality since in most other testimonies, I have found the experts were very careful in describing the tests without articulating such assumptions before interpreting the results.

41 "Wodurch augenbliklich die schwarzgraue Farbe in eine schön gelbe sich umänderte." Franken and Schmidt, "Expert Testimony Ruthardt."

42 Samuel Hahnemann, *Ueber die Arsenikvergiftung, ihre Hülfe und gerichtliche Ausmittelung* (Leipzig: Siegfried Leberecht Crusius, 1786).

43 In modern chemical terms: arsenic pentasulfide.

44 "Daß die fraglichen in dem Leichnam ... aufgefunden ... festen Körnchen Arsenik sind." Franken and Schmidt, "Expert Testimony Ruthardt."

45 "Es liegt in der Natur des Gegenstands [d. h. der forensischen Chemie], dass in einem Buche, welches ausschliessend [sic] judiciellen Zwecken dient, jede theoretische Schulansicht ausgeschlossen ist. ... Wenn der Vertreter des Rechtes Aufschlüsse von Sachkundigen fordert, so richtet er seine Frage nicht so sehr an die Person des Sachverständigen, als an die Wissenschaft, die letzterer vertritt, man soll daher nicht sein subjectives Meinen statt der objectiven Erkenntnis geben." Franz Schneider, *Die Gerichtliche Chemie für Gerichtsaerzte und Juristen* (Wien: Wilhelm Braumüller, 1852), 1–2.

46 For example, Marsh, "Account," 207–208.

47 William H. Brock, *The Fontana History of Chemistry* (London: Fontana Press, 1992), 210–240; Hasok Chang, *Is Water H$_2$O? Evidence, Realism and Pluralism* (Dordrecht: Springer, 2012), 133–201.

48 Burney, *Poison*, 87.

49 Burney, *Poison*, 87.

50 Carrier, "Evident Expertise."

51 See also Lissa Roberts, "The Death of the Sensuous Chemist: The New Chemistry and the Transformation of Sensuous Technology," *Studies in History and Philosophy of Science Part A* 26, no. 4 (1995).

52 For an example of a case in which the experts relied completely on smell as evidence for arsenic see Eblen and Halz, "Expert testimony in the Maier case, Weil der Stadt, 5 June 1809," LABW, StA LB, D 70 Bü 27.

53 See, for example, José Ramón Bertomeu-Sánchez, "Chemistry, Microscopy and Smell: Bloodstains and Nineteenth-Century Legal Medicine," *Annals of Science* 72, no. 4 (2015).

54 Anonymous, "Death Sentence in the Ruthardt Case, Esslingen, 23 December 1844," LABW, StA LB, E 319, Bü 159, Qu. 84.

55 Anonymous, "Record of the Execution of Christiane Ruthardt, Esslingen, 27 June 1845," LABW, StA LB, E 319, Bü 159, Qu. 111.

56 The following description of the case is based on the bill of indictment: Prosecution, "Bill of Indictment in the Flandern Case," LABW, StA LB, E 319 Bü 161, Qu. 5.

57 It may well be that this was the very same pharmacist as in the Ruthardt case. However, except for the name and the pharmacy being in Stuttgart, there is nothing else to support this theory.

58 Tomic, "Alkaloids," 283.

59 Anonymous and Franken, "First Expert Testimony in the Flandern Case," LABW, StA LB, E 319 Bü 16, Qu. 39.

60 "Einen ekelhaften, salzigen, scharfen Geschmack." Haidlen, "Second Expert Testimony in the Flandern Case, Stuttgart, 14 November 1852," LABW, StA LB, E319 Bü 161, Qu. 29.

61 "Vorstehender Gang der Untersuchung weist auf die ursprüngliche Absicht der Experten hin in dem Kartoffelgemüse einen etwa vorhandenen *nicht flüchtigen* giftigen Stoff, insbesondere ein *nicht flüchtiges* Alkaloid zu finden. Denn hätten sie schon bei Beginn der Untersuchung an das mögliche Vorhandensein eines *flüchtigen* Giftstoffes [wie Coniin] gedacht, so hätten sie nicht 3 aufeinanderfolgende Abdampfungen der Masse ... vorgenommen." Haidlen, "Second expert testimony Flandern." (Emphasis in the original)

62 "Diese Vermutung kommt aber zu spät, denn wenn ja ein flüchtiges Alkaloid u. vermuthlich Coniin in dem Gemüse war, so war nichts geeigneter als die Versuche der Herren Experten dasselbe zum größten Theile in die Luft zu jagen." Haidlen, "Second Expert Testimony Flandern."

63 "Was die Sachverständigen in dem Kartoffel-Gemüse gefunden hätten, sei er [Gmelin] nicht im Stande zu entscheiden. Coniin sei es nicht gewesen." Anonymous, "Record of the Flandern Trial, Esslingen, 19 November 1852," LABW, StA LB, E319 Bü 161, Qu. 36, Addendum 3.

64 For a similar case where this was not the case and where the expert successfully defended (at first) his practical claims against theoretical impossibility see James C. Mohr, *Doctors and the Law. Medical Jurisprudence in Nineteenth-Century America* (Baltimore: Johns Hopkins University Press, 1993), 122–139.

65 Burney, *Poison*, 78–115.

66 Carrier, "Value(s)," 50–52.

67 See, for example, Theodor Husemann, and August Husemann, *Handbuch der Toxicologie* (Berlin: Verlag Georg Reimer, 1862), 113.

68 Schneider, *Gerichtliche Chemie*, 2.

4 No "Mere Accumulation of Material": Fieldwork Practices and Embedded Evidence in Early (Latin) Americanist Anthropology

Julia E. Rodriguez

In 1888, Adolph Bandelier, the Swiss-American anthropologist known for his pioneering work in Mexico, the Andes, and the US Southwest, addressed a group of scholars at a meeting of the International Congress of Americanists (ICA) in Berlin. Describing the results of the Hemenway Southwestern archaeological expedition, Bandelier introduced his own eclectic methodology, noting that he was putting together material "to serve as guides for certain parts of the explorations, inasmuch as they embody whatever knowledge had of the regions to be investigated."[1] He urged his colleagues to cast a wide net in the search for traces of American civilizations:

> A mere accumulation of material for use of future generations, however laudable it is, could not satisfy the immediate practical wants of the Expedition. *It is obvious that there should be an intimate connection between the Archives and the field-work proper.*

This connection—developed and institutionalized in the twentieth century—would be facilitated by scientists' practices of mapping, contextualization, and first-hand observation decades earlier.

In the last decades of the nineteenth century, a diverse group of scientists, including archaeologists, ethnologists, and natural historians, built the expectations of being in place; presenting and analyzing their evidence in context; and finally having intimate sensory contact with objects and informants. In the process, Latin America was transformed from a site of extraction to a site of embeddedness; from a venue of collection and display to one of evidence.[2] This shift had implications for anthropology (and later, other social scientists) as an intellectual project. It was also consequential on the ground, as it brought foreign and local scientists into intimate contact with indigenous people and their present-day situations in new ways.

Gaining access to the expansive and rich landscapes of the Americas—especially in Latin America—scientists like Bandelier initiated

DOI: 10.4324/9781003188612-5

and then deepened a set of research practices in the field, creating a feedback loop between production and the application of knowledge.[3] This dynamic process encouraged these early anthropologists of the Americas to seek and rely on local knowledge in particular as both authentic and authoritative. Embedded evidence was both the goal and the result of Americanists' emergent fieldwork practices. Local knowledge was their most valuable capital.

The venue for Bandelier's speech was the seventh meeting of the ICA, a new professional organization which had been formed thirteen years earlier by a group of (mostly) European naturalists, historians, archeologists, linguists, and ethnologists as the first international organization devoted to—in their words—the "scientific" study of American civilizations. At their founding meeting in 1875, held in Nancy, France, the meeting's president, French archeologist Baron de Dumast, praised the participants as "benign luminaries ... an army of universal *peaceful progress* ... for the increase of knowledge."[4] These explorers saw themselves at the vanguard of an endeavor which, according to their bylaws, would carry out "ethnographic, linguistic, and historical" studies of American populations, with a focus on the remnants of pre-Columbian peoples' lives.[5]

Over the course of the following decades, anthropologists increasingly advocated for expeditions such as Bandelier's to uncover new discoveries and findings. In 1888, Sylvester Baxter, a Massachusetts journalist and public parks pioneer, published an account of the Hemenway Expedition entitled *The Old New World*.[6] This era saw more calls for in-person visits to the field. At the ICA meeting that same year, Baxter also extolled North American scientific expeditions, calling for "the investigation of the primitive people occupying a given section of country by ethnological, anthropological, and historical researches coordinately conducted under common auspices," and naming men like Bandelier, the US anthropologist Frank Cushing, and Dutchman Herman Ten Kate as examples of the new model of explorer-scientists.[7]

Americanism—a forerunner of modern four-field anthropology (a term applied after around 1900 recognizing the distinct approaches of archaeology, linguistics, physical anthropology, and cultural anthropology) and Latin American Studies—was a transnational project whose participants engaged with a bewildering array of scientific methods, discoveries, standards, languages, and cultures. The biannual meetings of the ICA were places where geographers, classicists, naturalists, physicians, linguists, and archaeologists mingled, shared ideas, and fought. A key characteristic of the Americanism of this moment was, in fact, the *lack* of consensus about the best type of evidence to solve the mysteries of the rise and fall of civilizations. Thus, the struggle over proper sources and methods was at the forefront of the first Americanist debates. Over time, it became clear that land, and evidence embedded in it, were central to the production of anthropological knowledge.

For much of the nineteenth century, natural historians interested in America had been divided between those whose examined objects in the laboratory or museum, and those who did so in the field. In the 1870s, despite a steady stream of collectors who had ventured to all corners of the earth, most European Americanists typically had little to no face-to-face knowledge of the places or people they wrote about. Many of them relied on collectors and informants in the field to ship artifacts to their offices. Further professional tension existed between those anthropologists who believed first and foremost in assembling abundant facts, and those who theorized from available data. This conflict, described by Kuklick with reference to early twentieth century anthropology, was present among early Americanists as well, but nonetheless the hunt for scientific treasure proceeded on various fronts, and there was a general tendency toward gathering as much "evidence" (both material objects and cultural knowledge) as possible.[8]

Up to that point, there had been a stable of competing ideas about human physical and cultural evolution with only a trickle of new evidence. The "Old New World" promised abundant new material to help solve these universal puzzles. Westward expansion in North America made available new land for scientific expeditions. At the same time, the newly accessible nations and regions in Latin America, with their diversity, wealth in natural material, and centuries of collected local knowledge, had the potential to change the state of universal knowledge.[9] Early Americanists were explicit in their view that the tools of the old sciences, combined with a wide-open field of largely untouched natural raw material could provide a virtually infinite amount of new knowledge and truths about the nature of humanity.[10] As Mexican scholar José Ramírez stated in 1895, the material found there, including "land, plants, animals, people, the remains (*restos*) of other flora and fauna, the arts, sciences, customs, and institutions" was "exclusively American." He continued, "None of this was sent in cargo from China or Carthage."[11]

Latin America had in fact been a rich source of ethnographic knowledge production as early as the sixteenth and seventeenth centuries, in the form of *relaciones*, or official reports to the King, as well as missionary documents. Then, in the late eighteenth and early nineteenth centuries, naturalists traveling through Spanish and Portuguese America, including Alexander von Humboldt and Charles Darwin, produced first-hand accounts of human physiology, language, and behavior, in addition to flora and fauna.[12] In the late nineteenth century, a new wave of nationally sponsored expeditions to study raw material in Latin America commenced, thanks to a few factors: newfound stability after decades of postcolonial wars and instability; a European scientific "arms race," leading to investment in expeditions; and new levels of consolidation and centralization of state power in many Latin American countries, which prompted intensified projects of territorial control and new forms of contact with Indigenous peoples.

Practices of finding, observing, gathering, and analyzing evidence in this setting were key to the professionalization and development of norms in anthropology. In the period under consideration here, approximately 1880–1910, Americanists changed "what counts" as anthropological evidence, albeit not in a straightforward way. Examining the complex interactions between historical actors and the substances of land, landscape, and soil illustrates the importance of context and the contingent nature of measures to establish scientific authority. At the same time, these encounters reflect the changing local, national, and geopolitical power relations in this period. Like other professional and academic fields, anthropology constructed a system for the authentication of evidence—including specialized knowledge and first-hand verification; a process that began long before the twentieth century, the heyday of modern anthropology. The practices generated consensus ideas, not the other way around.

This chapter pulls at one thread of this multifaceted process of consensus building around authority of evidence, namely, the significance of American land as a central channel to gold standard anthropological methods. While Americanists hungered after multiple sources of knowledge (including fossils, human remains, artifacts, monuments, languages, music, and stories), land was connected in one way or another to all of them. Among the many types of potential evidence available to early anthropologists, land was at once accessible and arguably, the richest medium, literally presenting multiple layers to be mined for data. Land along with some of the visible human transformations of the landscape and objects buried in soil—including caves, mounds, fossils, shards, and monuments—revealed the vestiges of human activity that, separately or together, promised to provide the keys to some of the most profound mysteries of human existence. I will argue that anthropologists' eventual emphasis on *place* plus *context* plus *contact* resulted in new definitions of scientific authority in their field and beyond. By 1910, the soil itself could confer authenticity not just to artifacts, but also scientific legitimacy to practitioners and their findings.

Most historians of anthropology locate fieldwork practices in the twentieth century, and in the British and US schools of anthropology, especially in the iconic fieldwork carried out in the South Pacific. Here I will shift our attention to the Americas, and in particular Latin America, as a crucial site in the development of fieldwork practices. A close examination of early Americanists' work in the field reveals that fieldwork practices were (and are) mediated by dynamics of place. Place is layered, with actors cutting through layers of space, from the local to the national to the transnational. Power dynamics infuse human relations through domination of land, not just through seizure and ownership, but also in the process of extracting local knowledge. The ability of outsiders to claim authentic knowledge in place increased their scientific authority.

Americanists approached land in three different ways: as landmass, landscape, and soil. In this chapter, I first take on landmass writ large: maps and geography, which at once reflected scientists' working out of abstract theories of human migration and culture, and provided a practical means of longer and more direct engagement with the land. Next, I turn to scientific analyses of landscape, which led to contextualization and close examination on the ground. Finally, I examine the third level of scientists' land practices, with buried evidence in layers of soil, including the earth itself and objects embedded in it. These practices built upon themselves, leading scientists to emphasize place, context, and eventually, personal contact with their evidence. This pattern was the result of complex interactions between people and objects, and between multiple actors of varying positions. But from the scientists' point of view, without a doubt, those with access to local *materia* had an advantage.

"Let's Go Marching on American Soil": Landmass and Mapping

At the most abstract level, late nineteenth-century human scientists beheld the American continent as a landmass that could yield new insight into the story of long-term cultural developments. Americanists in the 1870s and 1880s tirelessly reexamined various aspects of First Contact, scientific discovery, mapping, place naming, as well as speculation about pre-Columbian human migration; a preoccupation that reflected the Eurocentric nature of the Americanist project in its formative years. In this endeavor, they took on the largest time frame imaginable: the Indigenous settlement of the continent. This attempt to pin down the timeline and course of hemispheric migration had a large scope, going back to ancient, even prehistoric times.

A rich array of mapmaking precedents existed in the Americas, but mapping the American continent was also complicated by recognition of nations and tribes with sovereignty.[13] Colonial Mexican cartographers included Indigenous people on their maps, whereas North Americans did not. Historians have established the extent to which Indigenous maps influenced nineteenth-century Mexican geography and history. This process began immediately after representatives of the Spanish crown set foot in Mexico; not only did the invaders rely on local knowledge of the landscape for their own ends, but figures like Hernán Cortés included Indigenous maps in the bounty shipped back to the King. They also valued maps as artifacts, prominently displaying them at national and international scientific exhibitions.[14] Scientists relied on Indigenous mapmaking practices as well as content. Art historian Barbara Mundy has shown how modern Mexican cartographers borrowed Indigenous practices, such as conveying narratives, often portraying human agents and depicting scenes from history.[15]

Eventually, maps transformed from something abstract to concrete tools for evidence-gathering. Local geography professionals and organizations had been surveying landscapes long before European expeditions. And

Latin American scientists had been active in mapping, geography, and geological work—with their own agendas—since the beginning of independent Republics in the early nineteenth century.[16] But both national scientific elites and foreigners needed to be in place, and both prioritized surveying and mapmaking.

Two distinct national approaches to mapping illustrate the extent of on-the-ground geographical work in the timeframe in question. First, in Mexico, the postcolonial Mexican state in this period sought to increase state capacity and authority in order to solidify state power to tax or administer over the entire country. Mexican geography was important to this state expansion. The nation had established numerous institutions from the 1850s onward that produced surveys of land, people, and resources.[17] One of these early, and widely distributed, geographic projects was the *Atlas geográfico, estadístico é histórico de la República Mexicana*, published in 1858 by Antonio García Cubas. His work, like many Latin American geographers, built on Humboldtian visualizations, especially his depiction of mountains and major geological features, as well as the use of graphs on maps. Mexican geography was then an applied science in service to national goals. Creating detailed images and maps with scientific authority was key to that process.

In 1885, he produced another atlas, this one even more detailed. The *Picturesque and Historical Atlas of Mexico* provided 13 detailed maps of the Republic of Mexico organized and illustrated along themes such as ethnography, transportation networks, natural resources, archaeological sites, and historical figures. By layering types of information and images and incorporating types of data simultaneously, García Cubas forwarded a "holistic" idea of the nation. Mexico is literally at the center of each page, emphasizing the importance of the nation.[18] García Cubas dedicated a signed copy of his 1885 *Atlas* to the Smithsonian Institution and deposited it at the US Library of Congress.[19] But above all the atlas was intended for Mexicans. Combining images of various aspects of the nation, he looked to the future of Mexico as a nation on the move, destined for progress and embracing tropes of modernity. But this developing modernity was only possible with the natural landscape, and the historical and archaeological past.

While in nineteenth-century Mexico, the state used mapping to manage a settled, populous, and diverse nation, in Argentina, scientists were more likely to engage with large areas they frequently (and disingenuously) described as "empty."[20] In Argentina, scientists modeled themselves on the rich traditions of naturalists and geographers in the region, but their work took place in the context of Argentine state expansion and the ongoing frontier wars. Economic motivations dovetailed with the cultural agenda of national scientists. Unlike much of the rest of Latin America, in Argentina, borders were not yet fully established. By the 1870s, centralized state governments were attempting more systematic scientific surveys of borders, territories, and inhabitants that had been previously limited and incomplete. Sometimes the main goal was to stake out national boundaries to avert (or provoke) conflict with a neighboring country. At other times the explicit goal of the expedition was to reach remote tribes or populations and assess their level of threat to plans for national development.

There were border excursions throughout the last three decades of the nineteenth century to establish boundaries between Argentina and Chile. In December 1892, Francisco P. Moreno wrote to the Minister of Foreign Relations, Tomás de Anchorena, emphasizing the importance of careful surveys of the national landscape. Moreno wrote, "I think that every true Nation must know thoroughly the ground on which it develops .... It must, by all possible means [share with] its children knowledge of the native soil...." He continued, "Our civil strife has the same origin because we have ignored the physical environment in which men have developed."[21] Many Argentines called the region south of Buenos Aires, including Patagonia, "the desert"—implying sterility and emptiness.[22] However, the landscape was anything but empty. Recurrent military campaigns contained embedded scientists' intent on capturing and preserving ethnographic and archaeological artifacts, if not living specimens. The new expeditions had a national purpose, to gather symbols of the nation, to incorporate native cultures as relics of the past. At the same time, it was clearly advantageous for scientists in Argentina to regard Indigenous peoples' land as open territory. Their drive for evidence in place led them to claim Indigenous people as absent from or abstract to national space.[23]

Scientists from outside Argentina had a long-standing interest in the Southern territories' wealth of natural historical and paleographic material. At first, Americanists were preoccupied with large scale geographic and climate studies to prove theories of human origins. The abstract approach to American landmass was exemplified by John H. Beeker in 1877 in his work on the migrations of the Nahuas in the context of large-scale historical events. He argued that, as "in all cases of conquest," examining large-scale land formation could provide evidence of human cultural evolution:

> The dominant force 1) ... moves from a country of severer climate into a country of milder climate; 2) from a high or mountain land into the lowland; 3) from the dry steppe into the humid and fertile plain; 4) from the home of a barbarous nation of equal natural brain and physical power towards the seats of higher civilization and refinement; 5) from the home of a nation of superior average brain capacity towards the lands of nations of less average brain capacity.[24]

Mapping practices also had practical applications, as they facilitated firsthand presence on the land. Bandelier wrote in 1880 to Morgan that accurate, up to date maps were vital to his fieldwork:

> Now let me speak again of some very necessary details. I do not want to start all alone, if possible, but if you cannot come along yourself,

then give me a trustworthy boy who can *draw or photograph*, at least. I should insist upon such a companion at all events, since I am no draughtsman myself .... Another point of still greater importance yet is that I must have maps or charts of the country which I shall have to explore, in order to map down at once every ruin which is met with, and its character.[25]

Like their Northern colleagues, Mexican scholars took the continental landmass as the unit of analysis, seeing the region in abstract terms as a site of long-term mass migration. But they too noted the importance of having feet on the ground, and that maps could bring them to the right place. In 1895, Abraham Castellanos (of Oaxaca) suggested that in order to solve the origin of American populations [*pueblos*],

Let's go marching on American soil to expose our theory. From the arctic regions to the final southern crags of America, we can distinguish two essentially distinct races. One, daughter of a race crossed with the indigenous [*naturales*] of the land; the other, encountered as pure in the environment of the virgin jungles of America, where time has been unable to make them lose the traces of an unknown past.[26]

While he issued a general call for scientific studies of his homeland, Castellanos also made it clear that local scientists had inside knowledge of the land.

"A Beautiful Piece of Land, Fertile and Exuberant": Scientists and American Landscape

While they debated the continent in broad, general terms, Americanists also zoomed in at ground level. The result was more detailed descriptions of landscape and geology in the spirit of eighteenth-century naturalists, but shaped by late nineteenth-century sensibilities. It dawned on scientists that they needed to be in place to understand the material they worked with and presented to the world. At the same time, an influx of scientists engaged with the landscape went beyond an appreciation of being place. It also made them realize the importance of viewing their evidence in context. Thus, evidentiary practices in expeditions led to new forms of collecting, not just for display in cabinets but for making arguments with con-textualized data related to its specific location.

At the same time, geopolitical dynamics provided the backdrop to Americanists' absorption in landscape. There was a long history of colonial and neocolonial incursions into American landscape. Early nineteenth-century naturalists in particular were one of the main inspirations and sources for Americanists a few generations later. The best-known European naturalist to travel in Latin America was Alexander von Humboldt, who traveled to the New World in 1799. His time in America resulted in a multivolume set of publications with his detailed observations of South America, Mexico, and the Caribbean islands. (His importance to Latin Americans is reflected in the numerous sites named for him.) Humboldt's holistic approach to the natural world encompassed not just geology and botany, but also political and social description. Summarizing the contents of his report on New Spain (later Mexico), he listed "considerations on the extent and natural appearance of Mexico, on the population, on the manners of the inhabitants, their ancient civilization, and the political division of their territory."[27] Similarly, his study of the Cordilleras would "throw some light on the ancient civilizations of the Americans, through the study of their monuments of architecture, their hieroglyphics, their religious rites, and their astrological reveries."[28] A few years later Charles Darwin would spend about five years traveling through South America, the site of his most consequential observations, including observation of finches in the Galápagos that were the foundation of his theory of natural selection.

Throughout the nineteenth century, the Spanish and French governments sponsored scientific missions in the nineteenth century, and the new American republics sponsored their own naturalists, geography, and surveying expeditions.[29] Spanish monarchs had for centuries sent emissaries to their colonies (and later, their former colonies) to collect natural specimens for museums and other collections back home. Long before the better known journeys of Humboldt and Darwin, scores of Spanish (and Portuguese) scientists mapped the landscape, gathered, described, and drew plants and animals, and studied Indigenous languages. In the mid-nineteenth century, these expeditions contributed to the shift from cabinets of curiosity to scientific museums.[30]

A four-year Spanish scientific expedition to Latin America in 1862 predated the founding of the ICA.[31] One of the scientists on this mission was Marcos Jiménez de Espada, a zoologist at the University of Madrid, who was later active in the ICA. It was on the Spanish expedition that Jiménez de Espada shifted his focus to anthropology, eventually becoming a major scholar of Inca civilization. The 1866 expedition also had a Cuban member (at this point, Cuba was technically part of the Spanish Empire), Manuel Almagro y Vega; he was the ethnologist on board, and had trained in medicine in both Havana, Madrid, and Paris. Miller claims that

> Almagro was one of the first trained professional anthropologists to do field work in America, where he excavated archaeological sites and studied aboriginal life from Patagonia to Panama. He was responsible for sending hundreds of Indian artifacts to Spain, along with skulls, skeletons, and mummies.[32]

At this time, France was a leader in European science, and had sponsored naturalist expeditions to benefit collections and French anthropology.[33]

It was moreover motivated by ongoing interests in both formal and informal imperial expansion in America.[34] The French invasion of Mexico in 1861, accompanied by a state sponsored scientific expedition in 1864 (the *Commission scientifique du Méxique*) gave French Americanists further political and practical momentum and incentives. Later organizations would specialize further, culminating in the 1895 *Société Américaine de Paris* that was founded by Ernest Hamy to finalize the split from Orientalism.[35]

As transnational Americanists from various countries began exploring the landscape in closer proximity, their insertion into the landscape took various forms: natural historical surveys, expeditions for museum collections, and national scientific expansions subsidized by local governments and land developers. To cite an early example of recording the landscape, in 1883, Vicente de Vera y López, a Spanish participant, described "The variations in physical geography of the American continent from the era of the discovery to our times," noting that "there have been great variations [in the geography] in a relatively short interval of four centuries .... These variations on American soil are very extensive and important, more than one would think on first glance."[36] He cited changes in sea level and features of coastal areas in various parts of the Americas, including Florida, the Delaware River basin, and the Caribbean.

Europeans were far from the only naturalists to measure, describe, and draw the landscape. In the late nineteenth century, they increasingly confronted local scientific competitors in the field, as well as nationalist views of landscape. In the late nineteenth century, scientists' work conditions in the field were increasingly limited by new forms of sovereignty, autonomy, and nationalism throughout Latin America. While government mapping and scientific expeditions led to detailed descriptions of landscape in Latin America, the processes and outcomes were shaped by particularities of national context. Years of civil wars, political instability, and a slow recovery from the devastation of independence wars in the early nineteenth century resolved by 1880 in favor of centralized, modernizing states and more robust economies, especially in large nations like Mexico and Argentina. State leaders believed that key to the modernization project was investment in scientific institutions, including domestically run geographical and anthropological expeditions. Again, these nations in Latin America illustrate distinct experiences with close description of landscape, shaped by particularities of place and national culture.

In Argentina, the government invested in scientific surveys to strengthen its hold on "unconquered" territories as well as the border with Chile. As noted earlier, state official believed it crucial to present the Argentine hinterland as "empty," or in the particular parlance of the day, a "desert." In 1876, the Argentine delegation to the Philadelphia exposition published a report on Argentina, including detailed measurements and descriptions of the topography, climate, geology, plants, animals, as well as descriptions of government and social institutions. The report stressed Argentina's size,

natural resources, and diverse but mostly unconquered indigenous communities. The sheer size and dimensions of the territories was a frequent
refrain, referred to as "immense" and "considerable distances," that, for
example, "make impossible any attempt to surprise the Pampas Indians."
They also described "great natural obstacles" such as rivers and mountains.
The landscape provided an obstacle to pacification, for example in the
fertile plains surrounding Buenos Aires, where

> the complete submission of the savages of the Pampa not only presents
> immense difficulties, but ... almost all the advantages remain for the
> Indian .... Thanks to the immensity of these uncultivated plains, their
> groups of [Puelche and Ranquelche] riders almost always manage to
> approach the lines of frontiers unnoticed.[37]

Argentines dubbed this region "the desert"—implying sterility and emptiness. However, the landscape was anything but that. "The Indian, in
general, knows how to make the military expeditions to the *tolderías* [in-
digenous settlements] equally unfruitful. Moreover, his great ally, *the desert*,
supports him in a powerful way."[38] As a result, it was necessary for scientists
in Argentina to regard Indigenous peoples' land as open territory. As in
other parts of the Americas, the drive for embedded evidence led them to
claim Indigenous people as absent from or abstract to national space.

An interesting counterpoint was seen in Mexico, where the late
nineteenth-century state was equally invested in mapping and geography, but
in a different national context. Unlike in Argentina, the Mexicans understood their landscape to be "full"—inhabited and shaped by Indigenous and
mestizo peoples in all their diversity. There are various traditions and habits of
depiction in landscapes; as historian Jorge Cañizares-Esguerra notes, while
many Western European and North American nineteenth-century landscapes were devoid of humans, in Mexico this is not the case.[39] Geographical
projects led to thick descriptions of landscape by Mexicans, such as García
Cubas's widely read *Picturesque and Historical Atlas* (1885).

At the same time, scientific nationalism and insistence on local expertise
coexisted with international engagement. Mexicans and other Latin
American scientists increasingly participated in the ICA meetings and
publications. At the 1910 meeting in Mexico City, local officials made
arrangements for excursions and field trips to sites like Teotihuacán and
Tepoztlán. Mexican experts emphasized the value of their proximity to the
land and intimate knowledge of their local landscape to augment their
authority on the international scientific stage. Mexican Francisco Flores
described the landscape in his report that year on the archaeology and
anthropology in Tepic:

> There exists, in the west of the Republic, bordered by the states of
> Jalisco, Sinaloa, Durango, and Zacatecas, and, bathed to the West by the

waters of the Pacific, a beautiful piece [girón] of land, fertile and exuberant like all things Mexican, crossed by the flowing Ameca, Acaponeta, San Pedro, and Santiago rivers, and populated by jungles with palms, rubber, mahogany, and other valuable trees of great value.[40]

Flores also speculatively considered Tepic's prehistoric setting too. "What was the territory of Tepic like in primitive times? Which races populated or passed through it, *leaving their traces in the soil?*"[41] In this case, investigations of American origins and civilizations required immersion in the local landscape.

The ability to connect pieces of evidence, that is, to contextualize them in place, was also increasingly seen as necessary to interpretation. Bandelier—who had early on advocated an eclectic and comprehensive approach to evidence, including texts, fieldwork, and local informants—stressed the importance of measuring and documenting objects in place. In 1880, he wrote to his mentor Henry Morgan:

> You can easily see that such ruins are valuable only as long as they are in their original places, that however many monoliths of Copan any U.S. museum may contain for instance, such columns are but misplaced there, because they must be studied in foreign surroundings or rather in environs foreign to their origin. This has been the great drawback with all "antiques," which can only be understood and rightly interpreted when examined where they were.[42]

Within a few decades, the contextualization of evidence was fully accepted as the norm. Correspondence between Smithsonian Institution bureaucrats in Washington DC and scientists in the field in Mexico reflect a new concern with provenance and place. In 1907, W.H. Holmes, head of the US Bureau of Ethnology, wrote to archaeologist Zelia Nuttall for confirmation about an ancient codex and map. Along with the items, Nuttall's collaborator Alice Fletcher added instructions to keep them together:

> A map upon native cloth was originally with this codex, and the person to whom this map now belongs ... will agree to donate the map, so that the two which were originally together, may remain together and so be preserved with one another.[43]

Nuttall, a long-term resident of Mexico and of Mexican descent, was in a strong position to provide information about provenance and context of artifacts. Her knowledge of local sites, the Spanish language, and Mexican mores made her a valuable collaborator for well-funded foreign scientists. By this time, it had become clear that a scientist's deep knowledge of place, familiarity with local culture, and connection to local people did not just facilitate access to sites, but also endowed them with the tacit knowledge necessary to identify and gather evidence itself.

"There Is Much to Exhume": Evidence Buried in Layers of Soil

As the number and intensity of geographic and archaeological expeditions in Latin America increased, scientists drew ever nearer to the land, even below it, as they sifted through geographical strata. This close proximity led them to particular sites, to examine ever smaller patches of land, and to penetrate the surface. These practices increasingly required extended stays and interaction with local *materia*, and eventually, people. After 1900, two shifts in the larger global context shaped developments; first, national scientific institutions intensified their competition for anthropological evidence, with some scientists better funded than others. Secondly, the appearance of new technologies engendered new visualizations of land, including measurements, photos, and drawings. The result was a privileging of state money and new technologies, both of which bestowed prestige and authority selectively.

Americanists centered in North Atlantic institutions sought to insert themselves in the field in Latin American settings. Regardless of nationality, many saw themselves as the next generation of Humboldtian scientists. On both sides of the Atlantic, those creators of landscapes took as their touchstone Humboldt's visual representations of American sites. One scientist characterized their best work in 1910 as resulting from "the intimate relationship between the name Humboldt and Americanist studies."[44] Like the earlier generations of naturalists, they realized that their study of human-nature interactions required direct contact with evidence in the field; an embrace of landscape detail built on naturalist methodologies honed over centuries by these international figures.[45]

In light of the relative wealth and power of foreign (especially American and German) anthropologists, scientists' interactions were characterized by tensions, with local scientists both advantaged and disadvantaged. Again, their practices were situated in specific places. Aside from the importance of contextualizing evidence, gaining admission to sites often depended on local officials, labor, expertise, and hospitality. While foreigners flooded to archaeological sites, Latin American scientists leveraged the outsiders' reliance on local scientists, laborers, government officials, and merchants.

Local intellectuals served as conduits to foreign scientists. On a group outing in 1895 to Tepoztlán, south of Mexico City, during the ICA meeting held there, local architect Francisco M. Rodriguez described the pyramid's appearance and construction in detail, including materials, design, appearance and layout of the exterior and interior. He noted in addition that,

In the area around this pyramid there is much to exhume, much that still remains hidden in the depths of the forest, in the crags of the mountains ... which will complete a great picture and shed light on many obscure points in our history ... that the sages throughout the centuries and especially in our day, are charged to decipher.[46]

Argentina, too, was a key, if contentious, site for the working out of these practices in the late nineteenth century. The previous century had seen sporadic instances of collecting of animal fossils as oddities and curiosities to fill European cabinets; the precursors to natural history collections in state museums. The rich fossil fields south of Buenos Aires attracted local and foreign scientists alike. To European scientists racing to compile complete and persuasive classifications of prehistoric fauna, and to chart the geological age and formation of the earth, Argentina was a major and promising site, rich with mammalian fossil evidence. Throughout this period, articles and reports on Argentine geographical formations and fossils appeared in North American journals like *Science* and *The American Naturalist*. British, French, and Spanish scientific explorers had taken to the systematic exhumation and analysis of nonhuman animal fossils in Argentina beginning in the 1840s.[47]

With fieldwork increasingly valuable for scientific prestige, transnational scientists spent more time in the field, some even settling permanently or semi-permanently in the region. Particularly after 1900, European and Latin American scientists alike were increasingly outnumbered and out-spent by US anthropologists from the Smithsonian, Harvard, and Berkeley. In 1908, Marshall Saville described the geology of the area surrounding his work site—a hacienda in the tiny village of Tomsupa. Exemplifying the new style of close on-site observation, Saville painted a detailed picture of landscape, including the area surrounding a mound:

The whole mound is about four feet high and twenty feet in diameter. [It] is a solid mass of pot-sherds [shards] and sea shells with many whole vessels and small stones intermingled in a compact deposit. The mass of material so far as the excavation shows, extends downward at least five feet below the level of the surrounding plain....[48]

Latin American scientists—already ensconced in the field—were increasingly ambivalent about their relatively wealthy competition from Northern institutions. At the ICA meeting in New York in 1902, the powerful Mexican bureaucrat (and first Inspector of Monuments) Leopoldo Batres complained that he had to defend his turf from foreign, especially US incursions into Mexican space. He noted that he was one of only two Mexicans present at the ICA meeting in New York. In a letter to Batres, Minister of Education Justo Sierra echoed his nationalistic concerns about archaeological patrimony, with "general instructions" for the Inspector. He asked Batres to take photographs of items sent to the National Museum, with annotations, to document the removal of "objects that by your judgement came from an archaeological monument and have been removed from the same; and make known to the proper Secretary your opinion about *the authenticity, origin, and best probable destination of each object.*"[49]

US-based anthropologists may have held the upper hand in financing Americanist research projects, but increasingly they could not deny that Latin American sites were where one could "obtain the best results," as Gamio uttered in 1915. Moreover, only via full access to both horizontal and vertical dimensions of land would Americanists be able to determine—per Batres in 1902—the "authenticity, origin, and probable destination" of objects found there.

The emphasis on contextualization of evidence required scientists to be in place—although it is important to note that objects could still be, and were, removed. Also, then, it required close examination of objects on and below the surface. Thus it became increasingly clear that long-term contact was required to generate the authority necessary. If not always intended as cultural investigations, archaeological digs nonetheless brought an increasing number of scientists into the field and therefore in contact with people living there. Plainly put, scientists relied on local sources and knowledge, if not always acknowledged. This imperative facilitated more intimate exposure with living people and direct observation of the social dynamics of present-day communities.

"Among the Living Peoples": The Future of Fieldwork

By the turn of the new century, Americanist anthropologists' scientific authority increasingly relied on access to specific places, the ability to contextualize evidence from that place, and finally, on personal and intimate contact with it and its inhabitants. In the 1930s and thereafter, these expectations would solidify into the norms we now understand as anthropology's "gold standard" of participant-observer fieldwork. Americanists from the 1870s on had changed "what counted" as scientific, anthropological, and ethnographic evidence through very specific practices that were ultimately dependent on their relationships with both material and human subjects of study. These scientists—foreign and local alike, in constant negotiation with each other and with the land, material, and peoples they studied—*created* scientific authority with their evidence practices. Close, multisensory examination of material in context led to long-term immersion in the field. Like their evidence, scientists themselves recognized the need to embed in place.

As the number of expeditions increased, it became increasingly clear that new anthropological and archaeological studies would rise and fall not just on the authority of objects in the field, but on collaboration with local governments, laborers, and informants. The prominent US archeologist and geographer W.H. Holmes recognized in 1904 that the Americas were not just populated by artifacts, but with people:

In America all the steps of culture from the highest to the lowest within the native range are to be observed *among the living peoples*, and we are thus able to avoid many of the snares of speculation with respect to what men have thought and men have done under the greatly diversified conditions of primitive existence …. In America the past of man, for the most part at least, *connects directly with the present and with the living*. Each step backward along the course of culture development proceeds from a well-established and fully understood base, and *there is thus no baffling gap between history and prehistory, as in the Old World*.[50]

The search for land, the evidence buried in it, and the need for close and intimate knowledge of the local context had led Americanists to living people as authoritative sources of data about the past.

Observing living Indigenous peoples as a conduit to the past, however, did not mean Americanists looked at events from their subjects' points of view. The Salish, Mexica, Pueblo, and Fuegian people were means to scientific ends, not ends unto themselves.[51] Ironically, while in places like Argentina and Mexico, scientific evidence was used by the state to marginalize and even destroy native communities, anthropologists' fieldwork practices ultimately opened conduits to Indigenous land claims and demands for repatriation in the 1990s.[52] Moreover, after a period of soul-searching in the late twentieth century, many anthropologists began to recognize that legitimate (and ethical) fieldwork requires engagement with informants, and ultimately, reciprocity. Today, social scientists increasingly recognize collaboration as a generative manner of knowledge production. Some of the strongest voices in the field are those of Indigenous anthropologists themselves, who influence the profession in numerous ways, not least by teaching us that knowledge is relational, and that wisdom is rooted in place.[53]

Notes

1 Adolf Bandelier, "The Historical Archives of the Hemenway Southwestern Archaeological Expedition," *Congrès International des Américanistes. Compte-Rendu de la Septième Session. Berlin 1888* (Nendeln/Liechtenstein: Kraus, 1968), 458 (emphasis mine); hereafter cited as *ICA Proceedings*.

2 As historian David Livingstone has pointed out: "Scientific findings … are both local and global; they are both particular and universal; they are both provincial and transcendent." He further notes that: "Place is essential to the *generation* of knowledge. It is no less significant in its *consumption*. Ideas and images travel from place to place as they move from person to person, from culture to culture. But migration is not the same as replication. As ideas circulate, they undergo translation and transformation because people encounter representations differently in different circumstances." See David Livingstone, *Putting Science in Its Place: Geographies of Scientific Knowledge* (Chicago: University of Chicago Press, 2003), xi, 11. See also Julia Rodriguez, "Beyond Prejudice and Pride: The Human Sciences in Latin America in the 20th Century," *Isis* 104, no. 4 (2013), 807–817.

3 See the introduction to this volume.
4 Baron de Dumast, "Allocution," *ICA Proceedings*, 1875, vol. I, 31 (unless otherwise noted, all translations are mine).
5 Lucien Adam, speech, *ICA Proceedings*, 1875, vol. I, 7.
6 Sylvester Baxter, *The Old New World: An Account of the Explorations of the Hemenway Southwestern Archaeological Expedition* (Salem: Salem Press, 1888).
7 Sylvester Baxter, Introduction to Adolf Bandelier, "The Historical Archives of the Hemenway Southwestern Archaeological Expedition," *ICA Proceedings* 1888, 450. Cushing worked among the Zuni, Cheyenne, and other groups from the 1880s on, and was later hired by the Smithsonian Institution. Ten Kate conducted numerous anthropological surveys in the same time period in North America and South America, especially Argentina.
8 Henrika Kuklick, "Personal Equations: Reflections on the History of Fieldwork, with Special Reference to Sociocultural Anthropology," *Isis* 102, no. 1 (2011): 1–33. See also Efram Sara-Shriar, *The Making of British Anthropology, 1813–1871* (Pittsburgh: University of Pittsburgh Press, 2013); on fieldwork in biology, see Robert Kohler, *Landscapes and Labscapes: Exploring the Lab-Field Border in Biology* (Chicago: University of Chicago Press, 2002).
9 In spite of the rich historical records available from Latin American scientific expeditions, it has gone largely unnoticed in studies of the history of anthropology that the region was a major site of modern scientific anthropological fieldwork. On anthropology in nineteenth century Latin America, see Cristina Bueno, *The Pursuit of Ruins: Archaeology, History, and the Making of Modern Mexico* (Albuquerque: University of New Mexico Press, 2017); Mechthild Rutsch, *Entre el campo y el gabinete. Nacionales y extranjeros en la profesionalización de la antropología mexicana (1877–1920)* (Mexico City: INAH, 2007); Carolyne R. Larson, *Our Indigenous Ancestors: A Cultural History of Museums, Science, and Identity in Argentina, 1877–1943* (University Park:Pennsylvania State University Press, 2015).
10 On the precursors to Latin American Studies, see Helen Delpar, *Looking South: The Evolution of Latin Americanist Scholarship in the United States, 1850–1975* (Tuscaloosa: University of Alabama Press, 2008), and Ricardo D. Salvatore, *Disciplinary Conquest U.S. Scholars in South America, 1900–1945* (Durham: Duke University Press, 2016).
11 José Ramírez, "Los leyes biológicas permiten asegurar que las razas primitivas de América son autóctonas," *ICA Proceedings*, 1895 (México meeting), 363.
12 See extensive work on naturalists in Latin America, e.g. Cañizarres-Esguerra.
13 Martin Brücker, "Introduction," in *Early American Cartographies*, ed. Martin Brücker (Chapel Hill: University of North Carolina Press, 2012), 7. It was only later that a racial geography of the US-Mexico border developed in which a dichotomy between North and South was (and still is) seen as a racialized divide. Saldaña-Portillo thus calls the United States "as nonindigenous space atop Mexico as indigenous space, arguing argues that the "racial geographies" of both nations were "mutually constituted and imbricated in their colonial legacies." Moreover, the "colonial, generic constructs of Indians and indios … were derived in part from the perception of Indians and indios in an as landscape," María Josefina Saldaña-Portillo, *Indian Given: Racial Geographies across Mexico and the United States* (Durham, NC: Duke University Press, 2016), 6–8.
14 Barbara E. Mundy, "National Cartography and Indigenous Space in Mexico," in *Early American Cartographies,* ed. Martin Brücker (Chapel Hill: University of North Carolina Press, 2003), 364, 370, 376.
15 Mundy, "National Cartography," 380.
16 On the history of geography and maps in Latin America, see Raymond Craib, *Cartographic Mexico: A History of State Fixations and Fugitive Landscapes* (Durham: Duke University Press, 2004); Lina del Castillo, *Crafting a Republic for the World:*

Scientific, Geographic, and Historiographic Inventions of Colombia (Lincoln: University of Nebraska Press, 2018); Nancy Appelbaum, *Mapping the Country of Regions: The Chorographic Commission of Nineteenth-Century Colombia* (Chapel Hill: The University of North Carolina Press, 2016).

17 Geographic studies of Mexico included Juan de Dios Domínguez's *Catecismo Elemental de Geografía y Estadística del Estado de Querétaro* (1873), and Alberto Correa's *Geografía de México* (1885).

18 According to Magali Carrera, García Cubas's work "focused on the exposition of the diverse spaces of Mexico, making them visible and available for visiting and investment." Magali M. Carrera, *Traveling From New Spain to Mexico* (Durham: Duke University Press, 2011), 186; see also 202–203.

19 Carrera, *Traveling*, 225.

20 Argentina fits a model of territorial boundary-making seen in many other sites, as theorized by Charles S. Maier, *Once within Borders: Territories of Power, Wealth, and Belonging since 1500* (Cambridge: Belknap Press, 2016).

21 F. Moreno to Tomas de Anchorena, 5 December 1892, Archivo General de la Nación, Fondo Francisco P. Moreno, legajo 3097, no. 2, 1–81.

22 Argentina shared some features with other settler colonial societies, such as the United States, which had begun its expansionist project nearly a century earlier. The young Northern nation embraced a romantic myth of virtuous farmer-settlers, in contrast to an overcrowded and morally bankrupt England. But in Argentina in the 1870s, in contrast to the early US Republic's priorities, territorial expansion and boundary-making served the goal of creating a modern industrial export economy. For the United States, see Drew McCoy, *The Elusive Republic: Political Economy in Jeffersonian America* (Chapel Hill: UNC Press, 1980).

23 Recent insights of critical Latin American Studies scholars have for some time identified processes of *internal colonialism* as a way to frame ongoing patterns of violent and annihilating encounters. For example, in his theorizing of what he describes as the "coloniality of power," Aníbal Quijano pointed out the contribution of science early on in the establishment of Iberian colonies as a new way of thinking that codified "the relations between conquering and conquered populations," as-cribing to biological theories of race and naturalizing "hierarchical differences been the dominant and dominated." Aníbal Quijano, "Coloniality of Power and Eurocentrism in Latin America," *International Sociology* 15, no. 2 (2000): 216.

24 John H. Beeker, "On the Migration of the Nahuas," *ICA Proceedings* 1877, vol. 1, 325.

25 Adolph Bandelier to Henry Morgan, 31 March 1880, in *Pioneers in American Anthropology: The Bandelier-Morgan-Letters, 1873–1883*, ed. Leslie A. White (New York: AMS Press, 1940), 174–175.

26 Abraham Castellanos, "Plan general sobre 'Procedencia de los pueblos americanos' y 'Cuenta Cronológica'," *ICA Proceedings* 1895, 304.

27 Alexander von Humboldt, *Personal Narrative of Travels to the Equinoctial Regions of America During the Years 1799–1804* (New York: Benjamin Bloom, 1971), xvi.

28 Humboldt, *Personal Narrative*, xvii.

29 Neil Safier, Jorge Cañizares-Esguerra, and others have documented not just French and Iberian leadership (whose histories were long overshadowed by European and Scandinavians) of seventeenth and eighteenth century naturalist expeditions, but also their local collaboration with and reliance on Indigenous informants. Neil Safier, *Measuring the New World: Enlightenment Science and South America* (Chicago: University of Chicago Press, 2008); Jorge Cañizares-Esguerra, "How Derivative Was Humboldt? Microcosmic Narratives in Early Modern Spanish America and the (Other) Origins of Humboldt's Ecological Sensibilities," in *Nature, Empire, and Nation: Explorations of the History of Science in the Iberian World* (Stanford: Stanford University Press, 2006).

30 Daniela Bleichmar, *Visible Empire: Botanical Expeditions and Visual Culture in the Hispanic Enlightenment* (Chicago: The University of Chicago Press, 2012); according to Bleichmar, "These were lengthy, state-sponsored enterprises to gather information about plants and animals from the Spanish realms, assess economic possibilities, and provide plants and seeds for the Royal Botanical Garden in Madrid (founded 1755)."

31 Robert Ryal Miller, *For Science and National Glory: The Spanish Scientific Expedition to America, 1862–1866* (Norman: University of Oklahoma Press, 1968).

32 Ryal Miller, *Science and National Glory*, 17. Miller points out that Almagro's expedition notes are lost.

33 On the history of ICA meetings in France, see Paul N. Edison, "Latinizing America: The French Scientific Study of Mexico, 1830–1930" (PhD Dissertation, Columbia University, 1999); Christine Laurière, "La discipline s'acquiert en s'internationalisant. L'exemple des congrès internationaux des americanistes" (1875–1947), *Revue germanique internationale* 12 (2010): PAGES; Étienne Logie and Pascal Rivale, "Le Congrès des américanistes de Nancy en 1875: entre succès et désillusions," *Journal de la Société des Américanistes* 95, no. 2 (2009): 151–171; Leoncio López-Océn, Jean Pierre Chaumeil and Ana Verde Casanova, eds., *Los americanistas del siglo XIX. La construcción de una comunidad científica internacional* (Madrid: Iberomericana, 2005).

34 Edison, "Latinizing America. Since the 1990s, anthropologists have critically examined the ways in which their discipline served and was served by colonialism"; see, for example, the work of Talal Asad, Ann Laura Stoler, Warwick Anderson, and Claudio Lomnitz.

35 Christine Laurière, "La Société des Américanistes de Paris: une société au service de l'américanisme," *Journal de la Société des Américanistes* 95, no. 2 (2009): 93. According to Edison, the *Société Américaine de Paris* also signaled a turn toward biological racism in French anthropology: "embracing racial anthropology and museum-oriented ethnography, French Americanism sacralized the 19th century natural-historical paradigm for understanding non-European mankind." Edison, "Latinizing America," 406.

36 Vicente de Vera y López, "Sobre las variaciones ocurridas en la geografia fisica del continente americano desde la epoca del descubrimienta hasta nuestros dias," *ICA Proceedings* 1883, 369.

37 Napp, *La República argentina*, 401–402.

38 Napp, *La República argentina*, 402.

39 Jorge Cañizares-Esguerra, "Landscapes and Identities: Mexico 1850–1900." In *Nature, Empire, and Nation: Explorations of the History of Science in the Iberian World* (Stanford: Stanford University Press, 2006).

40 Francisco A. Flores, "Historia, Arqueología, y Etnogenia del territorio de Tepic," *ICA Proceedings* 1910, vol. II, 337.

41 Flores, "Historia, Arqueología, y Etnogenia," 337 (emphasis mine).

42 Bandelier to Morgan, 25 March 1880, in *Pioneers in American Anthropology*, 174.

43 Alice Fletcher note, letter from W.H. Holmes to Zelia Nuttall, 22 March 1907, National Anthropology Archive, Smithsonian Institution (A Fletcher correspondence).

44 Eugène Oberhummer, "L'oevre géographique d'Alexandre de Humboldt au Mexique," *ICA Proceedings* 1910, vol. II, 229.

45 Nancy Leys Stepan has described the process of visualizing landscape as a European concept, "a manner of perceiving space in terms of a scene situated at a distance from the observer ... it is rooted, then, in a Western way of organizing the visual field"; Nancy Leys Stepan, *Picturing Tropical Nature* (Ithaca: Cornell University Press, 2001), 25.

46 Francisco M. Rodriguez, "Descripción de la Pirámide llamada 'Casa del Tepozteco'..." *ICA Proceedings* 1895, 237.

47 Irina Podgorny, "De ángeles, gigantes y megaterios. El intercambio de fósiles de las provincias del Plata en la primera mitad del siglo XIX," in *Los lugares del saber*, ed. Ricardo D. Salvatore (Rosario: Beatriz Viterbo, 2007,), 125–157.

48 Marshall Saville, "Archaeological Researches on the Coasts of Esmeraldas, Ecuador," *ICA Proceedings* 1908, 336–337.

49 Justo Sierra to Batres, 7 November 1902, Intituto Nacional de Antropología y Historia, Mexico City, archive serie "Leopoldo Batres" (emphasis added).

50 W.H. Holmes, "Contributions of American Archaeology to Human History," *ICA Proceedings* 1904, 345 (emphasis added).

51 Anthropologists' subsequent study of "traditional" cultural forms would go on to perpetuate, and perhaps even accentuate, these divisions. Insight into this process would only emerge in the late twentieth century with the application of postcolonial theory: first, the anthropologists' reification of "traditional" culture of diverse American peoples; and second, the continuation of colonizing practices (including, and even especially, internal colonization). For a contemporary discussion of this process, see, among others, Johannes Fabian, *Time and the Other: How Anthropology Makes Its Object* (New York: Columbia University Press, 1983).

52 In 1994, after years of litigation, the Argentine government returned the first Indigenous human remains to the Tehuelche community. The bones belonged to Inakayal, who had been captured by the Argentine military and forced to live in the Museo de la Plata until his death in 1888; for the next century, his bones remained in the MLP's collection. See also Walter Delrio, et al., "Discussing Indigenous Genocide in Argentina: Past, Present, and Consequences of Argentinean State Policies toward Native Peoples," *Genocide Studies and Prevention* 5, no. 2 (Summer 2010): 138–159.

53 See, for example, the work of Margaret Bruchac and Kim TallBear.

Part II

Innovating Evidence: Contemporary Technoscientific Approaches

5 Prototyping Evidence: How Artifacts Demonstrate Technological Futures

Sascha Dickel

Present Futures

How can we get access to the future? In the modern ontology of time, the answer to this question is, "never."[1] The future is not an object that modern subjects may investigate like any other aspect of the world, because the future has not yet arrived. Luhmann famously argued that "the future cannot begin," because we will always be anchored in a specific present.[2] The *future presents* ahead cannot be known. The more modern subjects accept and reflect this aspect of their temporal existence, the greater the urge to construct *present futures*; constructions of how their future may be, will probably be, or should be.[3] Hence, a proliferation of present futures is a continuing trend associated with modernity.[4] As Adam and Groves observe, "Wherever we care to look we cannot fail but notice that the contemporary capacity and competence to produce futures is phenomenal."[5] Present futures represent the absent future presents. These representations *can* be accessed because they are a part of our present world like any other social construct.

Following an extensive and ongoing discussion about expectations and temporality,[6] the body of literature on present futures has grown considerably in recent years. This renewed interest in futures is linked to a rise in technological visions, and reflexive analysis of them in the field of science and technology studies.[7] Papers on technological visions "analyse how images of desirable or dystopic socio-technical futures affect the beliefs and perceptions of the public, and they look at how today's political, economic and societal decisions concerning technological change are influenced by such images of the future."[8] Reflexive studies on futures share the epistemological and methodological assumption that *the future* (in the sense of the future presents) cannot be known. Hence, their analytical strategy neither aims to predict the future, nor do they aspire to construct possible, probable, or desirable scenarios themselves. Rather, the reflexive approach to the future acknowledges the *immanence of the present* inherent in any construction of the future[9]—be it a fictional story, a scenario, or a prediction.

DOI: 10.4324/9781003188612-7

The majority of this line of work approaches futures as products of language; as "stories and images" of the future.[10] Armin Grunwald, one of the leading scholars on this topic, reaffirms that "the role of language is fundamental in discourses on the future, because *the future only exists in language*."[11] The value of an analysis of futures in the medium of language can hardly be contested. However, the recent turn to materiality suggests that the focus on language needs to be complemented by approaches that emphasize the role of artifacts in the constitution of present futures, and thus move beyond established modes of social constructivism in science and technology studies.[12] Present futures are not only linguistic constructs but may also appear to us as material artifacts.

Prototyping, described by Buchli as a material technology of "presencing the immaterial,"[13] is an illuminating case in this respect. Contemporary usage of the term prototype refers less to the past (to an original form), and more to an "objectivation" of present futures.[14] Prototypes are objects "in which an idea is manifested in different stages of development—in part only in its selected properties and components."[15] Prototypes are "socio-material devices for ordering the future in the present."[16] The *UXL Encyclopedia of Science* provides the following definition:

> A prototype is an initial model of an object built to test a design .... Prototypes are widely used in design and engineering to perfect items and processes before implementing them on a large scale. Automobile designers, for example, typically build prototypes of new cars to see if their ideas work in practice. A prototype is a vital part of the design process because it allows designers to see the product in action, so they can see what works and what does not. It is also useful for showing designs to corporate executives or investors to persuade them to support a project.[17]

Practices of prototyping are nothing new. Prototyping *avant la lettre* "has always existed and probably, for most of human history, has been more important than it's [sic] opposite, orderly science and planning."[18] In his seminal work on *thing knowledge* Davis Baird argues that "thought and design are not restricted to processes conducted in language. And working with models and prototypes is not more primitive ... than working with words and equations."[19]

Prototyping represents a crucial mode of knowledge production in technoscientific cultures. In professional fields like engineering and design, it is considered an essential practice in order to develop new products and technologies. Suchman, Trigg, and Blomberg, as well as Guggenheim, however, suggest that prototyping is no longer an activity reserved exclusively for experts but is increasingly performed as a participatory social practice.[20] Usually this involves the integration of prospective users who are invited into engineering labs and design studios to contribute to an optimization of the design process.

For advocates of "user innovation,"[21] prototyping represents a strategy to elicit latent user "needs," and to make them explicit and thus available for product design.[22]

However, prototyping today does not only take place in high-tech laboratories and professional design contexts. Prototypes are also increasingly produced, tested, and staged in places and events dedicated to open innovation, civic technoscience and science communication. Public workshops like FabLabs and makerspaces have been established in recent years, providing an infrastructure to make prototyping more accessible and affordable. Civic Hackathons include publics in the creation and presentation of prototypes. Videos on YouTube and various crowdfunding platforms show prototypes in action, increasing the public visibility of unfinished artifacts. Moreover, contemporary notions of design expand the meaning of prototyping,[23] and promote it as an everyday practice of "futuring."[24] Accessing the future with prototypes is turning into a public practice.

This chapter is based on several "focused ethnographies" of public prototyping.[25] In the last few years, I have studied makerspaces, design practices, hackathons, public beta tests, and demonstrations of prototypes in various media.[26] These studies involved participant observation and interviews, as well as document and video analysis. However, this chapter does not constitute a sociological analysis of technology development,[27] nor does it provide a thick description of design processes.[28] Rather, it draws upon diverse empirical studies to formulate a theoretical diagnosis how prototypes generate evidence for technoscientific innovation. Furthermore, it introduces public prototyping as a symptomatic "objectual practice" to render the future evident in contexts of public participation and communication.[29]

Prototypes as Epistemic Objects

Why are prototypes needed at all? Why do designers not just imagine a future state of an artifact and develop it accordingly? Schütz, for example, considered the anticipation of future actions as a mental activity of "motivated phantasying, motivated by the anticipated supervening intention of carrying out the project."[30] In his view, a design process would be based upon a mental construction of a finished product or project:

> I have to visualize the state of affairs to be brought about by my future action before I can draft the single steps of my future acting from which that state of affairs will result. Metaphorically speaking, I have to have some idea of the structure to be erected before I can draft the blueprint.[31]

According to Schütz, these "blueprints" are aspects of a mental journey into a future. Yet this very form of anticipation seems inadequate when it comes to more complex design processes in general, and the design of technical artifacts in particular. The step from the idea (present) to the finished and fully functional product (future) can seem too far, the imagination too vague, the plan inadequate. This is precisely where the prototype takes to the stage.

Prototypes are materializations of designs. With a prototype, the contingency of a designed future is combined with the contingency of present materialities. In dealing with a prototype, the anticipation of a future is translated into material design possibilities.[32] Prototypes thus stand between the intention of the design and what Schütz considers the action that has already taken place: the product that has already been realized. As a material artifact, the prototype translates fiction into fact, expectation into experience.[33]

Once the anticipation is objectivated, it is possible to do something with it. In this way, one's own idea can on the one hand solidify an anticipation. On the other hand, the anticipation might become alien to the design subject itself. The idea objectified then appears as a source of potential irritation that can shape the design process and modify the course of action. Through externalization, the design appears as an epistemic object that can be interacted with.

The notion of epistemic objects builds upon on Rheinberger's notion of epistemic things in scientific research.[34] Epistemic objects are symptomatic of the epistemic cultures of laboratories, in which objects are not only used as black boxes in an instrumental or commodified way, but are actively explored. In contrast to technical instruments that are ready-at-hand, stable, and unproblematic, epistemic objects are instable objects of problematization.

> [Epistemic objects] appear to have the capacity to unfold indefinitely. They are more like open drawers filled with folders extending indefinitely into the depth of a dark closet. Since epistemic objects are always in the process of being materially defined, they continually acquire new properties and change the ones they have. But this also means that objects of knowledge can never be fully attained, that they are, if you wish, never quite themselves. ... The *lack in completeness of being is crucial*: Objects of knowledge ... have material instantiations, but they must simultaneously be conceived of as unfolding structures of absences: as things that continually "explode" and "mutate" into something else, and that are as much defined by what they are not (but will, at some point have become) than by what they are.... The unfolding ontology of objects foregrounds the temporal structure.[35]

Hence, prototypes are epistemic objects *par excellence*. They illustrate certain socio-technical mechanisms of a desired (but absent) product. Prototypes do not serve a direct instrumental use. While "technical constructions are designed to secure the present,"[36] prototypes are used to access the future and to overcome the present. Prototypes promise to make the possibilities of the future testable in the present. If necessary, objects can be modified and/or new prototypes may be designed in an iterative process of error identification and correction.[37] What exactly may be tested depends on the complexity and "fidelity" of the prototype—in other words, on the question of how "far away" the prototype is from the intended end product. A paper prototype can be used to check whether a certain shape works. The mock-up of an app can show whether a certain interface makes it convenient to use. When the prototypical care robot lifts a dummy into bed, one may observe the pitfalls of an application involving the living.

The prototype thus appears as a material element of a feedback loop, which in turn has an effect on the conceptual design process and, if necessary, can give it a new direction. The irritating alienation caused by the present future has turned into externalized materiality, and can be an occasion for learning. Prototypes

> are created to debunk, complicate or reinforce ... ideas. ... Prototypes are tools to think with. They are particular and telling objects in that they represent at once the idealization of what is to be built as well as the rudimentary, necessarily incomplete experimental processes in which such building occurs. There is an iterative quality to the prototype that implies intrinsic plasticity and rapid exchange between the world of concepts and the world of things, a rhythmic and temporal flux that separates it from mass-produced artifacts.[38]

This iterative process of creating and testing prototypes is described in an interview by a member of a makerspace as an essential step of product design:

> Prototyping is an incredibly important step to simply determine what is doable, what is possible. But also in the area of user experience this is an incredibly important step. Well, I visited a few of the start-ups in [city anonymized for privacy] and there you start with the cardboards and walk around for a week with your phone cut out of cardboard in your pocket then an iteration made of wood. Then comes iteration two with a functional prototype to test the user interface. And in the industrial process—from the idea to the product—you can't avoid these iterations. Every product you hold in your hand goes through this prototyping process.

In design and engineering practices, prototypes mediate and translate between ideas and products. Prototyping allows the world to "object" to an idea: "Because they are physically implemented, and thus more than (more or less rigid) thought experiments, they take place not only in the mental realm but confront the imagined future with the resistances of present real-world components."[39]

As part of an ethnographic study, I accompanied the director of a makerspace, a public workshop for prototyping and product design, to a meeting with a group of young engineers. At the beginning of the meeting, he juxtaposed the habitus of a "prototyper" with the assumed habitus of a "perfectionist" who spent years in his workshop brooding over his plan for a product. This is an extract from my field notes:

> The director explains that his makerspace is pioneering a new culture of innovation which promises to revolutionize the supposedly slow mode of engineering in Germany. The typical German engineer would spend three years brooding abstractly over an idea and then trying to implement it directly—his misguided ideal would be the construction of an already finished product. He calls out to the crowd: "Build your prototype first from cardboard, then from wood, then go to the 3D printer. Test out what works and what doesn't." Instead of a tedious filing of a perfect end product, engineers should embrace an iterative generation of prototypes. He points out that it is "really important" to quickly build prototypes and test ideas early. The key to this: "Rapid prototyping," for which the equipment of the maker space would be optimal. He also promotes the makerspace as a place where different actors (engineers from large companies, start-ups, tinkerers and hackers) can meet and give feedback on prototypes.

During an interview, the director described the construction of his makerspace also as an incremental and iterative developing process. He narrated the history of the workshop as a process of learning through prototyping:

> You build a house and then you realize what you've forgotten or where something's not tight, ok. That is, at the opening you first think that you are finished—everything according to plan, everything according to specification. And then you realize, that is missing, that is missing, that is missing, that is missing. In other words, since the opening we have been constantly working on improvements and customer-oriented solutions. Customers ask you, "Where can I get a piece of wood?" and you think, "Oh shit, where is our wood?" Another says, "I need a screw" and you ask yourself, "Oh, how does my customer get a screw?" And that's really super important, knowing that, and actually always a key learning that you can take with you if you have to do it again at some point. But—and this is the nice thing— you can only experience this if you use your prototype, ok? When you build your new products, the idea you bring along is always flawless, then you build the prototype and you realize something is not possible.

And that's what we got to know with our makerspace. Construction and opening were prototypes. And now we are at prototype version 15. Ah, the nice thing is, I think a makerspace will never be an end product.

This extract shows that the notion of prototype may also be applied in retrospect, if actors realize that something has been a prototype for them because it served the epistemic functions of testing and evaluating. The interview also suggests that the notion is not restricted to specific categories of objects. Even whole organizations can be conceived as prototypes. Other interviews with makers and designers and observations of everyday practices in makerspaces confirmed that nearly everything can be understood as a prototype. The actors themselves not only use the term to describe objects with specific material properties; they use it to make sense of objects that fulfill a specific epistemic function. In all these cases, prototyping is distinguished from the alternative approach of planning the future. The epistemological advantage of prototypes is tied to their ability to let actors experience a future in the present.

Several interviews with members of the start-up and maker culture indicate this devaluation of long-term planning. They frame planning as a rather outmoded way of product development that no longer corresponds to the accelerated innovation society of today. Instead, they suggest that one should gather as much experience as possible with the epistemic object "prototype."

In this sense, prototyping is embraced as a promising way to generate *evidence* for actors involved in technoscientific design practices. The notion of evidence has many connotations. I use it in the sense of a present indicator that refers to something absent.[40] This is the way evidence is understood in police work and in court. A criminal act itself may be inaccessible, since it took place in the past; nevertheless, there may be traces that can be interpreted as evidence that the crime actually took place. We might also, for example, consider some (present and observable) smoke as an indication of a not yet observable fire.[41] The assumptions about what is absent may turn out to be completely wrong: we may learn that the cause of the smoke is not a fire, but a fog machine in a club. In order for the smoke to act as evidence, the fire (or machine) must be absent. If one could see the fire or the fog machine, the notion of smoke as evidence would be absurd, or at least redundant. In order for phenomenon A to function as evidence for phenomenon B, phenomenon B needs to be out of reach.

Prototypes fulfill a similar function. At first glance, a prototype refers to something that lies in the future. But here, one has to be more precise: prototypes refer to something that does not exist yet, but could manifest itself in future presents. Showing, sharing, and demonstrating prototypes can thus be interpreted as a make-believe game in which the respective artifacts function as props, whose presence creates an opportunity to experience something that is depicted as a representation of a future artifact.[42]

Prototypes as Demonstrative Evidence

When considered in this way, prototyping is also a mode of claiming epistemic authority.[43] In order to reconstruct this function of prototypes, I now turn to a civic hackathon that aspired to create alternative urban futures.[44]

> Two men enter a stage. They introduce themselves as Zachery and Sebastian. Sebastian reveals what the workshop is supposed to be about. We are to build prototypes for urban life in 2050. One participant asks if we should design utopias or dystopias. Zachery's prompt answer is: "Neither nor. Build prototypes! Prototypes are there to show something and learn something from it! This works better with positive scenarios." I'm working with others on a kind of autonomous vehicle service where the city itself acts as a platform operator and sets the prices. The tools available to us for building our prototypes are manifold: pens and notepads, modelling clay and LEGO bricks, cardboard and wood. 3D printers are also available (Sebastian: "If you can operate them"). We can see the other workshop participants doing handicrafts, drawing, gesturing and talking. Everyone works eagerly. Whenever we go over to discussing "too much," Sebastian and Zachery approach us with a cardboard sign. The sign reads: "Do, don't talk". In a short break I talk a little with them. I ask why talking is "forbidden." They tell us: anyone can talk about innovations. But today it should be about something new, about a competence that we do not yet have sufficiently, namely the rapid testing of project ideas. We should therefore distance ourselves from the mode of speech for a while and move on to a mode of physical creation.

This ideal of *design without discussion* was conveyed implicitly by the entire arrangement of the workshop, and expressed explicitly by the organizers. The tests of designs were meant to substitute discourses that would first have to make judgments about the feasibility and desirability of designs. At those points where a decision between two (or more) options was necessary, we were urged not to resolve the crisis through discussions, but through the realization of alternative prototypes, which were then compared and tested against each other. If verbal communication was necessary, it should be kept to a minimum (e.g. short explanations of the prototype in a feedback round), if possible it should be written in a concise form (e.g. one or two short sentences on a sticky note), but ideally language should be avoided—because the prototypes should be designed in such ways that they were able to speak for themselves. Language was positioned as a suboptimal functional equivalent for creating and demonstrating futures. Consequently, the processing of feedback should not lead to a long discussion. Instead, it was expected to channel the creativity of the group into building new prototypes.

Such an attribution of epistemic authority corresponds with a diagnosis of technoscientific knowledge cultures that valorize technical functionalities that can be materially evaluated. Epistemic authority in such cultures may be achieved less by carefully developed arguments in academic discourses that aspire to evaluate truth claims, and more through demonstrable functionalities.[45] Prototyping may be used to construct an opposition between the "hard"/tangible evidence of artifacts and the "soft"/intangible evidence of narrative scenarios. Thus, in addition to their epistemic function, the demonstrative function of prototypes is to provide evidence for artifacts which do not exist yet.

By transforming expectation into experience, and fiction into fact, prototypes may be used to convince others that a technology is possible (or even probable and/or desirable). Technological development and design are no solitary processes. Rather, a large number of actors typically have to be persuaded that a certain artifact should be constructed or that a certain technology path may be pursued further. For example, colleagues who are to participate in the development, sponsors that might invest money, or bureaucrats who want to know whether something can even be done in a certain way.[46] All this persuasive work requires communication. It requires a medium via which a message can be inscribed and read. Such media could be gestures, words, texts, or artifacts. As objectivations of present futures, also prototypes may function as media to present futures to others.[47] They function as demonstrations that are supposed to be more convincing than language and images due to their "testability."

In order to reconstruct the communicative role of prototypes, I will now turn to situations in which prototypes are presented to an audience. In these situations, the epistemic function may take a back seat, while the demonstrative function of prototypes takes center stage. Schulz-Schaeffer and Meister suggest that a prototype is more convincing than a narrative scenario of the future "because it seems to work in the real-world. In contrast to a narrative representation of a scenario, it is the working technology that makes the difference."[48] Prototypes present futures as something that can actually be *experienced*. The future becomes tangible and gives others an idea of the design goal as well as the design possibilities.

It is precisely these two options—narration versus prototypical demonstration—that a member of a makerspace unfolded in a thought experiment in which he explained the role that prototyping can play in fostering innovation:

> Imagine you're an [automobile company] employee, you come in and say, "Boss, boss, I've got a cool idea. It'll save the company." He says, "Hey, super good. Please make a presentation, because in six weeks the 'Great Ideas Committee' will meet and decide which idea will be followed up here." You go back and you do your presentation, wait six weeks, you present it to the committee. The committee says, "Super, we will decide, we will inform you." … Your idea is now gone. … OK, the new idea at [the company] is, you come and say, "Boss, boss,

boss." He says, "Cool, I can do without you for a week, I have someone who'll do your work, you go out to the Makerspace and make me a prototype of your idea, see if that's as cool as you think". You come back after a week and say one of two things, "Boss, look, there it is." And he says, "Now we go to the committee and present your prototype." OK. Or you come back and say, "You know, the idea in my head was good, but it doesn't work." OK, then the idea died, you don't need to pursue an idea that doesn't work because you've already tested it. OK.

In this extract, our interviewee presented three scenarios: In scenario one he presents a "revolutionary" idea to a decision-making body (the "Great Ideas Committee"). However, this committee would likely never be convinced of the idea immediately, but would first let the engineer waiting. Our interviewee contrasted this with another variant of the narrative—scenario two. In this scenario, the superior sends the engineer to the local makerspace to make a prototype. Now, two different possibilities could arise. In scenario two A, the prototype is (by change of context) transformed from epistemic object into demonstrative evidence. In scenario two B, the evaluation of the prototype has revealed that the idea doesn't work.

I now turn to an analysis of a YouTube video showcasing a high-speed transport system called Hyperloop, to discuss how prototypes are used as demonstrative evidence in front of larger publics.[49] As a technical object, the Hyperloop transport system has not yet become a reality, but the prototype already demonstrates its function in the present.

The video starts with an image of a desert (Figure 5.1). A text appears that informs the viewers about the time and place: "May 11th, 2016. 25 miles North of Las Vegas, Nevada."

Three men in casual business attire walk in slow motion in front of a larger object with the logo of the Hyperloop One company. Over a series of further images, a man begins to speak. We see him a few seconds later. He is one of the three men seen before. He wears a baseball cap with the logo of the company. "There are many engineering milestones to bring Hyperloop to reality. And this is one of the bigger, more tangible ones." In the following frame, the testbed (a track through the desert) is presented, along with a description of what is about to happen, a "propulsion open air test (Figure 5.2)."

The man continues his speech, "We developed a custom propulsion system. And we built a very specific test to showcase the trajectory we're on, which is to build a lot of hardware and absolutely prove to everyone that Hyperloop is coming, and coming very soon." While the man talks, a construction site and several hardware elements are shown in a quick montage, including a rendering of the propulsion test vehicle (Figure 5.3) as well as the physical prototype.

The three men in casual business attire walk toward a terrace, next to the testbed. Some people have already gathered there. Many people with large cameras are shown. In the next frame, one of the businessmen addresses the

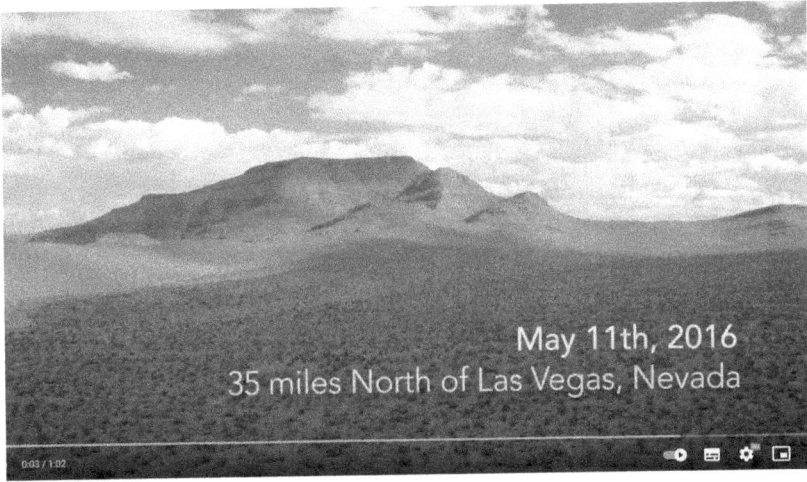

Figure 5.1 Time and location, Virgin Hyperloop, "Hyperloop One—Propulsion Open Air Test," uploaded 11 May 2016. https://www.youtube.com/watch?v=xiokghLXFYM

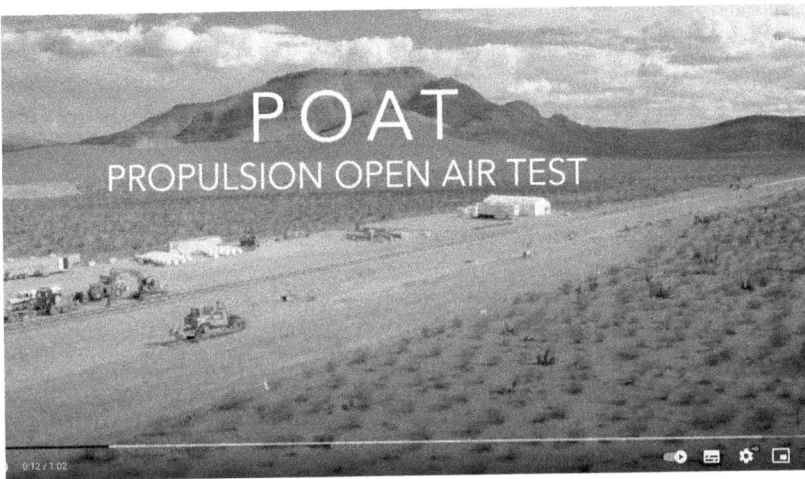

Figure 5.2 Propulsion air test, Virgin Hyperloop, "Hyperloop One—Propulsion Open Air Test," uploaded 11 May 2016. https://www.youtube.com/watch?v=xiokghLXFYM

viewers (and, as we may assume, a local audience). During these images we hear a voice say, "We are standing on hallowed ground." We see the terrace filled with people, recorded from above, and we hear the speech continue. "This is a significant moment for us." Another man in a business suit talks to the viewers. "When you think of Hyperloop you think maybe

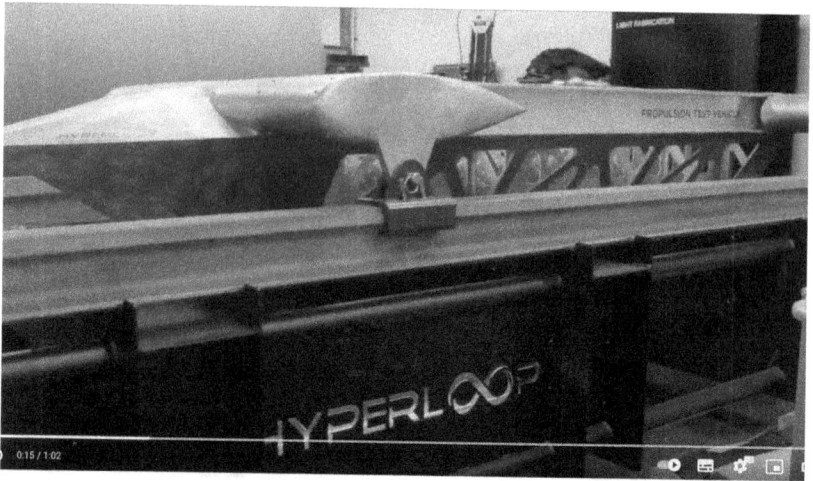

Figure 5.3 Model, Virgin Hyperloop, "Hyperloop One—Propulsion Open Air Test," uploaded 11 May 2016. https://www.youtube.com/watch?v=xiokghLXFYM

this is gonna happen years from now. It's gonna happen much quicker than anyone imagines. When it does, the world will never be the same." A mechanical sounding voice counts down from five to one. Each number is combined with a picture. Five: the prototype of the propulsion system on the tracks. Four: people in front of computer screens. Three: the live audience on the terrace. Two: people with cameras filming from the terrace. One: the three men in suits whom we already know. Finally, the prototype launches and runs across the tracks (Figure 5.4).

A few seconds later it decelerates, creating a cloud of dust. We see and hear the applauding crowd (in the front, we also see the three men in business suits applauding). The video ends with the prototype in motion, recorded from two different angles. The last frame is the logo of the company, filling the whole screen in front of a white background.

In the narrative of this video, a future in which the world is supposed to be transformed by technology (and "will never be the same" again) is linked with the test of a specific artifact. The successful test should demonstrate that this present future will be realized in future presents. The prototypical test is described as a "tangible" milestone. It is exactly the transformation of expectation into experience, of fiction into fact, in order to convince an audience, that this video demonstrates.

Indeed, the example has three distinct characteristics which are typical for the contemporary regime of mediatized technoscientific promises.[50] Firstly the video does not just suggest that a technology is possible; it also suggests that this technological future will happen sooner than the viewer might have thought. Thus, the prototypical demonstration moves the future closer

Figure 5.4 Demonstration, Virgin Hyperloop, "Hyperloop One—Propulsion Open Air Test," uploaded 11 May 2016. https://www.youtube.com/watch?v= xiokghLXFYM

to the present by setting the prototype in motion. Secondly, the test shows how a specific part of a technological system (the propulsion system), actually represents the whole; in this case, the complete Hyperloop transportation system. The people in the video do not just claim that the propulsion system is "coming very soon" but that Hyperloop itself is "coming very soon." Thirdly, the video is a public performance, a publicly staged "proof" of the arrival of Hyperloop. There are two publics assembled here: first, the local public shown in the video, and second, the public addressed by the YouTube video, which is much larger than the local audience. For the public assembled by digital media, the enthusiastic reaction of the local crowd is part of the apparatus of evidence generation. The applauding local crowd confirms that the technology has worked, and that this working technology is as spectacular as the people involved with the project claim it to be.

This example also demonstrates that it is not possible to criticize a future encoded in a prototype in the same way as a linguistic representation of the future. The established tools of discourse analysis and similar methods can of course be used to criticize the mode of presentation deployed in the video. However, it is harder to criticize the prototypical test itself. A textual narrative scenario of the future may result in disbelief, or may be deconstructed as part of a hype discourse. A prototype, in contrast, strengthens and solidifies an expectation of a specific future by demonstrating a functioning materiality. In this regard, the relation of written scenario and prototype is similar to the relation of scientific paper and laboratory infrastructure as described by Latour. In his monograph *Science in Action* the (fictional) reader in doubt of an author's true claims is led into a scientist's laboratory. "You doubt what I wrote? Let me show you."[51] The demonstration of the laboratory transforms the task of the disbeliever. She now needs to prove that the laboratory instruments are not producing true results, or are producing results that do not justify the claims made in the paper. In a similar way, disbelievers in a specific future must either show that the prototype does not perform the function that it claims to perform, or that the function of the prototype does not justify a suggested link between the prototype and the promised future. The disbeliever may also build a prototype herself to demonstrate the plausibility or desirability of an alternative design.

Public Evaluation of Prototypes

While the video discussed above involves its audience in an indirect way, local events of prototype testing may involve publics in a more direct manner as evaluators. During my fieldwork I visited a so-called "Prototypenparty" ("prototype party"). On its website, the party was called a "matching event" for different social groups, including,

> Professional product developers and designers from start-ups, universities and companies, students with their own product ideas and prototypes, makers, hardware hackers, up-cyclers and inventors, … competent "everyday experts," business angels, and investors in search of new investment opportunities, networkers who would like to support innovative ideas, companies looking for creative solutions, new markets and talents.[52]

The Prototypenparty's self-proclaimed goal is to close "the gap between the first implementation of ideas, detailed product tests, and market research."[53] In principle, anyone could take part in the event. Participants were divided into two groups, "prototypers" and "feedbackers," people who present and people who evaluate prototypes. There were no requirements for the participation of feedbackers, only the purchase of a ticket. The prototypers did not have to meet any specific criteria in order to participate either. However, requirements were formulated for the objects that they brought to the party. Only objects that were "in a feedback-capable" status should be presented.

> Those could be, for example, first implementations of hardware or software that can be "touched" and tested by the participants. Services might also be presented if they can be tested. A presentation of pitchdecks (e.g. as Powerpoint presentations), sketches or renderings is not enough.[54]

In other words, objects needed be in a not too unfinished state but—as was implicitly expressed—also not too finished, because a finished product would no longer be "feedback-capable."[55]

At the entrance I get three small gears pressed into my hand. "Fresh from the 3D printer! You'll need them later for the evaluation," says the young woman at the counter. Inside, a kind of club room awaits me in which there are about 100 people. It's a bit cramped because the room is full of stands where the prototyper teams present their prototypes. Some of them represent start-ups, some describe themselves as tinkerers who are curious if their inventions might be useful for others. Some people stand together in small groups and talk. Some roam the room and already look at the stands. Finally, a woman and a man step onto an improvised stage. They welcome prototypers and feedbackers. They emphasize that the event is something between a usability lab and a trade fair—but "much looser" than these established formats. Here, ideas could "see the light of day early." In the next hours, we (the feedbackers) evaluate different kinds of artifacts: an app for blind customers in supermarkets, a food replacement, designer clothes, a fitness tool, and a web platform for management purposes. We talk to the prototypers, we try things out, we rate the prototypes.

During the Prototypenparty the prototypes were tested in two ways. On the one hand, feedbackers evaluated the product which the prototype represented. On the other, they evaluated whether the prototype indeed evidenced the feasibility of the proposed product. In this regard, the demonstrative capacities of the prototypes were put to the test.

We have seen that prototyping involves at least two categories of actors; those who build and demonstrate prototypes, and those who serve as an evaluative audience for prototypical designs. Both roles are usually performed under laboratory conditions. The objectivation of futures that blur the borders of fact and fiction usually depend upon the stabilizing features of a laboratory setting.[56] Events like the "Prototypenparty," however, suggest that every member of society can take part in prototyping, if not as a "prototyper" who creates prototypes than at least as a "feedbacker" who evaluates them. Thus, prototyping is presented as a potentially all-inclusive practice, not something that requires specialist knowledge and is restricted to professional fields. The event also demonstrates how the epistemic and demonstrative functions of prototypes may be entangled: On the one hand, the event provided input for designers to improve their products, on other hand it enabled the start-up the showcase their projects.

Making Futures Evident

Prototypes make futures evident. They are present artifacts, that refer to absent objects, that are to be realized in the future. As epistemic objects, prototypes materialize a present future into a tangible object that may answer questions and also provoke new ones. In this regard, they produce *subjective evidence* for designers and engineers. As demonstrations prototypes represent *intersubjective evidence* that may convince others that a specific future might be realized. The creation and demonstration of prototypes constitutes an experimental way of dealing with social time, staging expected possibilities as material realities that can be experienced in the present. They are thus temporally paradoxical objects: prototypes represent a future object with which they themselves, as a present object, are not identical. They are props in make-believe games of futurity.[57]

According to the famous aphorism of science fiction author William Gibson, "the future is already here—it's just not evenly distributed." Prototyping redistributes the future, either by demonstrating the prototype to an audience outside the lab or by creating quasi laboratory conditions in settings like makerspaces and during events like the "Prototypenparty." Public prototyping extends the practice of prototyping by inviting the public to participate in this practice.

I suggest that the construction and evaluation of prototypes is currently promoted as an apparatus to render the future accessible. In our present culture of innovation, the chronopolitics of prototyping provide a pre-emptive strategy to govern the unknown future.[58] Prototyping celebrates the iterative irritation of the new. It does not offer a stable and fixed vision of the future; it favors incrementalism instead of abstract utopianism, and yet is still woven into the technoscientific narrative of a world shaped by technology and design.[59] The socio-material apparatus of prototyping poses the implicit question of whether a future can already be demonstrated in the present.

Public prototyping does not only promote certain prototypes, it also fulfills an ideological function: Public prototyping promotes the practice of prototyping itself. It endorses material technologies of constructing and evidencing present futures and it implies an epistemic superiority of practices that objectivate futures and makes them "testable," instead of painting a future "merely" by using words and images. When prototyping is staged as a public issue, the epistemic authority of tangible artifacts becomes a social affordance, and constitutes an implicit criterion of including or excluding present futures. A future that cannot be demonstrated in public may be discarded from public discourse. However, a reflexive analysis of prototyping reveals that the "hard" evidence of material objects needs to be combined with the "soft" evidence of a narrative in order to enact demonstrative evidence. Objects need to be framed as prototypes to fulfill their functions.

Notes

1 For alternative ontologies of time see, for example, Barbara Adam, *Timewatch: The Social Analysis of Time*; Barbara Adam and Chris Groves, *Future Matters: Action, Knowledge, Ethics*.
2 Niklas Luhmann, "The Future Cannot Begin: Temporal Structures in Modern Society," *Social Research* 43, no. 1 (1976).
3 Barbara Adam and Chris Groves, *Future Matters: Action, Knowledge, Ethics*, Volume 3 (Leiden, Boston: Brill, 2007).

4 Reinhart Koselleck, "Vergangene Zukunft der frühen Neuzeit," in *Vergangene Zukunft: Zur Semantik geschichtlicher Zeiten* (Frankfurt am Main: Suhrkamp, 1979).

5 Adam and Groves, *Future Matters*, xiv.

6 Wendell Bell, ed., *The Sociology of the Future. Theory, Cases, and Annotated Bibliography* (New York: Russell Sage Foundation, 1971); Luhmann, "Future"; Hans Ulrich Gumbrecht, "How Is Our Future Contingent? Reading Luhmann against Luhmann," *Theory, Culture & Society* 18, no. 1 (2001): 18–59; Lucian Hölscher, *Die Entdeckung der Zukunft* (Frankfurt am Main: Fischer, 1999); Reinhart Koselleck, *Vergangene Zukunft: Zur Semantik geschichtlicher Zeiten*(Frankfurt am Main: Suhrkamp, 1979).

7 Cynthia Selin, "The Sociology of the Future. Tracing Stories of Technology and Time," *Sociology Compass* 2, no. 6 (2008): 1878–1895; Armin Grunwald, "The Hermeneutic Side of Responsible Research and Innovation," *Journal of Responsible Innovation* 1, no. 3 (2014): 274–291; Nik Brown and Mike Michael, "A Sociology of Expectations. Retrospecting Prospects and Prospecting Retrospects," *Technology Analysis & Strategic Management* 15, no. 1 (2003): 3–18.

8 Ingo Schulz-Schaeffer and Martin Meister, "Laboratory Settings as Built Anticipations—Prototype Scenarios as Negotiation Arenas between the Present and Imagined Futures," *Journal of Responsible Innovation* 9, no. 1 (2017): 3.

9 Armin Grunwald, "Nanotechnologie als Chiffre der Zukunft," in *Nanotechnologien im Kontext: Philosophische, ethische und gesellschaftliche Perspektiven,* eds. Alfred Nordmann, Joachim Schummer, and Astrid Schwarz (Berlin: Akademische Verlagsgesellschaft AKA GmbH., 2006).

10 Grunwald, "The Hermeneutic Side," 279.

11 Grunwald, "Nanotechnologie," 56 (Translation mine, emphasis in original).

12 For materiality in general, see Bruno Latour, *Reassembling the Social: An Introduction to Actor-Network-Theory* (Oxford: Oxford University Press, 2005); Karen Barad, "Posthumanist Performativity: Toward an Understanding of How Matter Comes to Matter," *Journal of Women in Culture and Society* 28, no. 3 (2003): 801–831. For objects in particular, see Karin Knorr-Cetina, "Sociality with Objects: Social Relations in Postsocial Knowledge Societies," *Theory, Culture and Society* 14, no. 4 (1997): 1–30; Herbert Kalthoff, "Practices of Calculation," *Theory, Culture & Society* 22, no. 2 (2005): 69–97.

13 Victor Buchli, "The Prototype: Presencing the Immaterial," *Visual Communication* 9, no. 3 (2010): 273–286.

14 Objectivation is an essential term in the "new sociology of knowledge" of Berger and Luckmann. Objectivations are "products of human activity that are available both to their producers and to other men as elements of a common world." Peter L. Berger and Thomas Luckmann, *The Social Construction of Reality: A Treatise in the Sociology of Knowledge* (New York: Anchor Books, 1966), 49.

15 Christoph Gengnagel, Emilia Nagy, and Rainer Stark, *Rethink! Prototyping: Transdisciplinary Concepts of Prototyping* (Heidelberg: Springer Cham, 2016), 5.

16 Wilke 2010.

17 Amy Hackney Blackwell and Elizabeth P. Manar, eds. *UXL Encyclopedia of Science* (Farmington Hills: UXL, 2015).

18 Michael Guggenheim, "The Long History of Prototypes," *Limn* 0 (2010), https://limn.it/articles/the-long-history-of-prototypes/.

19 Davis Baird, *Thing Knowledge: A Philosophy of Scientific Instruments* (Berkeley: University of California Press, 2004), 149.

20 Lucy Suchman, Randall Trigg, and Jeanette Blomberg, "Working Artefacts: Ethnomethods of the Prototype," *The British Journal of Sociology* 53, no. 2 (2002): 163–179; Guggenheim, "Long History."

21 Erich von Hippel, *Democratizing Innovation* (Cambridge: MIT Press, 2005).

22 Suchman, Trigg, and Blomberg, "Working Artefacts," 166.

23 Tim Seitz, *Design Thinking und der neue Geist des Kapitalismus: Soziologische Betrachtungen einer Innovationskultur* (Bielefeld: transcipt, 2017).

24 Edward Cornish, *Futuring: The Exploration of the Future* (Bethesda: World Future Society, 2005).

25 Hubert Knoblauch, "Focused Ethnography," *Forum Qualitative Sozialforschung/ Forum: Qualitative Social Research* 6, no. 3 (2005): Art. 44.

26 Sascha Dickel, *Prototyping Society. Zur vorauseilenden Technologisierung der Zukunft* (Bielefeld: transcript, 2019).

27 Trevor J. Pinch and Wiebe E. Bijker, "The Social Construction of Facts and Artefacts: Or How the Sociology of Science and the Sociology of Technology Might Benefit Each Other," *Social Studies of Science* 14, no. 3 (1984): 399–441; Bruno Latour, *Aramis, or the Love of Technology* (Cambridge: Harvard University Press, 1996).

28 Valentin Janda, *Die Praxis des Designs: Zur Soziologie arrangierter Ungewissheit* (Bielefeld: transcript, 2018).

29 Karin Knorr-Cetina, "Objectual Practice," in *The Practice Turn in Contemporary Theory,* ed. Eike Savigny, Karin Knorr-Cetina, and Theodore R. Schatzki (London: Routledge, 2001).

30 Alfred Schütz, "Tiresias, or Our Knowledge of Future Events," in *Collected Papers II: Studies in Social Theory,* ed. A. Brodersen (The Hague: Martinus Nijhoff, 1976), 289.

31 Schütz, "Tiresias," 289–290.

32 Janda, "Praxis," 208–209.

33 Schulz-Schaeffer and Meister, "Laboratory Settings," 11.

34 Knorr-Cetina, "Objectual Practice," 181.

35 Knorr-Cetina, "Objectual Practice," 181–182.

36 Rheinberger 1992, 33.

37 Janda, "Praxis," 200–201.

38 Lina Dib, "Of Promises and Prototypes: The Archaeology of the Future," *Limn* 0 (2010), https://limn.it/articles/of-promises-and-prototypes-the-archeology-of-the-future/.

39 Schulz-Schaeffer and Meister 2017, "Laboratory Settings," 13.

40 Ian Hacking, *The Emergence of Probability: A Philosophical Study of Early Ideas about Probability Induction and Statistical Inference* (Cambridge: Cambridge University Press, 1975), 34–37; see also Thomas Kelly, "Evidence," in *The Stanford Encyclopedia of Philosophy,* ed. Edward Zalta (Stanford: Stanford University, 2016).

41 Kelly, "Evidence."

42 Adam Toon, *Models as Make-Believe: Imagination, Fiction and Scientific Representation, New Directions in the Philosophy of Science* (Basingstoke: Palgrave Macmillan, 2012).

43 Thomas F. Gieryn, *Cultural Boundaries of Science: Credibility on the Line* (Chicago: University of Chicago Press, 1999).

44 Hackathons are events in which hardware and software developers build prototypes collaboratively within a predefined time frame. In the case of civic hackathons, citizens are encouraged to develop technical solutions to address local or global issues and challenges. Peter Johnson and Pamela Robinson, "Civic Hackathons: Innovation, Procurement, or Civic Engagement?" *Review of Policy Research* 31, no. 4 (2014): 349–357. Johnson and Robinson (2014).

45 Alfred Nordmann, "A Forensics of Wishing: Technology Assessment in the Age of Technoscience," *Poiesis and Praxis* 7, no. 1–2 (2010): 5–15; Alfred Nordmann, Bernadette Bensaude-Vincent, and Astrid Schwarz, "Science vs. Technoscience: A Primer, Version 2.0," (PDF) modified December 2011, https://www.philosophie.tu-darmstadt.de/media/philosophie___goto/text_1/Primer_Science-Technoscience.pdf.

46 Bruno Latour, *Science in Action: How to Follow Scientists and Engineers through Society* (Cambridge: Harvard University Press, 1987).

47 Arie Rip, "Technology as Prospective Ontology," *Synthese* 168, no. 3 (2009): 405–422.

48 Schulz-Schaeffer and Meister, "Laboratory Settings," 16.

49 Virgin Hyperloop, "Hyperloop One—Propulsion Open Air Test," uploaded 11 May 2016. https://www.youtube.com/watch?v=xiokghLXFYM.

50 Sascha Dickel and Jan-Felix Schrape, "The Renaissance of Techno-Utopianism as a Challenge for Responsible Innovation," *Journal of Responsible Innovation* 19, no. 2 (2017): 1–6.

51 Latour, *Science*, 64.

52 "Prototypenparty? Was ist denn das?," *Prototypenparty*, accessed 3 February 2019, https://prototypenparty.com/informationen/ (my translation).

53 "Prototypenparty," (my translation).

54 *"Prototypenparty,"* (my translation).

55 Bröckling argues that feedback is a social technique that rests on the expectation that subjects are willing and capable of change ("Und…Wie War Ich? Über Feedback," *Mittelweg 36*, no. 15 (2006): 27–44).

56 Schulz-Schaeffer and Meister, "Laboratory Settings," 16.

57 Toon, *Models.*

58 Mario Kaiser, "Reactions to the Future: The Chronopolitics of Prevention and Preemption," *NanoEthics* 9, no. 2 (2015): 165–177.

59 Nordmann, Bensaude-Vincent, and Schwarz, "Science vs. Technoscience."

6 On Top of the Hierarchy: How Guidelines Shape Systematic Reviewing in Biomedicine

Alexander Schniedermann, Clemens Blümel, and Arno Simons

Certifying Biomedical Evidence

Evidence practices in biomedicine have changed profoundly since the postwar era. After medical treatments ceased to reveal "blockbuster" effects at the beginning of the twentieth century, experts developed and promoted more systematized attempts to determine the most effective "therapeutic interventions."[1] They redefined the ways of how knowledge can be legitimately produced and claimed. In this manner, the randomized-controlled-trial (RCT) combined different methods and techniques and has become the main template for generating biomedical knowledge. Full of ethical and epistemic promises, proponents turned this research design into a marker for "evidence" in biomedical policy and practice, and one which allows for certificating knowledge claims irrespective of the various complexities in their making.[2]

But not only experimental designs became part of the evidence movement in biomedicine. Also, research syntheses summarizing and synthesizing primary research were highly demanded. By using systematic reviewing to find and appraise the relevant trials, as well as meta-analyses as a reliable statistical technique to pool trial data, researchers started to aggregate multiple trial results into overall conclusions about treatment effectivity.[3] Facing the growing output of the "clinical trial industry,"[4] experts hoped that these methods would help to cope with a huge load of information. Further, it was hoped that the aggregation of studies would solve contradictive results and sufficiently represent scientific consensus. These steps toward fact making would then ultimately allow for drawing conclusions and recommendations as to what knowledge should be perceived as "evidence" and which interventions are to be considered effective.[5]

Like clinical trials, systematic reviews were ascribed evidencing qualities based on the intended purpose of the genre as identifying and presenting the most convincing research. Displaying such authority claims, meta-analysis and systematic review have been coined the "platinum standard of evidence,"[6] in comparison to RCT as the "the gold standard."[7] In addition, synthesis formats have been placed at the top of what was called the

DOI: 10.4324/9781003188612-8

"hierarchy of evidence."[8] They have become the main input to "evidence-based medicine," and one of the core areas of international organizations that deal with quality assurance in health care, such as the Cochrane Collaboration which essentially produces and disseminates systematic reviews about medical interventions.[9]

Likewise, the roles of meta-analysis and systematic review in biomedicine became a benchmark for other agoras where research and intervention meet. With generous references to biomedicine, researchers and practitioners from social policy, education or climate science discuss the benefits and potentials of these formats for their areas of expertise.[10] As a result, systematic research syntheses also shaped concepts like "evidence-based policy" and informed the work of the Campbell Collaboration or the Collaboration for Environmental Evidence, organizations that are very similar to Cochrane. Despite the efforts to establish systematic review and meta-analysis in other fields, their diffusion was rather limited which is why biomedicine remained the only research community with a wider adoption of these standardized research formats.[11] For this reason, we consider in more detail what makes systematic reviewing an evidence practice in biomedicine, and the ways in which this practice was established and stabilized.

In order to shed light on how systematic reviews are attributed to represent evidence, we focus on the exploration of methods and guidelines that play a crucial role during the writing of such reviews. Today, systematic reviews are a widely accepted subtype of review articles, the latter often being perceived as a very diverse and little standardized genre in scientific literature.[12] Criticism related to the writing of review articles in science argued that those would suffer from little methodological control and a lack of transparency (e.g. how primary research is included, how to evaluate research, etc.).[13] Against this background, the systematic review article was positioned as a more controlled format of research synthesis, where strict and transparent criteria assure the quality of conduct and reporting. Method experts have developed distinct conceptions of what systematic reviews are, and how they can be separated from other forms of synthesizing research.[14]

As systematic reviews spread among the sciences, their status hinges on the agreed upon standard ways of writing and doing such reviews, that is, guidelines.[15] The epistemic authority of systematic reviews is based on the idea that various forms of biases can be ruled out by providing a strict set of procedures to follow. Since meta-analysis alone was perceived to be incapable of ruling out various biases, systematic reviewing as a higher-order method became pervasive.[16] It is against this background that we intend to explore the systematic review guideline documents in more detail. In steering the author through various stages in the conduct of a systematic review, the method promises to reduce the influence of the author's individual preferences and viewpoints. Rather than defining or certifying the outcome, the systematic reviewing method defines the practices and steps taken within the process. Therefore, the epistemic authority is based on a procedural conception of objectivity, that is, a set of procedural rules which authors need to follow if they want their article to be certified as systematic review.[17]

We are focusing our analysis on more recent guidelines for systematic reviewing, as these can be best understood as solutions for what was called the "new worries of science."[18] In addition to the sometimes detrimental influence of the pharmaceutical industry on the outcome of trials,[19] experts also criticized how the pressure to publish influences which knowledge is reported how and where.[20] Reporting guidelines such as the "preferred reporting items for systematic reviews and meta-analyses,"[21] (PRISMA) inform the methodologies for *reporting* systematic reviews, thus the writing of the reports. We have therefore chosen to focus our analysis on these guidelines, in order to explore the ways in which the renegotiating of evidence practice took place in the biomedical field.

Effectively being a checklist for review authors, PRISMA represents the claims for procedural conception of objectivity that is common to systematic review methodologies. Yet, at the same time PRISMA seems to be different from the rather extensive methodological frameworks for systematic reviewing, for example, defined by the Cochrane Collaboration which guides the whole process of registering, writing and publishing reviews. Different to these large infrastructures, PRISMA consists of a small checklist of rules that, if followed by the author, promise to ensure the credibility of the decisions made during the systematic review process.

In contrast to clinical treatment guidelines, PRISMA influences the practices of researchers rather than doctors. As such, its successful dissemination and implementation seems to depend on the voluntary acceptance of a wider community, rather than the enforcement of a clinical director or a healthcare system. Treatment guidelines may not be accepted, and instead create lines of conflict between the social contexts of the guideline developers and the guided individuals.[22] In addition, because the making of treatment guidelines incorporates various actors—such as researchers, practitioners, industry agents, or healthcare officers—the resulting definition of a medical practice is shaped by a multitude of interests and values correspondingly.[23] Likewise, actors and corresponding values define the evidence practice of systematic reviewing when issuing guidance for systematic reviews. Thus, the making and dissemination of PRISMA entangles characteristics of scholarly communication, evidence practices, and practices from clinical guideline making. To understand the processes leading to the production of PRISMA guidelines and the justification of their status as evidence practices, we address the following questions.

First, how and by whom was PRISMA constructed, and which factors and arguments were negotiated to make it persuasive and acceptable? Second, how was PRISMA disseminated in biomedical communities, and how has it become so pervasive for authors of systematic reviews? Third, if the social and practical configurations that led to the construction of PRISMA mirror the making of medical practice guidelines, how do procedural and interventionist paradigms interfere with the prevailing modes of evidence practices or academic cultures?[24]

We believe that understanding these processes can shed light on how evidencing practices are constructed. Interestingly, the case of guidelines for systematic reviews provides an example of how instruments initially designed to guide practitioners outside academia (guidelines) were taken back and adapted to what is central for the research enterprise, the art of presenting evidence in scientific writing. In the next section, we present our approach to tackle the questions mentioned above. Presenting and discussing our results, we provide five sections that are based on different aspects in constructing PRISMA: the creation of a narrative, making the guidelines credible, applicable, and explaining how the guidelines were disseminated, and implemented. Finally, we present a conclusion based on our research questions.

Methods and Theoretical Background

To tackle the research questions above, we conceptualized reporting guidelines as a specific genre of literature, that lies between typically academic and also more performative modes of narrating.[25] Based on this conception, we performed a document analysis to reveal the guideline's rationales and aims as well as the reported methodology that led to its construction. To further deepen our understanding, we developed and performed expert interviews based on the findings.

In a first document search, we explored the environment of the PRISMA guidelines by identifying relevant publications listed on the guideline's website as well as the website of the EQUATOR network, which collects and disseminates various reporting guidelines.[26] We further used the Web of Science bibliometric database to identify all guideline updates, translations, and guideline forks. Focusing on the main PRISMA publications, we performed an exploratory document analysis to study not only its bibliographic but also the textual characteristics.[27] This analysis was guided by a framework which focused on issues of scholarly communication, such as publication formats, authors, referencing, or number of pages.

Key to understanding guidelines such as PRISMA as a genre of their own is the idea that they offer a specific and shared set of communicative purposes which are different from that of traditional research articles.[28] First of all, guidelines appear to be more technical than other forms of scientific writing, providing a list of rules to follow. Yet for these rules to be followed and have an impact, guideline publications needed to be granted legitimacy by community members.[29] Therefore, guidelines also contain textual elements designed for persuasion, that is, a narrative or rationale why guidelines and standards are needed and why the concrete effort appears to be plausible. For this reason, the document analysis focused on exploring the ways in which the guidelines and the process of creating them were designed to create academic credibility.

To further deepen our understanding of what has been found in the document analysis, we planned and performed semi-structured interviews with six guideline experts during the first quarter of 2021. Interview participants were sampled from authors, translators, or workgroup members of one or more of the PRISMA publications or its updates. While interviews have been anonymized for ethical reasons, the interviewees have been asked to provide substantial information about their role in the development process. Analyzing the interviews, we used deductive coding based on the key themes identified in the document analysis as well as theoretical accounts of standardization, the construction of clinical practice guidelines, and communication in academic communities.

Each of the interview participants had different roles in the making of the earliest version of the guidelines—QUOROM—or one of its later updates. Two were involved as authors in the 1999 version (Participants B and C), and its 2020 version (Participants E and F), while only one authored the 2009 version of the guidelines' explanatory document (Participant A). Most interviewees were part of the expert group that developed PRISMA, yet all of them were researchers active in various fields, such as information retrieval, research ethics, clinical research, or research design. In the 2009 version of PRISMA, two participated in the guidelines (Participants A and B) and one also contributed to the explanatory document (Participant B). Although many workgroup members were listed as authors in the 2020 update, two interviewees remained mere workgroup members (Participants A and B). While all of the interviewees have at least some background in the biomedical sciences, one participant is especially known for his work as an editor of an academic journal (Participant B). Several are also associated with the Cochrane Collaboration (Participants A, B, and F). In addition, one interviewee has substantial experience in the medicine related industry (Participant C).

Our analysis is guided by a framework that focuses on "the multiple 'worlds' of a guideline."[30] It consists of four "worlds," or repertoires, that express the discursive forces in the construction of guidelines. First, the repertoire of science consists of argumentative strategies about the identification and appraisal of relevant evidence and mainly comes in the form of systematic reviews or meta-analyses.[31] Second, the repertoire of practice, which attempts to link a guideline text to the practices in a clinic in order to evaluate its usefulness. Third, in the repertoire of politics, guideline developers envision different stakeholders to discuss the acceptability of the included statements, and how the guideline embodies political positions that may affect power relations. Last, the repertoire of process understands the group as an apparatus of knowledge generation, which constructs guidelines by employing a reliable methodology.[32]

By identifying such dimensions in the creation of PRISMA, we can analyze substantial similarities between the construction of clinical practice guidelines and reporting guidelines for researchers.

However, we must consider the substantial differences between clinical practices and evidence practices in scholarly communication, and inform the analytical framework proposed above accordingly. Different to other fields, biomedicine has a well-established set of research techniques which lead to determinable and replicable outcomes. When facing new research problems, biomedical researchers can rely on an agreed upon set of practices and tasks to establish knowledge claims. This low level of task uncertainty is characteristic of what Richard Whitley has termed "professional ad-hocracy,"[33] an organizational configuration of professions which are characterized by their degree of division of labor. We are interested in how these task uncertainties are reduced and accepted due to technical stan-dardization and formal training. Such standards enable the communication between distant communities by changing local disciplinary practices in relation to overarching goals and aims.[34] But how does such standard-setting take place?

Scientific communities are structured by disciplinary authorities that define standards and practices in their local domain.[35] In order to retain their status, such authorities eventually become more resistant to overarching standardi-zations. In comparison to other scientific fields, biomedical communities in particular are more fragmented into local centers of authority that have to negotiate new forms of standards on a dynamic level.[36] In recent years, biomedicine and its subfields have also been influenced by the strengthening of role models to act more autonomously in negotiating priorities between research and application—or bench and bedside.[37] For this reason, estab-lishing standards has become more complicated and requires more rhetoric and discursive effort, in order to not only to be accepted by researchers, but also to compete against other potential forms of standardization, for example, provided by the Cochrane Collaboration. In addition, standard conflicts may not be avoided or resolved by overarching or central authorities, as in the case of clinical practice in which checklists and treatment guidelines are im-plemented by clinical directors or healthcare policies.[38] Rather, the standard must be made highly compatible with existing and agreed upon disciplinary as well as local regulatory authorities, for example editors of academic journals as the gatekeepers of scientific fields.[39]

Given these theoretical considerations, we extended the "multiple worlds framework," proposed by Tiago Moreira, to fit the specific char-acteristics that can be found in academic practices and scholarly commu-nication. First, we put a much stronger emphasis on how the standard employs a narrative for the profession to provide rationales for application and build the "repertoire of science." Second, we focused on the role of academic journals in disseminating the guideline, for example publishing the guidelines as well as enforcing it by implementation. And we further investigated how a "repertoire of journals" also influenced the creation of the guidelines in order to fit it into the role of the gatekeepers of science.

Results and Discussion: Invoking the Crisis

PRISMA attempts to influence the practice of writing systematic reviews. It is dependent on making authors aware of the genre's shortcomings, and creating acceptance for change. To understand how PRISMA constructs such acceptance and calls for change, we analyzed its argumentative strategies and their evolution. Essentially, in order to persuade readers, the guidelines establish a narrative which makes the practice of systematic reviewing problematic.

PRISMA creates a story about the current state of systematic reviewing, the problems, and the potential solutions—of which the PRISMA guidelines are only one. As such, they not only construct a profession and "enroll" the reader into their reasoning,[40] but also employ an operational mode that calls for action.[41] The professional story unfolds with the role of systematic reviews and meta-analyses for contemporary biomedicine:

> Systematic reviews and meta-analyses have become increasingly important in health care. Clinicians read them to keep up to date with their specialty, and they are often used as a starting point for developing clinical practice guidelines.[42]

Beyond iterating the functions and epistemic promises of systematic reviews, this story also creates narrative links to the actors who value and promote them to shape a community.[43] While guideline developers, doctors or healthcare systems are mentioned in all versions of PRISMA, in its first iteration (the QUOROM guideline), it also mentions the Cochrane Collaboration.[44]

In a second step, the guideline texts explain the various flaws in the reporting of systematic reviews. In doing so, the text refers to the common variance in the quality of scientific publications, rather than accusing systematic reviewing or review authors more explicitly.

"As with other publications, the reporting quality of systematic reviews varies, limiting readers' ability to assess the strengths and weaknesses of those reviews."[45] In addition, the text refers to several observations to support these claims, and persuades the reader that applying the guidelines will provide a viable solution to the presented problems. In citing studies that either prove the lacking reporting quality of reviews, or evaluate how a guideline can improve reporting, the PRISMA document combines two strands of research into a new story.[46] Utilizing the repertoire of science, it gathers the support of several researchers and their studies to become a technical document itself. This makes rejecting its statements and conclusions more tedious, since critics have to reject all supporting references and claims.[47]

Each version—QUOROM in 1999, and PRISMA in 2009 and 2020—witnessed roughly a decade of development in systematic reviewing. While the main rationale and several arguments can be found in each version in a similar way, their differences relate to the genre's role for scientific communication. With QUOROM as its first version,[48] the narrative of the responsible profession was built particularly around meta-analyses—the statistical aggregation technique—even though the guideline mentions the conceptual differences between meta-analysis and systematic reviewing by referring to the Potsdam consultation held in 1994, which was one of the earlier international gatherings to discuss the state of research syntheses:[49]

> Several queries addressed the distinction between the meta-analysis and systematic review. As we indicate in the introduction, and throughout the statement, the QUOROM group agreed to observe the distinction as defined by the Potsdam consultation on meta-analysis.[50]

In the later versions this was changed.[51] A more inclusive wording was used to explicate a wider methodological scope and applicability. Besides renaming the guideline from QUOROM to PRISMA to actually include the words "systematic review," its narrative sections now mentioned how the systematic review has become important to actors other than healthcare decision makers or researchers. It adds guideline makers, clinicians, funders, and even journal editors, and thereby claims that the guidelines' role in changing practices is relevant to a wider array of publics and communities.

The updates in 2020 renew the original intentions and target audiences but still mention wide applicability, even for nonmedical practices such as "social or educational interventions."[52] Although its narrative employs an overall more neutral tone, it distinctively establishes links to other actors that gather around the guidelines. Most notably, it mentions the wide array of PRISMA extensions that modify the 2009 version in order to account for specific review methods, study types, or disciplinary specialties. In addition, the document refers to the EQUATOR network of reporting guidelines, and explains its compliance with the network's guidance on creating reporting guidelines.[53] It thereby stabilizes the PRISMA guidelines by situating them in a network of standardization organizations.

Similar to how the narratives within clinical guidelines evolved from healthcare decision making to other uses of knowledge syntheses, guideline developers aligned PRISMA's story with current topics in the research community. Most notably, several participants elaborated on the relation between PRISMA and the problem of reproducibility or replicability of systematic reviews.[54] Subordinating the problems of systematic reviews under those "new worries of science,"[55] and with PRISMA as a weapon in the "credibility revolution,"[56] the overall endeavor is equipped with substantial intellectual value and societal legitimization.

The stories, and their evolution during the development of PRISMA and its updates, construct multiple professions. In the early versions, a diverse set of actors such as researchers, practitioners, or policymakers are narrated into a community of users and producers of systematic reviews. Furthermore, PRISMA's terminology and selection of items for inclusion shapes this community's genuine understanding of what a systematic review is. Framing the epistemic crises in and around systematic reviewing, the story fuels the quality assurance movement in contemporary biomedicine. This consists of academic researchers and publication experts, who address the problems of reviews by scientifically evaluating the genre. Similar to how traditional review articles narrate topics, assumptions, and results into a scientific field,[57] the guidelines connect topics, arguments, studies, and even institutions.[58] How this profession was shaped into an active discourse community will be elaborated in the next section, in more detail.

Guidelines in the Making

In order to be granted legitimacy, the guidelines had to present the process by which the different items relevant for the practice of systematic reviewing, the contributing guideline authors, and the description of the tasks were articulated. As such, the content of the PRISMA guidelines and its updates were drafted and discussed in several meetings, surveys, and conferences. Based on interviews with the guideline authors and document analysis, we will now further elaborate on the processes that shaped the making of PRISMA. Beside the role of the steering committee, the formation of expert groups, and formal consensus practices, the enrollment of various actors effectively fostered the dissemination of the guidelines and initiated a network of what might be considered as PRISMA's own professional community.

We first analyzed how, and in what ways, experts were invited, and what role they played in convincing communities to change their practices. The core of the guidelines was formulated by a small group of actors. A central role was taken by the steering committee—as the method sections of the guidelines and our interviews have revealed. As the first iteration of PRISMA was built from scratch, this committee collected available knowledge about reporting, and drafted the first items. It was this group which also coordinated later revisions or versions of the guidelines. Although the updates were built upon each other and involved additional actors, a few core experts still led the development by writing substantial parts of the texts, and dealing with comments. Recalling experiences with other guidelines, interviewees stressed how the steering committee's efforts can benefit or harm the overall workflow:

When the team is very capable, they would have done a lot of prep (aration) work ahead of time, before they engage with you to solicit input on the specific things. When the team is less capable … it's going around and around and you never get to what needs to be done or how this can be done in a way that not only incorporate most people's feedback but also efficiently … in terms of time-wise and how many rounds of revision we need.[59]

Crucial to the construction of PRISMA were the involved experts. Starting with 29 contributors in 1999, growing to 42 in 2009, and finally to 139 in 2020, PRISMA and its network grew substantially over time. Similar to what Latour has called "bringing friends in" to explain the argumentative force of referencing,[60] the number and status of the involved experts provides the guidelines with intellectual and social authority. Since most intended recipients and readers of PRISMA are authors of systematic reviews, the group must achieve a proper representation of disciplinary and methodological plurality to avoid conflicts with local authorities, such as prolific specialists or groups that usually establish disciplinary practices and standards:

So we did the first guideline and it was an interdisciplinary group of people. There were statisticians. There were trialists, people who run randomized controlled trials. And perhaps most importantly, we had some influential journal editors.[61]

Authoring a guideline document such as PRISMA can boost one's academic impact, due to the high citation rates of this genre.[62] In addition to selecting experts based on their skills and roles of contribution, selection criteria must warrant the expert's proper motivation to participate: "As you probably know, these guidelines are highly cited. People want to have them on their CVs. It's beneficial to your career. Sure, if you have something that's been cited a thousand times."[63]

Analyses of each individual author's affiliation reveal that these were carefully selected, each representing different fields of expertise. Negotiating and defending their boundaries against other experts, prolific researchers take a central role and make local authority claims, for example, about valid methods or canonical interpretations. By agreeing or disagreeing on the narratives, other disciplinary researchers gather around these new cores, and build social structures and networks.[64] The disciplinary experts who contributed to the new standard can transform the perception of who (or what) the local authorities are. The local integration of PRISMA devalues systematic reviews that do not comply with this standard. Therefore, as with other standardizations, the negotiation and integration of PRISMA may reshuffle authority claims in biomedical disciplines.[65] Yet, according to our interviewees, the effects of guidelines are often less glamorous than expected though we have indication from our fieldwork that PRISMA has achieved to become a particularly reputed way of reporting systematic reviews, especially when it is planned as a collaborative endeavor.

But the selection of experts focused not only on the intellectual contribution of experienced researchers but also on the effective dissemination of the guidelines. By bringing experts from different fields into the process of formulating the guidelines, the guidelines become related to these different fields of which these authors were part of. Therefore, the representation of targeted groups is crucial for the acceptance and dissemination of guidelines, as it creates awareness and supports "marketing."[66] The involvement of journal editors particularly served this role, as one interviewee pointed out:

> I remember we would be looking at who were the key stakeholders and journal editors were key with the view that both you've got the voice of the journal editor influencing [the] guideline, but you've (also) got the ability for other journal editors to say, "Oh, somebody of my type was involved. Maybe I should pay more attention to it." And then it was a bit of lobbying of the journals to say, "This is a good thing to be doing. It will help your transparency."[67]

By establishing cross-disciplinary narratives and enrolling various disciplinary authorities, PRISMA can be disseminated and implemented into several biomedical subcommunities. Furthermore, since there is no central authority that issues regulations and standards—setups that can be found, for instance, in clinical practice—guidelines must enroll and cooperate with local authorities in order to become pervasive.[68] In representing the shared efforts of such connected local authorities, the PRISMA documents establish the network of group members and their institutional affiliations. To display this expert group, members were turned into a composite author called "the PRISMA group," that was listed as last author.[69] In contrast to merely listing the contributors' names in the acknowledgement, this rather uncommon type of authorship stresses that members contributed in multiple ways in addition to the writing of text. Yet at the same time, the text employs a personal tone which makes not only the authors visible but also the wider group.[70]

How the different experts collaborated with each other on PRISMA is outlined as a method section in the guideline document. This establishes a textual link between the procedure and the resulting guideline items. Additionally, the resemblance to other reports of empirical research presents guideline making as a research process itself. As such, PRISMA attempts to create causal links between the group meetings and the resulting rules for reporting.[71] By unfolding and communicating this link, the text of the guidelines provides the repertoire of process, and attempts to understand the making of PRISMA as a reliable methodology to negotiate and consent on a proper set of rules.[72] Therefore, the social configuration of the expert group as well as its communicative processes become an abstract recipe that contains mechanisms that build trust and acceptance by the biomedical community.

Turning the construction of PRISMA into an abstract procedure, its techniques were related to other agreed upon methods or standards. This increases the transparency and acceptability of the PRISMA standard, since authors can understand the standard's connections to their local practices.[73] For example, the guidelines and the interview respondents both mentioned the prominent role of the Delphi methodology to achieve formal and reliable consensus. Promoted by the United States' National Institutes of Health (NIH) during the 1970s and 1980s, it has become an established procedure to organize expert consensus and mitigate individual biases.[74] In another example, the PRISMA authors had to comply with the authorship standards of the International Committee of Medical Journal Editors, by omitting "the PRISMA group" as an author in the 2020 version.

Unsurprisingly, guideline construction procedures became standardized themselves. Influential guideline experts have formalized this procedure into "guidance for developers of health research reporting guidelines,"[75] in essence a guideline for creating guidelines, or a practice that constructs evidence practices. As interview participants noted, the PRISMA group was eager to standardize their own communication procedures and contribute to such a guideline. Recalling efforts by one of the leading experts, one interviewee mentioned:

> I mean there is standardization at the root of the guideline, but then it's the process and getting to that guideline, that also is its own standardized process that I think he's trying to standardize because there are many different approaches.[76]

In turn, the latest update of the PRISMA guideline was based on this now standardized procedure and also reports its compliance with it. Such "network-building" provides a mutual legitimization of both standards—the guidelines for reporting reviews, and the guidelines for creating guidelines—and makes them more authoritative.[77] Turning its own construction even further into a procedurally regulated endeavor, PRISMA benefits from the same valuations as systematic reviews that comply with PRISMA. Since PRISMA turns practices into evidence-practices by the procedural inscription of values, guideline making has also been transformed into an evidence-practice. In other words, after redefining how knowledge turns into authoritative evidence, the guideline can become evidence by the very same definition.

Beyond the construction and interaction with guideline objects, developers further professionalized the role and authority of guidelines by systematically evaluating their effectiveness. Titles such as "Epidemiology of systematic reviews,"[78] witness how epidemiological methods and concepts were turned into a new professional narrative, coming from "meta-epidemiology,"[79] into more interdisciplinary conceptions of "meta-research."[80] Leaving the boundaries of epidemiology aside, interdisciplinary researchers, institutes, journals, and educational programs focused on providing the "the repertoire of 'science'" in guideline development.[81] Aligning these tasks with the overall goal of moving toward higher quality science and evidence-based medicine, the network became a constituency that not only benefits from various expertise, but also from the representation of different groups.[82]

As we have shown in this section, the steering group attempted to manage what has been called the "second tension" throughout this book. Although the definition and evaluation of guidelines is heavily influenced by scientific practices, the formed constituency provides a cross-professional space where review authors, journal editors, and guideline researchers are invited to contribute. This helped to make the issues of PRISMA visible and valuable to those various contexts. At the same time, the possibility to equally influence the logic of the guidelines avoids the impression that one group or field authoritatively translates its knowledge to another. Thus, potential frictions between the diverse groups were actively minimized throughout the process.

Applicability by Simplicity

In order to influence writing practices, a guideline needs to formulate concise rules which authors can follow. In this section, we deal with how these items are formulated and—based on the interview results—what motivated the guideline authors to do so.

PRISMA complies with the typical form of a journal publication. As such, it not only provides a structure consisting of a narrative and methodology section but also presents its regulatory items like research results. First, in a sample flow chart, it displays how authors of systematic reviews can properly report the stages of including and excluding primary studies. Second, the item list displays the various aspects and rules of proper reporting, which may be used as a checklist by authors. It grew from six overarching categories that roughly address the stages of reviewing,[83] to 27 separate and rather detailed items in PRISMA's 2009 version.[84] With the latest update, some of the items were given additional subitems so that the overall number did not change in the 2020 version. As interviewees have noted, much of the efforts during item formulation negotiations were related to stripping down the number of items and rules so that the guidelines would not only be publishable, but also easy to use. Not surprisingly, some respondents have explained that PRISMA represents only the minimum or mandatory reporting, rather than what could be considered as recommended or optimum.

> Because things were discussed (that) we couldn't possibly put everything in. So we had sort of an outline about which would be the required items in the checklists. And that's what the majority of the time the in-person meeting was involved with, then we all went away.[85]

In its focus on PRISMA's applicability, the group took a different approach to other attempts in standardizing the methodological quality of reporting or conduct. Especially in the case of systematic reviewing, potential guidelines compete against the epistemic and social authority of the Cochrane Collaboration. By providing extensive and strict methodological guidance for conduct and reporting, software infrastructures, detailed editorial supervision, and publication in its own database, the Cochrane Collaboration defines the overall process of systematic reviewing on a much more comprehensive level.[86] For this reason, interviewed participants not only mentioned the greater quality of Cochrane reviews in conducting and reporting but also the much higher efforts to perform and write them. Thus, the group also implicitly focused on applicability by juxtaposing their goals and efforts to those of Cochrane.

> Cochrane is very detailed, and there's Cochrane for every kind of thing that you want to do. It was clear from the get go, that we were not going to be a Cochrane group, we were not going to hang together for the rest of our careers doing this.[87]

Limiting the size and scope of guidelines is a common phenomenon in the construction of clinical treatment guidelines. Since the guideline has to fit into its applicatory context, the group tries to envision the "repertoire of 'practice,' which refers to the usefulness of the guideline."[88] Developers weigh the guidelines' knowledge claims against the circumstances of different contexts and usage cases in order to judge whether the guidelines will be useful and improve the overall outcome. In contrast, extraordinary situations may demand improvisation and are not covered by the guidelines' knowledge claims, rendering it inapplicable or useless. Envisioning such cases enables developers to configure the standard's complexity and scope of application.

A careful consideration of complexity and scope is particularly important for the dissemination of multidisciplinary guidelines, such as PRISMA.[89] Highly applicable guidelines provide substantial support in most situations and practitioners perceive them to be useful, so that they willingly accept the guideline. In addition, translations into different languages increased its applicability among nonacademics, especially medical practitioners, as one interviewee noted.[90] Therefore, what some interviewees perceived to be "the minimum," may be better interpreted as the most cost-effective level of reporting, in which the effectiveness of each rule is weighed against the necessary efforts to comply with it.[91]

Seen in this light, PRISMA's particularly restricted focus on reporting makes it applicable as a standard for systematic reviews of various study types and research topics. In contrast to Cochrane's standard, which provides guidance for the design and conduct of reviews, PRISMA explicitly regulates only the writing of reviews, and tells authors what details have to be included in their manuscript. Although the distinction between conduct and reporting is fuzzier in the case of reviews than in clinical trials, the focus on reporting avoids conflicts with methodological plurality, as participants explained. In that sense, the guideline limits its own scope and controls the number and size of targeted groups, which makes potential conflicts manageable.[92] However, the restriction to reporting also mobilizes the regulatory capabilities of academic journals, as we will explain in the next sections.

Dissemination by Journals

The guidelines were published in biomedical journals in order to effectively reach potential authors of systematic reviews. In this section, we discuss some of the unique characteristics of its publications, found during the document analysis and the interviews.

The PRISMA documents were published in multiple scientific periodicals. While QUOROM was published only once,[93] the 2009 version was published in seven different journals,[94] and was officially translated three times,[95] in order to quickly approach multiple biomedical sub-communities. Likewise, the explanation and elaboration document—an additional paper that offers a more detailed look into the guideline's items and also provides some examples—was published in five different journals in 2009 and translated once.[96] The latest update has been published as a pre-print, and in five different journals.[97] In addition, there was one additional explanatory document.[98] Since cross-publication is usually considered unethical for researchers, interviewees noted how it was suggested and discussed by the team's publication experts.

Beside the potential for faster communication, cross-publication also promised wider access, since researchers and their respective institutions may have different journal subscriptions. But it also means the involvement of multiple editorial offices and peer review processes, which demands huge efforts from the guideline developers. In addition, different editorial standards or peer reviews can lead to huge variations in the final documents, although all are intended to represent the same standard.[99] More strikingly, since the guidelines already incorporate the intellectual contributions of many experts, peer reviews may undermine the sophisticated consensus practices. As one participant noted:

> What I would call the aftermath of that ... was a general sense of, "we're not going to do this again." Why are we publishing in so many journals the same thing? And it is one of the challenges with publishing. That means what are the journal(s) supposed to do about peer review, given that so many dozens of people have been involved in reaching this consensus? How can any of that be changed by one or two peer reviewers?[100]

The guidelines were placed among some of the "big five,"[101] which are the most impactful journals in biomedicine. While QUOROM was the only document in *The Lancet*, PRISMA of 2009 and its 2020 update were published in the *British Medical Journal* and *Annals of Internal Medicine*. Other journals were *PloS Medicine*, the *Journal of Clinical Epidemiology*, the *International Journal of Surgery*, *Systematic Reviews*, *Open Medicine* (discontinued), *Physical Therapy*, and the *Italian Journal of Public Health*. In addition, four official translations are listed on the PRISMA website, of which three were also published in national journals.

In the interviews, participants argued that the group targeted high impact journals, as these have more resources for methodological quality assurance, and the necessary awareness for improving reporting. But more importantly, such journals improve the dissemination, since they are more central and authoritative within their respective communities. High impact can be achieved by specialized journals that have become local authorities within usually smaller sub-disciplines. On the other hand, generalized journals often reach high citation impacts too. Since such generalized journals target various audiences or even link many sub-disciplines, their published methods and standards are subject to intense criticism and academic competition.[102] Therefore, the successful enrollment of more generalized journals into PRISMA's narrative provides the guidelines with cross-field legitimization and authority.

In publishing and implementing PRISMA, a journal can make an individual commitment toward reporting quality in its domain. The journal employs the professional story of the guidelines, and aligns itself with the experts and networks that developed the guideline; whether they are from the same domain as the journal or not. But similar to the guideline developers that contributed mainly in order to boost their academic recognition, journals benefit from PRISMA's high impact too. Not surprisingly, interview participants noted that cross-publication was brought up by journal editors, and some of them might have been motivated by improving the impact metrics of their periodicals. Similar to scholars who can shape their epistemic practices in accordance with such metrics,[103] journal editors try to influence their metrics by inviting authors or to solicit review articles.[104] Likewise, journal editors can participate in the development and dissemination of highly cited guidelines and standards.

Enforcement by Gatekeeping

In contrast to other standard innovations, PRISMA would also be implemented into editorial processes, so that authors of systematic reviews would have to comply with PRISMA in order to get their manuscript accepted. In this section, we discuss the extent and role of this practice. Usually, new standards or methods for "doing science" become established by orchestrating and demonstrating their epistemic superiority, so as to finally persuade individual researchers. Whereas evaluation and demonstration of effectiveness ordinarily requires time and additional resources, as explained above, PRISMA was legitimized with the help of the participation of prolific researchers and, as will be shown, academic journals.

Traditionally, individual researchers demand more evidence for a standard's effectiveness before they comply with, and this can hinder quick acceptance and dissemination.[105] Many methods or standards become pervasive only due to traditional modes of academic quality assurance; largely, peer review. Since, in a peer review system, experienced and prolific scientists evaluate the intellectual contributions of other researchers, they can utilize whatever standard or guideline they find appropriate. Although this system generally ensures a certain level of academic quality control, it notably did not prevent the decrease in reporting quality in the first place.

However, peer reviewers may now consider PRISMA when they appraise the reporting quality of systematic reviews, or when editors ask them to do so. But this still means that the dissemination and application of guidelines is dependent on the awareness, and application by many individuals. So, varying expertise in relation to reporting does not only result in authors submitting incompletely reported manuscripts, but also peer reviewers not detecting the flaws.

"Unfortunately, you're assuming everybody is at (a) certain level. You are assuming people would know how to write a paper, but that's not the case."[106]

In addition to the role of peer reviewers in the evaluation of manuscripts, editorial offices also have substantial influence on deciding whether a submission will be published or not. Usually, editors decide which submission will be sent for review, and select appropriate peers, which is why they are called the "the gatekeepers of science."[107] They determine the intellectual contexts of a submitted manuscript and also heavily influence formal characteristics of the texts, such as writing styles or references.[108] In fact, interviewees have mentioned that editorial offices may even have aggravated the reporting crisis by putting limitations on word or reference counts. Nevertheless, guideline developers hoped that PRISMA would be implemented into journal processes and overseen by editorial offices or method editors.

Ideally, authors would fill out the checklist and submit it as additional material together with their manuscript. The editorial office or specifically trained methods experts would then appraise the manuscript and the checklist in order to judge the level of compliance. Due to the procedural nature of PRISMA, compliance can be checked on a per-rule basis, which turns the overall handling into a more formalized process that can be performed without proficiency in either the content or method of the review. Such a process could even be automated, similar to the checking of statistical reporting.[109] Utilizing editorial capacities in advance of the peer review system would provide the necessary centrality to ensure that every systematic review is judged by the same criteria. This increases the authority of PRISMA by extending its intellectual authority with a more formalized type of regulation enacted by the gatekeepers of science.

Beyond the publication in high impact journals, several journals officially endorsed the guideline. As the 2020 version concludes, PRISMA was endorsed by almost 200 journals or organizations publishing systematic reviews, which provide the guidelines with the necessary outreach to achieve a wide dissemination. However, the mere endorsement does not clarify the actual level of implementation which can vary a lot. So, while engaged editors verify the submitted checklists and occasionally ask authors for further clarifications, other endorsers may just publish them together with the manuscript.[110]

As our interviews have shown, the goal of making PRISMA implementable by academic journals played a crucial role in the making of the guidelines. As Tiago Moreira suggests, guideline makers discuss the acceptability of the guidelines by assessing the "repertoire of 'politics.'"[111] In doing so, they envision the guidelines' recommendations through the lenses of the various groups and identify potential lines of conflict. In interpreting the relations between authors and journals, the PRISMA group imagined the latters' regulatory capacity and decided what editors can demand from authors, before they withdraw their submissions and turn to a competitor. As already indicated above, the often emphasized boundary between conduct, as performing the review, and reporting, as writing down the results, reflects the careful appraisal and utilization of these regulatory capacities. In other words, PRISMA was tailor-made to be implemented on the journal level, as participants explained:

> Journals are able to implement a guideline by way of telling authors who must adhere to this guideline, it's a lot easier to get to authors and researchers that way. I think that's one of the key reasons from my understanding as to why reporting guidelines have been the focus over conduct guidelines.[112]

Reducing the guidelines to a very specific set of values and requirements enables the journal to establish it as a required standard. Although academic journals have achieved some level of authority independent of individual experts or the overall referee system, they are subject to various epistemic and social constraints.[113] As such they must navigate between the economic expectations of the publishers, and authors who might submit to a different periodical if the imposed formal requirements become too burdensome or not applicable to their research.[114] Similar to when clinical guidelines are neglected whenever anomalies occur during medical treatments, authors can turn to a different journal when their systematic reviews are of such a type that PRISMA is not applicable. In this respect, the spread of PRISMA is more like standardization in the industry, in which standards are issued by a variety of competing organizations that develop and market their standards as a form of decentralized regulation or soft law.[115]

Templating Evidence in Biomedicine

In this study, we tried to understand how biomedicine's most appraised evidence practice, the writing of systematic reviews, is shaped and defined by formal standards, most notably the "preferred reporting items for meta-analyses and systematic reviews,"[116] or PRISMA. Understanding this reporting guideline as an attempt to standardize evidence practices, we investigated how texts, researchers, methodologists, and journal editors got engaged in social configurations and practices to create a guideline that is applicable, and acceptable. To do so, we performed a document analysis of the PRISMA reporting guideline, and interviews with its developers. By using the "multiple worlds" framework suggested by Tiago Moreira,[117] we identified several similarities and differences between the construction of PRISMA and clinical treatment guidelines.

The emergence of PRISMA exemplifies the interaction of medical practice and biomedical research, described as the "second tension" throughout this book. In telling a coherent story about the problems of systematic reviewing and how those can be solved, PRISMA focuses on the practices of reviewing and how those can be improved procedurally. Conceived as a regulatory tool, PRISMA represents an intervention aimed at improving reports of systematic reviews so that they can be considered as biomedical evidence. This originates from what is often understood as the core of evidence-based medicine, in which doctors are provided checklists and standards to increase treatment uniformity and reduce errors.[118] Likewise, PRISMA tries to guide authors through the writing of systematic reviews to ensure that all reviews contain the necessary information and nothing is left out. So, the basic principles behind the improvements of medical practices were used to improve evidence practices such as systematic reviewing.

In becoming a new standard, PRISMA interferes with already established practices and local authorities that have defined the characteristics of systematic reviews. In our analyses, we found that some efforts were taken to make PRISMA applicable, as well as acceptable. Besides keeping the guidelines rather simple for easy application, it was designed with respect to the regulatory capabilities of academic journals. Since PRISMA focuses only on reporting, compliance can be enforced or supervised by editorial offices. This turned the guideline not only into a regulatory tool for journal editors, but also fostered a wider dissemination and application. In addition, its developers formed a professional community that evaluates endorsements and compliance with PRISMA. This community also provides updates and extensions in order to keep PRISMA relevant to contemporary trends in biomedical research.

Our analysis has shed some light on the configurations and decisions that enabled the wide dissemination and application of the PRISMA guideline in a diverse and global endeavor that lacks in central authorities. As such, this study provides insights into standardizations and the very mechanisms that shape evidence practices in science. Since disciplines such as the social sciences, psychology or climate science also proclaimed crises, our results can be used to inform similar efforts in those fields, or at least, explain if and when such forms of overarching standardizations are inapplicable.[119] However, more research is needed to better understand some factors affecting the dissemination and application of PRISMA, as well as the reflexive discourse that follows its implementation.

Notes

1 Marcia L. Meldrum, "A Brief History of the Randomized Controlled Trial. From Oranges and Lemons to the Gold Standard," *Hematology/Oncology Clinics of North America* 14, no. 4 (2000): 745.

2 Harry M. Marks, *The Progress of Experiment: Science and Therapeutic Reform in the United States, 1900–1990* (Cambridge: Cambridge University Press, 1997), 1–14; Heiner Raspe, "Eine Kurze Geschichte Der Evidenz-Basierten Medizin in Deutschland," *Medizinhistorisches Journal* 53, no. 1 (2018): 71–81; Stefan Timmermans and Aaron Mauck, "The Promises and Pitfalls of Evidence-Based Medicine," *Health Affairs (Project Hope)* 24, no. 1 (2005): 18–28.

3 D.J. Cook, D.L. Sackett, and W.O. Spitzer, "Methodologic Guidelines for Systematic Reviews of Randomized Control Trials in Health Care from the Potsdam Consultation on Meta-Analysis," *Journal of Clinical Epidemiology* 48, no. 1 (1995): 67–171. While often confused or used interchangeably, we follow the distinction stemming from the Potsdam Consultation. It defines "systematic review" as the overall and structured process of defining a question, searching for relevant studies and appraise their quality. In contrast, "meta-analyses" is a statistical aggregation technique that can be applied in systematic reviews.

4 Meldrum, "A Brief History of the Randomized Controlled Trial," 755.

5 Morton Hunt, *How Science Takes Stock: The Story of Meta-Analysis* (New York: Russell Sage Foundation, 1999), 1–19.

6 Jacob Stegenga, "Is Meta-Analysis the Platinum Standard of Evidence?" *Studies in History and Philosophy of Science Part C: Studies in History and Philosophy of Biological and Biomedical Sciences* 42, no. 4 (2011): 497.

7 Stefan Timmermans and Marc Berg, *The Gold Standard: The Challenge of Evidence-Based Medicine and Standardization in Health Care* (Philadelphia: Temple University Press, 2003), 26.

8 Maya Goldenberg, "Iconoclast or Creed?: Objectivism, Pragmatism, and the Hierarchy of Evidence," *Perspectives in Biology and Medicine* 52, no. 2 (2009): 168.

9 Miriam Solomon, *Making Medical Knowledge* (Oxford: Oxford University Press, 2015), 105–132.

10 See Neil R. Haddaway and Gary S. Bilotta, "Systematic reviews: Separating fact from fiction," *Environmental International* 92–93 (2016): 578–584.

11 Ann Oakley, David Gough, Sandy Oliver, and James Thomas, "The Politics of Evidence and Methodology: Lessons from the EPPI-Centre," *Evidence & Policy* 1, no. 1 (2005): 6–14.

12 Ali Sorayyaei Azar and Azirah Hashim, "Towards an Analysis of Review Article in Applied Linguistics: Its Classes, Purposes and Characteristics," *English Language Teaching* 7, no. 10 (2014): 76–88.

13 Judy Virgo, "The Review Article: Its Characteristics and Problems," *The Library Quarterly* 41, no. 4 (1971): 275–291.

14 Clemens Blümel and Alexander Schniedermann, "Studying Review Articles in Scientometrics and Beyond: A Research Agenda," *Scientometrics* 124, no. 1 (2020): 714–717.

15 David Moher et al., "Preferred Reporting Items for Systematic Reviews and Meta-Analyses: The PRISMA Statement," *BMJ* 339, no. 1 (2009): b2535. Hereafter referred to as "Moher et al., *BMJ* 339."

16 Victor Montori, Marek Smieja, and Gordon Guyatt, "Publication Bias: A Brief Review for Clinicians," *Mayo Clinic Proceedings* 75, no. 12 (2000): 1284–1288.

17 Saana Jukola, "Meta-Analysis, Ideals of Objectivity, and the Reliability of Medical Knowledge," *Science & Technology Studies* 8, no. 3 (2015): 102.

18 Janet Kourany, "The New Worries about Science," *Canadian Journal of Philosophy* 50, (2020): 1.

19 Marks, *Progress of Experiment,* 343–355.

20 See Sarah de Rijcke et al., "Evaluation Practices and Effects of Indicator Use—a Literature Review," *Research Evaluation* 25, no. 2 (2016): 161–169.

21 Moher et al., *BMJ* 339.

22 Stefan Timmermans and Steven Epstein, "A World of Standards but Not a Standard World: Toward a Sociology of Standards and Standardization," *Annual Review of Sociology* 36, no. 1 (2010): 69–89.

23 Tiago Moreira, "Diversity in Clinical Guidelines: The Role of Repertoires of Evaluation," *Social Science & Medicine* 60, no. 9 (2005): 1975–1985.

24 Joan Busfield, "'A Pill for Every Ill': Explaining the Expansion in Medicine Use," *Social Science & Medicine* 70, no. 6 (2010): 937; Sandra J. Tanenbaum, "Knowing and Acting in Medical Practice: The Epistemological Politics of Outcomes Research," *Journal of Health Politics, Policy and Law* 19, no. 1 (1994): 27–31.

25 Charles Bazerman and James Paradis, *Textual Dynamics of the Professions: Historical and Contemporary Studies of Writing in Professional Communities* (Madison: University of Wisconsin Press, 1991), 3–10; John Swales, *Research Genres: Explorations and Applications* (Cambridge: Cambridge University Press, 2004), 1–32.

26 "PRISMA statement," PRISMA, accessed May 4, 2021, www.prisma-statement. org; "The PRISMA 2020 statement," EQUATOR, accessed 4 May 2021, www. equator-network.org.

27 Glenn Bowen, "Document Analysis as a Qualitative Research Method," *Qualitative Research Journal* 9, no. 2 (2009): 27–40; Kalpana Shankar, David Hakken, and Carsten Osterlund, "Rethinking Documents," in *The Handbook of Science and Technology Studies,* ed. Ulrike Felt, Rayvon Fouche, Clark A. Miller, and Laurel Smith-Doerr (Cambridge: MIT Press, 2017), 59–85.

28 Brun Latour, *Science in Action: How to Follow Scientists and Engineers through Society* (Cambridge: Harvard University Press, 1987), 21–62; Swales, *Research Genres,* 1–32.

29 Timmermans and Epstein, "World of Standards," 94–99.

30 Moreira, "Diversity," 1976.

31 Solomon, *Making Medical Knowledge,* 105–132.

32 Moreira, "Diversity," 1982–1984.

33 Whitley, *Intellectual and Social,* 187.

34 Nils Brunsson and Bengt Jacobsson, *A World of Standards* (Oxford: Oxford University Press, 2002), 1–16; Timmermans and Epstein, "World of Standards," 69–89.

35 Timmermans and Epstein, "World of Standards," 83.

36 Whitley, *Intellectual and Social,* 187–193.

37 Clemens Blümel, "Translational Research in the Science Policy Debate: A Comparative Analysis of Documents," *Science and Public Policy* 45, no. 1 (2018): 24–35; Barbara Hendriks, Arno Simons, and Martin Reinhart, "What Are Clinician Scientists Expected to Do? The Undefined Space for Professionalizable Work in Translational Biomedicine," *Minerva* 57, no. 2 (2019): 219–237.

38 Tanenbaum, "Knowing and Acting in Medical Practice," 27–31.

39 Timmermans and Berg, *Gold Standard*, 55–81; Timmermans and Epstein, "World of Standards," 69–89.

40 Michel Callon, "Some Elements of a Sociology of Translation: Domestication of the Scallops and the Fishermen of St Brieuc Bay," *The Sociological Review* 32, no. 1 (1984): 211–214.

41 Bazerman and Paradis, *Textual Dynamics*, 8–10.

42 See Moher et al., *BMJ* 339, 1.

43 Greg Myers, "Stories and Styles in Two Molecular Biology Review Articles," in *Textual Dynamics of the Profession*, ed. Greg Myers and James G. Paradis (Wisconsin: University of Wisconsin Press, 1991), 64–70.

44 David Moher et al., "Improving the Quality of Reports of Meta-Analyses of Randomised Controlled Trials: The QUOROM Statement," *The Lancet* 354 (1999): 1896. Hereafter referred to as "Moher et al., *The Lancet* 354."

45 See Moher et al., *BMJ* 339, 1.

46 Ryan Boyd, Kate Blackburn, and James Pennebaker, "The Narrative Arc: Revealing Core Narrative Structures through Text Analysis," *Science Advances* 6, no. 32 (2020): 1–9.

47 Latour, *Science in Action* 32–33.

48 See Moher et al., *The Lancet* 354, 1896–1900.

49 see D.J. Cook et al., 67–171.

50 Moher et al., *The Lancet* 354, 1899.

51 See Moher et al., *BMJ* 339, 1–8.

52 See Matthew Page et al., "The PRISMA 2020 Statement: An Updated Guideline for Reporting Systematic Reviews," *BMJ* (March 29, 2021): 1. Hereafter referred to as "Page et al., *BMJ*."

53 See Iveta Simera et al., "A Catalogue of Reporting Guidelines for Health Research," *European Journal of Clinical Investigation* 40, no. 1 (2010): 35–53.

54 Steven Goodman, Daniele Fanelli, and John Ioannidis, "What Does Research Reproducibility Mean?" *Science Translational Medicine* 8, no. 341 (2016): 1–6.

55 Kourany, "New Worries," 1.

56 Simine Vazire, "Implications of the Credibility Revolution for Productivity, Creativity, and Progress," *Perspectives on Psychological Science* 13, no. 4 (2018): 411.

57 Clemens Blümel. "What Synthetic Biology Aims At," in *Community and Identity in Contemporary Technosciences*, ed. Karen Kastenhofer and Susan Molyneux-Hodgson (New York: Springer International Publishing, 2021), 65–84.

58 Eric Borg, "Discourse Community," *ELT Journal* 57, no. 4 (2003): 398–400; Myers, "Stories and Styles," 45–76.

59 Interview Participant F in discussion with the authors, March 2021.

60 Latour, *Science in Action*, 31.

61 Interview Participant C in discussion with the authors, March 2021.

62 Based on the 2019 version of the database at the German Kompetenzzentrum Bibliometrie, PRISMA and its explanation document have ~49k and ~14k citations in Scopus and ~32k and 10k~ citations in Web of Science.

63 Interviewing Participant E in discussion with the authors, March 2021.

64 Christoph Luetge, "Economics in Philosophy of Science: A Dismal Contribution?" *Synthese* 140, no. 3 (2004): 279–305; Kevin Zollman, "The Epistemic Benefit of Transient Diversity," *Erkenntnis* 72, no. 1 (2010): 17–35.

65 Whitley, *Intellectual and Social*, 144.
66 Anneke Francke et al., "Factors Influencing the Implementation of Clinical Guidelines for Health Care Professionals: A Systematic Meta-Review," *BMC Medical Informatics and Decision Making* 8, no. 1 (2008): 1–11.
67 Interview Participant A in discussion with the authors, March 2021.
68 Timmermans and Berg, *Gold Standard*, 55–81.
69 See Moher et al., *BMJ* 339, 1.
70 Myers, "Stories and Styles," 56–59.
71 Bazerman, *Shaping Written Knowledge*, 260; Karin Knorr-Cetina, *The Manufacture of Knowledge: An Essay on the Constructivist and Contextual Nature of Science* (Oxford: Pergamon Press, 1981), 114–120; Eddy Lang, Peter Wyer, and R. Brian Haynes, "Knowledge Translation: Closing the Evidence-to-Practice Gap," *Annals of Emergency Medicine* 49, no. 3 (2007): 355–363.
72 Moreira, "Diversity," 1982–1983.
73 Timmermans and Berg, "Standardization in Action," 273–305.
74 Solomon, *Making Medical Knowledge*, 99.
75 David Moher et al., "Guidance for Developers of Health Research Reporting Guidelines," *PLoS Medicine* 7, no. 2 (2010): 1.
76 Interview Participant E in discussion with the authors, March 2021.
77 Timmermans and Berg, "Standardization in Action," 273–305.
78 M. Alabousi et al., "Epidemiology of Systematic Reviews in Imaging Journals: Evaluation of Publication Trends and Sustainability," *European Radiology* 29, (2019): 517.
79 Jong-Myon Bae, "Meta-Epidemiology," *Epidemiology and Health* 36, (2014): e2014019.
80 John Ioannidis et al., "Meta-Research: Evaluation and Improvement of Research Methods and Practices," *PLOS Biology* 13, no. 10 (2015): 1.
81 Moreira, "Diversity," 1978.
82 Arno Simons and Alexander Schniedermann, "The Neglected Politics behind Evidence-Based Policy: Shedding Light on Instrument Constituency Dynamics," *Policy & Politics* 49, no. 4 (2021): 513–529; Arno Simons and Jan-Peter Voß, "The Concept of Instrument Constituencies: Accounting for Dynamics and Practices of Knowing Governance," *Policy and Society* 37, no. 1 (2018): 14–35.
83 See Moher et al., *The Lancet* 354, 1897.
84 Moher et al., *BMJ* 339: 5–6.
85 Interview Participant C in discussion with the authors, March, 2021.
86 Iain Chalmers, Larry Hedges, and Harris Cooper, "A Brief History of Research Synthesis," *Evaluation & the Health Professions* 25, no. 1 (2002): 29–30.
87 Interview Participant C in discussion with the authors, March 2021.
88 Moreira, "Diversity," 1980.
89 Francke et al., "Factors," 1–11.
90 Interview Participant D in discussion with the authors, March, 2021.
91 Timmermans and Berg, *Gold Standard*, 70–81. The reduced scope of PRISMA has also led to the emergence of various forks that attempt to slightly adjust the guideline to be more applicable for other data types or study designs. However, those have not been investigated in this study.
92 Kristina Tamm Hallström. "Organizing the Process of Standardization," in *A World of Standards*, ed. Nils Brunsson and Bengt Jacobsson (Oxford: Oxford Scholarship, 2002), 86–99; Cook Sackett and Spitzer, "Methodologic Guidelines," 167–171.
93 See Moher et al., *The Lancet* 354: 1896–1900.
94 See Moher et al., *BMJ* 339; David Moher et al., "Preferred Reporting Items for Systematic Reviews and Meta-Analyses: The PRISMA Statement," *Open Medicine* 3, no. 3 (2009): 123–130; David Moher et al., "Preferred Reporting Items for

Systematic Reviews and Meta-Analyses: The PRISMA Statement," *Annals of Internal Medicine* 151, no. 4 (2009): 264–269; Moher et al., *BMJ* 339; David Moher et al., "Preferred Reporting Items for Systematic Reviews and Meta-Analyses: The PRISMA Statement," *PLoS Medicine* 6, no. 7 (2009): e1000097; David Moher et al., "Preferred Reporting Items for Systematic Reviews and Meta-Analyses: The PRISMA Statement," *Journal of Clinical Epidemiology* 62, no. 10 (2009): 1006–1012; David Moher et al., "Preferred Reporting Items for Systematic Reviews and Meta-Analyses: The PRISMA Statement," *International Journal of Surgery* 8, no. 5 (2010): 336–341; David Moher et al., "Preferred Reporting Items for Systematic Reviews and Meta-Analyses: The PRISMA Statement," *Physical Therapy* 89, no. 9 (2009): 873–880.

95 See David Moher et al., "Principais Itens Para Relatar Revisões Sistemáticas e Meta-Análises: A Recomendação PRISMA," trans. Taís Freire Galvão and Thais Andrade, *Epidemiologia e Serviços de Saúde* 24, no. 2 (2015): 335–342; David Moher et al., "Bevorzugte Report Items für systematische Übersichten und Meta-Analysen: Das PRISMA Statement," trans. Andreas Ziegler, Gerd Antes, and Inke R. König, PRISMA, 2010; David Moher et al., "Linee guida per il reporting di revisioni sistematiche e meta-analisi: il PRISMA Statement," trans. Elena Cottafava and Marco Da Roit, *Evidence* 7, no. 6 (2015).

96 A. Liberati et al., "The PRISMA Statement for Reporting Systematic Reviews and Meta-Analyses of Studies That Evaluate Healthcare Interventions: Explanation and Elaboration," *Annals of Internal Medicine* 151, no. 4 (2009): 65–94; A. Liberati et al., "The PRISMA Statement for Reporting Systematic Reviews and Meta-Analyses of Studies That Evaluate Healthcare Interventions: Explanation and Elaboration," *BMJ* 339, no. 1 (2009); A. Liberati et al., "The PRISMA Statement for Reporting Systematic Reviews and Meta-Analyses of Studies That Evaluate Healthcare Interventions: Explanation and Elaboration," *PLoS Medicine* 6, no. 7 (2009); A. Liberati et al., "The PRISMA Statement for Reporting Systematic Reviews and Meta-Analyses of Studies That Evaluate Healthcare Interventions: Explanation and Elaboration," *Italian Journal of Public Health* 6, no. 4 (2009): 354–391; A. Liberati et al., "The PRISMA Statement for Reporting Systematic Reviews and Meta-Analyses of Studies That Evaluate Healthcare Interventions: Explanation and Elaboration," *Journal of Clinical Epidemiology* 62, no. 10 (2009): e1–34; A. Liberati et al., "PRISMA Statement per il reporting di revisioni sistematiche e meta-analisi degli studi che valutano gli interventi sanitari: spiegazione ed elaborazione," trans. Elena Cottafava, Fabio D'Allesandr, and Christiana Forni. *Evidence* 7, no. 6 (2015).

97 See Matthew Page et al., "The PRISMA 2020 Statement: An Updated Guideline for Reporting Systematic Reviews," MetaArXiv (2020); Page et al., *BMJ*; Matthew Page et al., "The PRISMA 2020 Statement: An Updated Guideline for Reporting Systematic Reviews," *Systematic Reviews* 10, no. 89 (2021); Matthew Page et al., "The PRISMA 2020 Statement: An Updated Guideline for Reporting Systematic Reviews," *International Journal of Surgery* 88 (2021): e105918; Matthew Page et al., "The PRISMA 2020 Statement: An Updated Guideline for Reporting Systematic Reviews," *PLoS Medicine* 18, no. 3 (2021): e1003583; Matthew Page et al., "The PRISMA 2020 Statement: An Updated Guideline for Reporting Systematic Reviews," *Journal of Clinical Epidemiology* 134, (2021): 103–112.

98 See Matthew Page et al., "PRISMA 2020 Explanation and Elaboration: Updated Guidance and Exemplars for Reporting Systematic Reviews," *BMJ* 372, (2021): n160.

99 See David Moher et al., "Guidance for Developers," 7. This was not systematically investigated in this analysis. However, differences in formatting and punctuation, as well as wording are visible. For instance, while the PloS version mentions "field [of research]," the BMJ version says "specialty."

100 Interview Participant A in discussion with the authors, March 2021.

101 Pascale Allotey, Caitlin Allotey-Reidpath, and Daniel Reidpath, "Gender Bias in Clinical Case Reports: A Cross-Sectional Study of the 'Big Five' Medical Journals," *PLoS ONE* 12, no. 5 (2017): 1.

102 Ann Forsyth, "In Defense of the Generalist Journal: Speaking Beyond Silos," *Journal of the American Planning Association* 86, no. 2 (2020): 139–141.

103 See Ruth Müller and Sarah de Rijcke, "Thinking with Indicators. Exploring the Epistemic Impacts of Academic Performance Indicators in the Life Sciences," *Research Evaluation* 26, no. 3 (2017): 157–168.

104 Blümel and Schniedermann, "Studying Review Articles," 722.

105 Kevin Zollman, "The Communication Structure of Epistemic Communities," *Philosophy of Science* 74, no. 5 (2007): 574–587.

106 Interview Participant F in discussion with the authors, March 2021.

107 Diana Crane, "The Gatekeepers of Science: Some Factors Affecting the Selection of Articles for Scientific Journals," *The American Sociologist* 4, no. 2 (1967): 195.

108 Crane, "Gatekeepers," 195–201; Bazerman, *Shaping Written Knowledge*, 136–137.

109 Mohammadreza Hojat, Joseph S. Gonnella, and Addeane S. Caelleigh. "Impartial Judgment by the 'Gatekeepers' of Science: Fallibility and Accountability in the Peer Review Process," *Advances in Health Sciences Education* 8 (2003): 75–96; See Halil Kilicoglu, "Biomedical Text Mining for Research Rigor and Integrity: Tasks, Challenges, Directions," *Briefings in Bioinformatics* 19, no. 6 (2017): 1400–1414.

110 Interview Participant E in discussion with the authors, March 2021.

111 Moreira, "Diversity," 1981.

112 Interview Participant E in discussion with the authors, March 2021.

113 Whitley, *Intellectual and Social,* 18.

114 J.A. García, Rosa Rodriguez-Sánchez, and J. Fdez-Valdivia, "Competition between Academic Journals for Scholars' Attention: The 'Nature Effect' in Scholarly Communication," *Scientometrics* 115, no. 3 (2018): 1413–1432; Vincent Larivière, Stefanie Haustein, and Philippe Mongeon, "The Oligopoly of Academic Publishers in the Digital Era," *PLoS ONE* 10, no. 6 (2015): 1–15.

115 Weisz et al., "Emergence," 691–727; Brunsson and Jacobbson, *World of Standards,* 3; Timmermans and Epstein, "World of Standards," 79–81.

116 Moher et al., *BMJ* 339, 1.

117 Moreira, "Diversity," 1976.

118 Tanenbaum, "Knowing and Acting in Medical Practice," 27–31; Timmermans and Mauck, "The Promises and Pitfalls," 20–21.

119 Vazire, "Implications of the Credibility Revolution," 412.

7   On the (Im)possibility of Identifying the Evidence Base of the Impact of Star Architecture Projects

Nadia Alaily-Mattar, Diane Arvanitakis, Martina Löw, and Alain Thierstein

Multidisciplinary Research of Star Architecture as a Case for Complicated Practice of Evidence

Since the opening of the Guggenheim Museum in Bilbao (GMB), designed by star architect Frank Gehry, the idea has been popularized that exceptional built artifacts can become agents of change for urban settings, wider social and cultural processes, and the economic trajectory of cities, to become "an urban policy in motion."[1] Although a number of factors played into the supposed success of GMB, it was largely attributed to the effects of what the mass media coins "star architecture." City officials became eager to duplicate such assumed effects by commissioning star architecture in their cities. But what evidence is there that such projects trigger effects which result in long term impact? The possibility of identifying any evidence base is complicated by the complex nature of such projects, and by the necessity of undertaking multidisciplinary research. What does such complexity indicate about the nature of evidence in the fields of urban studies, urban planning, and urban policymaking? Does such complexity render impossible the exercise of identifying an evidence base? What can we learn about negotiation processes in the validation of knowledge in these fields?

Based on the idea that multidisciplinary research is required to scientifically investigate the effectiveness of architectural and urban planning interventions, a research group consisting of scholars of architecture, planning, economics, economic geography, sociology, and cultural sociology undertook research into three case studies of star architecture projects. Faced with the challenges of a complex object of study and multidisciplinary research, the research group identified the need to develop and apply a communication tool to manage the complexity of the object of research—star architecture—and the integration of different disciplinary specific problematizations. What was developed was a conceptual impact model—hereafter "impact model"—that captures both the "outputs" of star architecture projects and the assumed effects. Furthermore, the

DOI: 10.4324/9781003188612-9

impact model was to serve as heuristic tool to assist a multidisciplinary group of scholars, with divergent evidence practices, and organize the assumptions that underlie the search for evidence. By visualizing complex relationships between actions, outputs and effects, the impact model assisted in organizing the chain of argumentation and identifying assumptions about causal links in the discussion of the impact of star architecture. In the following, the missing explanatory links for investigating the impact of star architecture projects are outlined. Then, using the case study of Phaeno in Wolfsburg, the usefulness of the impact model for organizing evidence and describing the evidence practices of various disciplines is illustrated. The conclusion discusses the limitations and opportunities of using such a model and its relevance for the understanding of evidence in the field of architecture.

The Missing Explanatory Links: Current Investigations into the Impact of Star Architecture Projects on Cities

The term star architecture is used to capture the nuanced—but not mutually exclusive—aspects of signature/branding, iconicity, and flagshipness[2] of architecture projects, whose recognition status is exceptional. Substantial scholarly research has been undertaken to describe, discuss, and evaluate specific aspects of star architecture projects. However, by focusing too narrowly either on the projects or on the accompanying socioeconomic and cultural changes within cities, much of this research fails to uncovering whether, and if so how, these projects actually trigger a transformative change to their respective cities.[3]

Within architecture itself, research into star architecture is restricted to the monographs of particular star architects, the architecture of cultural buildings, or the iconic aspect of star architecture. The discussion of whether iconic architecture is a new genre or merely a stylistic movement occupies much of the discourse;[4] the question of impact, however, is not addressed. A notable exception is literature by architectural scholars, which applies a political economy perspective to critique the underlying motivations and legitimization of the initiatives behind star architecture projects. Much of this literature builds on the work of Lawrence Vale, who has shown how the production of "landmark" buildings contributes to the communication and consolidation of state power.[5] For example, Deyan Sudjic argues that iconic buildings serve to soothe the "edifice complex" of the rich and the powerful.[6] Maria Kaika goes so far as to argue that these buildings do not just reflect but also "produce a new identity for elites or institutions in need of reinvention."[7] Such literature mostly discusses the political dimensions associated with architectural projects, and how architecture can be used as an instrument for asserting, legitimizing, and creating political power, and inventing new identities. It is concerned with revealing why such projects are developed rather than analyzing and synthesizing why, and how, such projects work the way they do.

Tourism, geography, urban policy, cultural, and political economy studies also rarely consider questions that link the specific characteristics of star architecture projects to effects and impact. For example, in tourism and urban policy studies, the problematization of star architecture projects revolves around measurement issues; particularly how effects can (or cannot) be attributed to a singular attraction. But by considering star architecture only in the context of flagship tourist destinations[8] such studies do not sufficiently flesh out the specificity of star architecture. Research within economics and economic geography focuses on investigating the effects on the local economy and land value capture on, for example, the creation of jobs, the local art scene, as a global networking engine, and as value creator of iconic design.[9] Some research discusses the effects on the local societal enhancement of culture and on hard branding the city.[10] However, as with tourism studies, such investigations do not sufficiently engage with the specific characteristics and capacities of star architecture to generate effects.

It is remarkable that architecture—as the discipline that is foremost in enabling such projects to take shape—seems unconcerned with investigating whether, and if so, how outputs produced by those who are highly recognized in the field meet such exaggerated expectations. A systematic literature review of the *Scopus* and *Avery* databases for peer reviewed articles published between 1997 and 2017 and on the topic of star architecture did not yield a single article by architects concerned with the question of the impact of star architecture.[11] Based on this finding, the question can be posed: Is investigating the impact of star architecture irrelevant to architecture research, or are the evidence practices of architects simply inadequate to describe such an impact? In addition, what does this mean in the light of the move to science- and evidence-based practice in multiple fields of public policy?

Architecture's silence vis-à-vis the scientific investigation of the impact of star architecture can be explained by the need to mobilize research methods outside the typical skill set of architects, who use only the power of their own expert knowledge as evidence. Tourism, urban policy studies, geography, economic geography, political economy, sociology, and cultural urban studies have attempted monodisciplinary analysis, each with their own evidence practices.[12] The disciplines of economics use indicators and statistics, or "hard/tangible evidence"; sociocultural studies make reference to theory and concepts to organize their findings derived from observing the world as "soft/intangible" evidence; as is the morphological analyses undertaken by architects who apply their expert skills to produce knowledge about the built environment. Clearly the practices of evidence differ from one discipline to the other and the application of different research methods generates a different nature of evidence. Varying practices of evidence result in discipline specific problematizations. Investigations into the effects of star architecture projects and the impacts on their respective cities would benefit from bringing into conversation different disciplines, which have generally remained too unspecific regarding the complex nature of the object of analysis.

The Impact Model

The interaction on multiple spatial scales, exemplified by star architecture, serves to magnify the multidisciplinary processes inherent in architecture; and in turn, the decoupling of various disciplinary effects is also possible with a project of this scale.[13] The impacts and effects of star architecture are multisectored[14] and therefore no one discipline alone—especially architecture—is able to comprehensively evaluate such projects and do such an evaluation justice. Multisector effects demand multidisciplinary analysis and thus the impact model was developed as a tool to address the challenges in the process of analysis and evaluation.

The basic building blocks of the impact model—to be read from bottom to top—are (1) the starting conditions out of which a project idea emerged, (2) the activities taken, (3) the resulting outputs, and (4) the effects generated by these outputs (see Figure 7.1).[15] This chapter will focus on describing the third and fourth elements of the impact model, namely, outputs and effects.[16] Outputs occur incrementally at each step, as activities are undertaken. Besides the usual, and tangible construction project outputs, as interest groups form novel alliances around the intention to reposition the city, new actor networks emerge within this process. Unlike most buildings that are simply unremarked upon, star architecture building projects "arouse public attention causing disputes and controversies";[17] an attention that increases throughout the process. Development usually takes many years and is more often than not accompanied by media exposure from its onset. In addition to these new actor networks, the impact model identifies three other main outputs, namely, the material building itself, the function, which is operated inside this building, and the complex and dynamic bundled "offerings" (indicated in the dashed rectangle of Figure 7.1).

The term "offerings" is borrowed from Pine and Gilmore, who use it to highlight the intangibility of outputs also offered alongside the building and its functions.[18] The offering of "spectacle" is often associated with avant-garde architectural style, exuberant costs, and the surrounding controversies. With the commission of a star architect, the spectacle begins, and with it comes media attention. The second offering is that of a "signature," which relates to the recognition value of an architect's signature style, and associated with the fame of the star architect's persona and the brand value of his/her other artifacts. The third offering is that of an "icon," and one which relates to the specific architectural form. Designed to become icons in their respective cities, Jencks argues that iconic buildings must successfully deform codes; they "must carry a negative charge, a paranoia that challenges contemporary taste, a disturbing value, and something new."[19] Both signature and icon offerings are related to the materiality of the

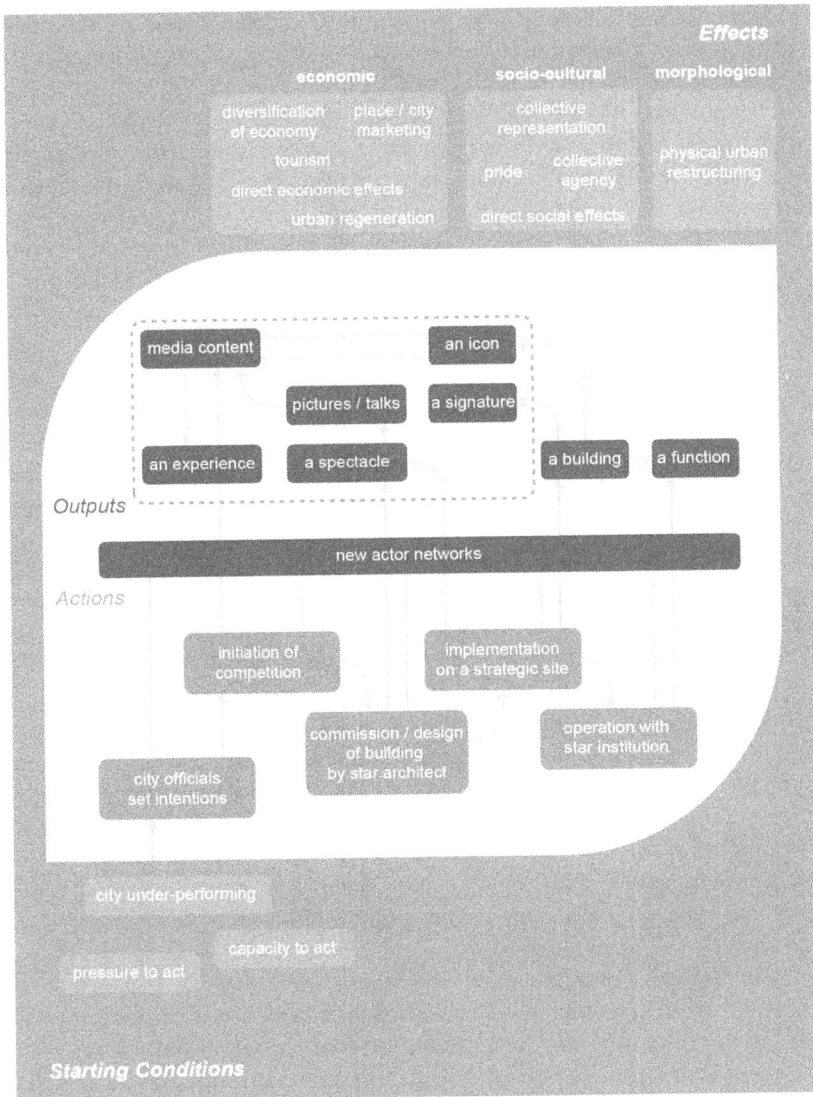

Figure 7.1 A conceptual impact model of how a star architecture project "works." It depicts the flow (bottom to top) from starting conditions to actions, which results in outputs and their respective effects. The particular "offerings" of star architecture—as a subset of outputs—are indicated by the dashed rectangle. Reproduced from Alain Thierstein, Nadia Alaily-Mattar, and Johannes Dreher, "Star Architecture's Interplays and Effects on Cities," in *About Star Architecture: Reflecting on Cities in Europe*, ed. Nadia Alaily-Mattar, Davide Ponzini, and Alain Thierstein (Basel: Springer International Publishing, 2020), 49.

building. With the next stage of action—implementation—an abstract architectural design becomes a physical object, a material reality. This signature of a well-known architect, an icon in the city, now contributes to the production of "pictures and talk"; the fourth offering of star architecture projects. Pictures and talk become "media content" that is circulated widely, and this media content packages and transforms a project into an experience. Pine and Gilmore argue that experiences have become new economic offerings for which demand exists just like for services and goods.[20] Hence, "experience" is another offering of star architecture projects. The dashed rectangle in the model with dynamic and bundled offerings describes the particularity of star architecture as one of the outputs of star architecture projects.

The fourth part of the impact model is labeled "effects," which are supposedly triggered by the outputs. Star architecture projects are developed by public authorities based on the hypothesis that they can deliver certain benchmarked effects, be it morphological, social, or economic. The impact model captures only these three fields of effects. This is not to say that these are the only possible effects of star architecture projects. For example, projects might have an effect on technological innovation in the field of architecture. However, such an effect will not likely play a role in the repositioning of a city. The impact model therefore only includes those effects which might contribute to the repositioning of cities; the model accounts also for adverse effects, which we term "unintended" or "unexpected" effects in contrast to the intended or expected effects. Urban morphological effects could refer to the physical reconfiguration in relation to direct surroundings. Economic effects could refer to the direct economic effects of these projects (the diversification of the economy, city marketing, boosting tourism, and urban competitiveness). Social effects could refer to the direct social effects of these projects, such as access to enhanced amenities in the city, enhancing citizen pride, collective representation, and agency.

In the model, the outputs and effects are deliberately not connected, in order to avoid prejudicing the analysis with prescribed relational flows. Rather the particularity of each case study would dictate investigations as to which output(s) generate which effect(s). In other words, the impact model calls for researchers to ask what outputs generate which effects, and, crucially what evidence can support claims that effects have been generated by a star architecture project.

The Impact Model in Action: The Phaeno Project in Wolfsburg

The Phaeno is a science center, designed by Zaha Hadid for the city of Wolfsburg, Germany, and inaugurated in 2005 (see Figure 7.2). Wolfsburg is a medium-sized city of around 124,000 inhabitants in 2015. The city was founded by the Nazis in the late 1930s as an industrial city adjacent to the Volkswagen (VW) factory. Since its inception, Wolfsburg's fate has been tied to that of VW. In the 1990s, the city found itself in a precarious situation.[21] As a result of the 1992/93 crisis in the automobile industry which affected VW, Wolfsburg faced unprecedented levels of unemployment. Uncertainty about the role of Wolfsburg as a production site and corporate headquarters of the VW group shook the foundations of this relationship between city and corporation. A city that had for decades been at the forefront of progress and prosperity became suddenly aware that it lagged behind, and that as a result, its future was in jeopardy. In response to the shock of the early 1990s, there was huge pressure to act, and the city of Wolfsburg tried to reorient itself. Its main objective became to diffuse the dominance of VW by increasing competitiveness and local attractiveness in relation to external stakeholders such as investors, companies, highly skilled workers, and regional tourists. Paradoxically, VW would actually be a close partner in Wolfsburg's endeavor to stand on its own feet, with a cornerstone of this move toward independence being "a shoulder-to-shoulder alliance" forged in the mid-1990s between VW and the city of Wolfsburg.[22] The result was a strategic concept, titled AutoVision, which laid the ground for a radical reorientation of urban development policy based on a wide-ranging of public-private partnerships.[23] The idea was to make Wolfsburg competitive at a regional level.

A range of measures were undertaken to diversify the economy and labor market structure of Wolfsburg, and the implementation of leisure facilities to boost the tourism sector was a key element. From this urban governance "à la Wolfsburg!"[24] the idea of the Phaeno emerged. In 1999, the city initiated an international architecture competition to develop a vacant lot strategically located along a major axis in Wolfsburg, adjacent to the city's main railway station and high-speed rail stop and VW's factory and its Autostadt, a 28-hectare automobile themed experience park. Opened in 2000, Autostadt proved to be a huge success, attracting about 2.2 million visitors annually in the first few years. The idea of the Phaeno emanated from this incentive of developing a valuable public addition in a strategic plot in Wolfsburg. An architectural design competition was held in January 2000, and the London-based architect Zaha Hadid won. The building was built over a period of five years at a total cost of 99 million euros. Around 250,000 visitors—mostly school children from the region—now visit the Phaeno annually.[25]

The study of morphological effects of the Phaeno project—in terms of the changes in the form of Wolfsburg as a city—has focused on the analysis and comparison of archival material, and has looked to historical figure-ground plans, architectural plans, competition briefs, and documents related to planning laws and urban development strategies as evidence. These were compared with testimonies from interviews, and a phenomenological analysis of visual relationships through photographic material and on-site

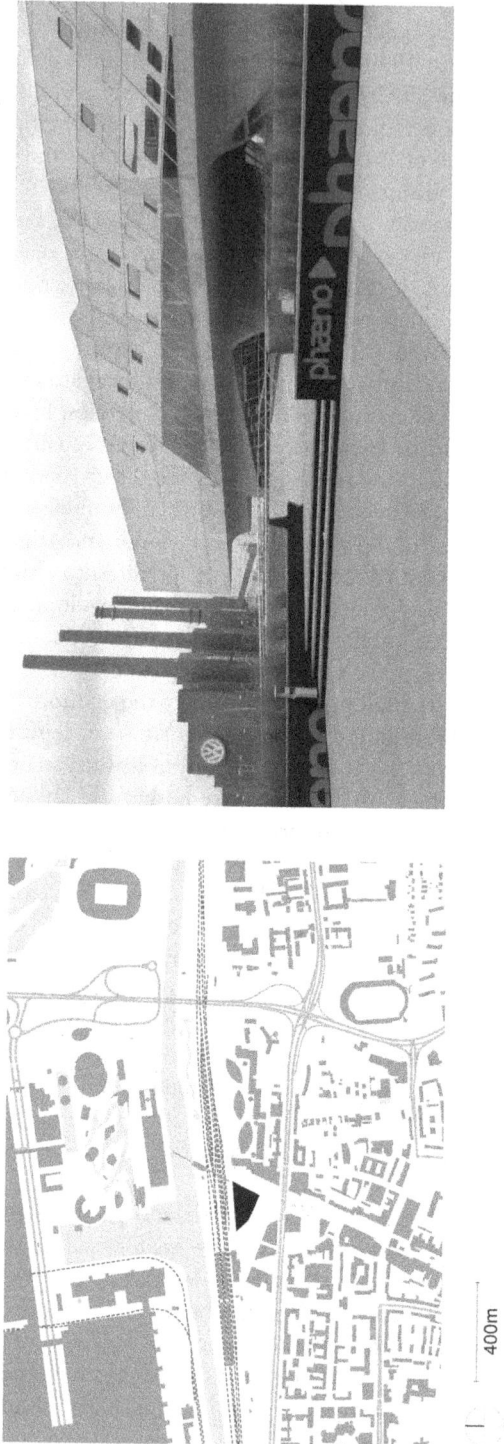

Figure 7.2 To the left, the location of the Phaeno in Wolfsburg (Source: authors, based on OpenStreet Map). To the right image of Phaeno with Volkswagen factory in the back (Source: photograph by Dominik Bartmanski).

observations. The morphological analysis indicates that being one of the first buildings in the undeveloped buffer zone, the Phaeno itself was exposed to physical restructurings, as the later addition of new buildings changed subsequent view sights onto the Phaeno. However, the Phaeno also had a morphological effect on its latterly realized direct surroundings. In other words, any argument of a unilateral morphological effect on its direct surrounding—positive or negative—would be a misunderstanding of the urban situation. Instead, the extravagant appearance of the Phaeno formally and materially integrates the wider urban structure and, in some ways, architecturally orchestrates Wolfsburg's process of repositioning. Surrounded by a morphologically introverted "commercial island urbanism,"[26] the Phaeno mediates between the factory, the train station and the historical Koller axis. In this sense, the Phaeno as a building embodies the key processes of urban restructuring and the city's changing self-perception: the consolidation of a city center, its new transregional connection and its growing self-confidence toward VW. Once seen as "an extension of an industrial zone"[27] rather than as a city in and of itself, and one suffering from a lack of a consolidated urban structure or center, beginning with the Phaeno building, the development of this central part of Wolfsburg in the 2000s radically transformed the physical structure of the city of Wolfsburg.

In the study of sociocultural effects, semi-structured interviews with the proponents of the Phaeno, interviews in the public space of Wolfsburg, and document analyses were undertaken. The evidence in this investigation indicates that the radical transformation of the physical structure of the city of Wolfsburg described also led—among other sociocultural effects—to a transformation of the city's self-image as a suburban residential agglomeration serving the VW factory. In other words, relevant sociocultural effects were triggered and increased by the morphological effects described earlier, a hypothesis confirmed by the interviewees.[28] Further evidence to support this are the findings of an analysis of the city image of Wolfsburg, conducted for the city by the consultancy form CIMA Beratung + Management GmbH. The 2009 analysis was based on telephone interviews with citizens of Wolfsburg (400 interviewees) and its region (600 interviewees). The results indicate that the complex and dynamic bundled "offerings" of the impact model can be linked to accentuated citizen pride and collective representation. Citizen pride effects should be understood in the context of Wolfsburg's recent past, lack of preexisting symbols of pride, and its identity as a showcase of Nazi-era industrial urban planning, a kind of "spoiled" identity.[29] The development of the Phaeno by a globally recognized architect—triggering media attention—signified to citizens a new confidence in the city's life by signaling the city's willingness and ability to take its technological character to a new level, beyond VW. Zaha Hadid's signature extravagant design counteracted the mundane narrative of Wolfsburg, and even if it was not unanimously and instantly praised by the inhabitants, it helped the city reassert itself vis-à-vis its main actor—the VW factory. Indeed, Hadid's confirmation as a star architect following her award of the Pritzker Prize in 2004 and *The Guardian's* listing of the Phaeno as one of the 12 most significant modern works of architecture in the world have a high symbolic value in the city of Wolfsburg.

For the study of economic effects, longitudinal socioeconomic data on the Phaeno and the city of Wolfsburg were analyzed. Statistical analysis was limited to Pearson's square correlation analysis,[30,31] which examines the dependency between two variables, such as overnight stays and visitor numbers. The Phaeno is a science center, a museum with interactive exhibits that encourage visitors to experiment and explore scientific phenomena. Due to its architecture, its function as a science center—which at the time of its creation constituted a unique feature in Germany—and its location it caters predominantly to regional and national audiences with educational interests. Conceived as a threshold to and from the railway station, directly on the ICE high speed rail route to Berlin, the Phaeno gained supra-regional attention and importance. It is the second most visited cultural institution in Wolfsburg after VW's Autostadt,[32] with an average of approximately 252,000 visitors a year (2006–2015).[33,34] The construction of the Phaeno has attracted further projects such as hotels or the Designer Outlet Center, and the proximity to the Phaeno was an important factor for the project initiators, with the Phaeno acting as the "nucleus for further projects,"[35] and contributing to the increasing attractiveness of the city center. The revitalization of the area led, for example, to a strong expansion of gastronomic offerings.[36] However, an increased attractiveness of the city of Wolfsburg and its surroundings as a place of residence cannot be proven quantitatively. An analysis of commuter flows in the period from 2002 to 2012 shows that both the proportion of commuters and the proportion of long-distance commuters increased.[37] In addition, a tourism effect of the Phaeno—using the number of overnight stays as indicator—cannot be identified. The effect is visible in conjunction with other attractions, but the isolated importance of the Phaeno cannot be quantified. The situation is similar with the effects on the labor market. The sixty-five full time jobs created by the Phaeno have had a negligible effect on the city's economy, except when considered together with the knock-on effects in gastronomy, design, and tourism sectors. An analysis of socioeconomic indicators thus shows the difficulty of isolating the effect of the Phaeno, let alone its individual outputs, as the project itself was developed within a larger strategic concept—AutoVision—which saw the development of 12 other attractions in the city of Wolfsburg within a very short time period.

Strengths and Limitations of the Impact Model

This chapter investigates star architecture projects beyond their specific architectural design and their visual appearances and aesthetics. For example, while an increased focus on the visuality of architecture is acknowledged, no investigation is undertaken on what enables architecture to perform as image (offering opportunities for future research). Rather the focus is on whether, and if so, how the performance of architecture as image generates effects. The discussion of aesthetics and unique design considerations does not yield a solid corpus of variables, which can then be scientifically interrogated for effects. So it becomes difficult to generate robust evidence that can be generalized to evaluate impact, especially when considering how individualized the public reception of such complex notions as aesthetics is. Instead, aspects such as design, visual appearance, and aesthetics are abstracted into what they enable a built artifact to "offer";[38] what we call "offerings" in the impact model. These can be treated as evidence (or lack thereof) and interrogated accordingly.

The impact model is a visual and didactic tool that compliments more traditional forms of communication grounded in speaking and writing. Because it assists in organizing findings and identifying the shortcomings of empirical research, the impact model helps identify the inconclusiveness of the evidence where it occurs. For example, Beatriz Plaza investigates the economic effects, but does not consider sociocultural effects and morphological effects.[39] Scholars such as Leslie Sklair and Maria Kaika address the question of the political motivations that drive the development of such projects and legitimize them, but leave aside the question of whether such projects contribute to urban regeneration or tourism.[40] In the case study of the Phaeno, the study of economic effects shows the difficulty of isolating the effects of the outputs of a project in general, let alone the outputs that relate to its star architecture specificity. Would a tin box with a banner rather than a signature architecture have delivered the same number of visitors to the Phaeno and created the same number of jobs? Likewise, the morphological analysis of the Phaeno illustrated the effects of the building on the physical structure of the city of Wolfsburg. But even this analysis does not take into consideration the specificity of star architecture. Only the sociocultural analysis has delivered evidence about the effects of such specificity. The challenge of tracing back effects to star architecture specificity is related to the limitations of research methods—e.g. the changing magnitude and composition of intended and unintended effects over time—and availability of quantitative and longitudinal data. The impact model assists in identifying where such a challenge persists and where it cannot be resolved. As such, using the impact model it is possible to identify where evidence-based arguments are possible or not. This challenge also points to the performative and dynamic nature of evidence practices and to the fact that different evidence cultures exist in different contexts.

The decoupling or analytic separation of effects is useful for operationalizing research. The impact model helps to organize this separation. However, the analytic separation of effects should not sideline the exercise of investigating the intertwining of effects, and how effects support each other to generate impact. In other words, it is important to acknowledge the cross referential and mutually constitutive nature of effects. In Wolfsburg, the realization of Phaeno did radically transform the physical structure of the city. This was initially a morphological effect; but one which in turn drew in the medium-term sociocultural effects in the city. Hence, while the impact model's isolation of effects serves the operationalization of research, the investigation of the impact of star architecture projects on their respective cities must draw on the changing interrelatedness of sociocultural, economic and morphological effects.

This highlights the potential of the using the impact model for urban analysis as a field of dynamic and intertwined relations. It helps us to see particular local developments, for example, that Phaeno has changed the city of Wolfsburg significantly in sociocultural and morphological terms. Using the impact model has the potential to structure viability studies of future high-profile architectural projects. Its conclusions would be drawn from a rich pool of findings, and not restricted by the boundaries or biased by the problematizations of a single discipline. Such an exercise is pressing given the call for evidence-based public policy, especially in the context of the widely celebrated success of such projects, often occurring in parallel to loud and vocal opposing views highlighting their failure.[41]

Scholars of architecture are often concerned with the critique of the physical output of the process; they lament the instrumentalization of their profession to produce outputs that serve political purposes rather than address spatial problems. For political scientists, the process is the object of their inquiry—a process in which there are winners and losers. Economic geographers are frequently concerned with the effectiveness of the outputs of the process, while sociologists are concerned with long-term impact in which different effects intertwine. The impact model, however, enables us to draw on the evidences presented by different disciplines and analyze how the parts are interrelated while not losing track of the bigger picture. For scholars of architecture, drawing on the findings of different disciplines can enable them to generate knowledge that pertains to the exaggerated expectations as to what architecture projects can or cannot deliver. The coexistence and complementary use of the apparently opposing "soft and hard" evidence becomes relevant for any scientifically grounded discussion of the failure or success of star architecture projects, which must first start with a common ground of identifying, describing, and explaining the object of inquiry, its relations, and managing its complexity. The impact model does that in a comprehensive manner, either proactively as a viability assessment, ex-ante, and or respectively for analytical purposes, ex-post, which can serve evidence-based urban planning.

Acknowledgments

This work was supported by the DFG Deutsche Forschungsgemeinschaft, grant number KO 5090/1-1, LO 1144/14-1, TH 1334/11-1. This chapter draws on the following publications emerging from the research group: Alain Thierstein, Nadia Alaily-Mattar, and Johannes Dreher, "Star Architecture's Interplays and Effects on Cities"; Nadia Alaily-Mattar, et al., "Unpacking the Effects of Star Architecture Projects"; Nadia Alaily-Mattar et al., "Situating Architectural Performance: 'Star Architecture' and Its Roles in Repositioning the Cities of Graz, Lucerne and Wolfsburg"; Nadia Alaily-Mattar and Alain Thierstein, "The Circulation of News and Images: Star Architecture and Its Media Effects"; Nadia Alaily-Mattar, Joelean Hall, and Alain Thierstein, "The Problematisation of 'Star Architecture' in Architecture Research"; and Johannes Dreher, Nadia Alaily-Mattar, and Alain Thierstein, "Star Architecture Projects and Their Effects: Tracing the Evidence."

Notes

1 Sara González, "Bilbao and Barcelona 'in Motion.' How Urban Regeneration 'Models' Travel and Mutate in the Global Flows of Policy Tourism," *Urban Studies* 48, no. 7 (2011): 1397–1418.
2 For signature and branding see Donald McNeill, *The Global Architect: Firms, Fame and Urban Form*; for iconicity, see Charles Jencks, "The Iconic Building Is Here to Stay"; for flagshipness, see Adi Weidenfeld, "Iconicity and 'Flagshipness' of Tourist Attractions."
3 Nadia Alaily-Mattar and Alain Thierstein, "The Circulation of News and Images: Star Architecture and Its Media Effects," in *About Star Architecture: Reflecting on Cities in Europe*, ed. Nadia Alaily-Mattar, Davide Ponzini, and Alain Thierstein (Basel: Springer International Publishing, 2020), 97–114.
4 Charles Jencks, "The Iconic Building Is Here to Stay," *CITY* 10, no. 1 (2006); William Saunders, ed., *Commodification and Spectacle in Architecture. A Harvard Design Magazine Reader* (Minneapolis: The University of Minnesota Press, 2005).
5 Lawrence J. Vale, *Architecture, Power, and National Identity* (New Haven: Yale University Press, 1992).
6 Deyan Sudjic, *The Edifice Complex. How the Rich and Powerful Shape the World* (London: Penguin Press, 2005).
7 Maria Kaika, "Architecture and Crisis: Re-inventing the Icon, Re-imag(in)ing London and Re-branding the City," *Transactions of the Institute of British Geographers* 35, no. 4 (2010): 454.
8 Adi Weidenfeld, "Iconicity and 'Flagshipness' of Tourist Attractions," *Annals of Tourism Research* 37, no. 3 (2010): 851.
9 For job creations, see Beatriz Plaza, "Valuing Museums as Economic Engines"; for the local art scene, see Beatriz Plaza, Manuel Tironi, and Silke Haarich, "Bilbao's Art Scene and the 'Guggenheim effect' Revisited"; for global networking, see Beatriz Plaza and Silke Haarich, "The Guggenheim Museum Bilbao: Between Regional Embeddedness and Global Networking"; for iconic design, see Gabriel Ahlfeldt and Alexandra Mastro, "Valuing Iconic Design: Frank Lloyd Wright Architecture in Oak Park, Illinois."
10 For culture, see Luciana Lazzeretti and Francesco Capone, "Museums as Societal Engines for Urban Renewal. The Event Strategy of the Museum of Natural History in Florence"; for societal enhancement, see Graeme Evans, "Measure for Measure: Evaluating the Evidence of Culture's Contribution to Regeneration."
11 Nadia Alaily-Mattar, Joelean Hall, and Alain Thierstein, "The Problematization of 'Star Architecture' in Architecture Research," *European Plannin Studies* 30, no. 1 (2022): 13-31.

12 Alaily-Mattar, Hall, and Thierstein, "Problematization of 'Star Architecture.'"
13 Arguably, multidisciplinary effects are implicit in even the most localized architectural intervention, but it is prohibitive in terms of economies of scale to give them the same degree of attention.
14 "Effect" refers to linear, presumably causal relationships, while "impact" is a process that involves the interplay of effects.
15 Alain Thierstein, Nadia Alaily-Mattar, and Johannes Dreher, "Star Architecture's Interplays and Effects on Cities," in *About Star Architecture: Reflecting on Cities in Europe*, ed. Nadia Alaily-Mattar, Davide Ponzini, and Alain Thierstein (Basel: Springer International Publishing, 2020), 45–53.
16 Other elements are described in published papers by Johannes Dreher, Nadia Alaily-Mattar, and Alain Thierstein, "Star Architecture Projects. The Assessment of Spatial Economic Effects by Means of a Spatial Incidence Analysis," *Raumforschung und Raumordnung | Spatial Research and Planning* 78, no. 5 (2020): 439–453.
17 Albena Yaneva, *The Making of a Building. A Pragmatist Approach to Architecture* (Oxford: Peter Lang AG, 2009), 8.
18 James Gilmore and Joseph Pine, "Beyond Goods and Services: Staging Experiences and Guiding Transformations," *Strategy & Leadership* 25, no. 3 (1997): 11.
19 Jencks, "Iconic Building," 12.
20 Gilmore and Pine, *Beyond Goods and Services*, 11.
21 Nadia Alaily-Mattar et al., "Situating Architectural Performance: 'Star Architecture' and Its Roles in Repositioning the Cities of Graz, Lucerne and Wolfsburg," *European Planning Studies* 26, no. 9 (2018): 1874–1900.
22 Annette Harth et al., *Stadt als Erlebnis: Wolfsburg. Zur stadtkulturellen Bedeutung von Großprojekten* (Wiesbaden: VS Verlag für Sozialwissenschaften, 2010), 191.
23 McKinsey & Company Inc., "*AutoVision – Stärkung der wirtschaftlichen Leistungsfähigkeit der Stadt Wolfsburg*, unpublished Report, 21–25, 1999.
24 Jürgen Pohl, "Urban Governance à la Wolfsburg," *Informationen zur Raumentwicklung* 10, no. 9 (2005): 637–647.
25 Alaily-Mattar and Thierstein, "*Circulation of News and Images*," 97–114.
26 Ulfert Herlyn et al., *Faszination Wolfsburg 1938–2012* (Wiesbaden: VS Verlag für Sozialwissenschaften, 2010), 125.
27 Till Briegleb, "Die Traurigkeit einer Reissbrettsiedlung," *Süddeutsche Zeitung*, 21 June 2016.
28 See for the composition of interviews with stakeholders and inhabitants but also contesting views: Alaily-Mattar, "Situating Architectural Performance: 'Star Architecture' and Its Roles in Repositioning the Cities of Graz, Lucerne and Wolfsburg," 1874–1900.
29 Lauren Rivera, "Managing 'Spoiled' National Identity: War, Tourism, and Memory in Croatia," *American Sociological Review* 73, no. 8 (2008): 613–634.
30 Walter Nägeli, "Situating Star Architecture: The Case of Phaeno in Wolfsburg," in *About Star Architecture: Reflecting on Cities in Europe*, ed. Nadia Alaily-Matta, Davide Ponzini and Alain Thierstein (Cham: Springer Nature Switzerland, 2020), 169–186.
31 Nadia Alaily-Mattar et al., "Unpacking the Effects of Star Architecture Projects," *Archnet-IJAR International Journal for Architectural Research*, (2021): ahead of print.
32 Johannes Dreher, Nadia Alaily-Mattar, and Alain Thierstein, "Star Architecture Projects and Their Effects: Tracing the Evidence," *Journal of Urbanism: International Research on Placemaking and Urban Sustainability*, (2021).
33 Since the first full year of operation (2006–2015), the number of visitors to the Phaeno has fluctuated between 275,000 (2006, 2011) and 225,000 (2007), by a maximum of 18%.
34 Analysis of the origin of visitors highlights the supra-regional importance of Phaeno. According to a report by Phaeno gGmbH 91% of visitors do not come from

Wolfsburg. According to the German newspaper *Die Welt*, two thirds of the Phaeno visitors are not from the greater region of Lower Saxony, and 15% have a journey of at least 250 km. A report of the regional economic effects of the Phaeno conducted by Carmen Kissling and published in 2010, documents the results of an investigation into the spending behavior of Phaeno visitors. The report shows that retail (1.1 million euros), the catering and hotel industry (2.1 million euros), and other service companies (0.9 million euros) benefited from a total of 4.1 million euros in expenditure from outside visitors since the opening of Phaeno in November 2005 to the end of October 2012. This corresponds to a total of approximately 585,000 euros annually. The analysis of the number of visitors to the Phaeno and the increasing number of overnight stays in Wolfsburg show only a medium statistical correlation ($R2 = 0.6$). On the other hand, if one looks at the increasing number of leisure facilities and the increasing number of overnight stays, a strong positive correlation ($R2 = 0.9$) with high significance can be identified. This supports the thesis that the AutoVision concept promoted tourism in Wolfsburg. According to information published by Wolfsburg AG, since 1998—the year in which the AutoVision was adopted—the number of overnight stays in Wolfsburg had doubled by 2014.

35 Dreher, Alaily-Mattar, and Thierstein, "Star Architecture Projects."
36 Herlyn et al., *Faszination Wolfsburg,* 173–174.
37 Bundesagentur für Arbeit, unpublished data, "Employees by Economic Sector," received 25 November 2016.
38 Dreher, Alaily-Mattar, and Thierstein, "Star Architecture Projects."
39 Plaza, "Valuing Museums," 155–162; Plaza and Haarich, "The Guggenheim Museum Bilbao," 1456–1475.
40 Sklair, "Iconic Architecture and Capitalist Globalization," 21–47; Kaika, "Architecture and Crisis," 453–474.
41 See chapters about contesting views on star architectural projects in the cities of Paris, Athens, Vienna, or UNESCO world heritage sites in general, in: Nadia Alaily-Mattar, Davide Ponzini, and Alain Thierstein, *About Star Architecture: Reflecting on Cities in Europe* (Basel: Springer International Publishing, 2020), 247–306.

Part III

Governing Evidence: Evidence-Based Practice and Politics

8 The Thing We Call Evidence: Toward a Situated Ontology of Evidence in Policy

Kari Lancaster and Tim Rhodes

Calls for policy to be "evidence-based" have proliferated within government bureaucracies, academic institutions, and media to the extent that evidence-based policymaking has "become a movement unto itself."[1] However, it is abundantly clear that contemporary governance and policymaking routinely falls short of this ideal. Evidence-based policymaking is in many ways a "hopeless illusion,"[2] and yet the question of what prevents policy being "evidence-based" has remained a perennial preoccupation.[3] Studies seeking to identify barriers and facilitators to the use of evidence for policy have commonly conceived of evidence and policy as two separate domains, with this characterization in turn stimulating efforts to "bridge the gap" and connect these "two communities" through knowledge brokering, transfer, and translation.[4] Although it is widely used, this metaphor of "bridging" through "translation" has increasingly been critiqued as "too linear and simplistic,"[5] constraining thinking about the link between knowledge and the policymaking process.[6] Several problematic premises underpin this idealized characterization of how evidence comes to bear on policy, including the assumption of a technical-rational process in which objective evidence is thought to sit outside of policy processes waiting to be "brought in." The work of policy is far messier, and more complex in practice, as is evidence-making itself. The metaphor of "bridging the gap" through translation has, arguably, served to "close our minds to alternative framings which could add to the illumination and analysis of this complex field" and has perhaps "outlived its usefulness."[7]

As noted in this volume, the translation metaphor tends to underappreciate the co-constitutive force of the field in which evidence is "received" and the mediating effects of evidence-making practices in different contexts. New ways of thinking about the relations between evidence and policy are needed. As Greenhalgh and Russell[8] have argued, acknowledging that the complex world of policy is not "exclusively (or even predominantly) concerned with 'what works'" requires a reconceptualization of how evidence is used for policy, doing away with the assumption that policy can ever be "evidence-based" as conventionally construed, with the selection and presentation of evidence perhaps instead "considered as moves in a rhetorical

DOI: 10.4324/9781003188612-11

argumentation game and not as the harvesting of objective facts to be fed into a logical decision-making sequence." Rather than seeing evidence and policy as two separate domains to be "bridged," there is a need to reframe evidence itself "as part of the 'stuff happening' of policy making."[9] There is a need to attend to evidence as a thing *inside* the process. To make this conceptual move requires a more performative vision which recognizes evidence not as separate, isolatable, or distinct but rather constituted in local practices,[10] entangling in an assemblage of situated relations.

The challenges thrown forth by the COVID-19 pandemic have laid bare the illusory promise of a technical-rational linear process of evidence in translation for policy, illuminating the need for more nuanced ways of thinking about evidence as situated and processual. Idealized conceptualizations of "evidence-based policy" are ill equipped to deal with the uncertainties arising in evolving situations of need.[11] In the early stages of the COVID-19 pandemic calls to "follow the evidence" became a "rallying cry" for action.[12] The limits of this cry were quickly felt as governments scrambled to respond to contain a novel coronavirus by implementing largely untested interventions that radically disrupted ways of life, with modeled projections of potential effect speculatively generating multiple scenarios and filling gaps in what could be known.[13] What it means to "follow the evidence," in the more conventional parlance of evidence-based policymaking, has in many ways been altered as research, deliberation, and implementation have proceeded simultaneously, in greatly compressed timeframes and under conditions of heightened social and political concern. Interventions to restrict movement across borders, close schools and businesses, mandate facemasks, or lockdown communities are not isolatable in their effects but each and all entangle, including socially and politically, with major impacts on work, family, economies, education, welfare systems, health services, cultural life, and citizenship. Urgent and evolving situations of need, such as we are facing in the current global public health crisis, provide a salutary reminder that there are always degrees of uncertainty and that outcomes are without guarantee. We have become aware that the relevant question is no longer simply "what works?" but rather "what constitutes 'evidence-enough'?" for intervening in the unknown,[14] the answer to which is always situated and relational, in-the-moment, value-laden, and political. In this context, to "follow the evidence" means watching and learning to iterate accordingly, in an adaptive mode.

Although there is a tendency to side-step the complexity of the issues which have been more demonstrably brought to bear in the COVID-19 crisis, and act as if evidence and policy can proceed according to the ideal of "evidence-based policy" as conventionally construed, or assume that there can be some return to the status quo, what if we recognized the inherent limits of this ideal characterization and instead adapted our approach? How might these profoundly uncertain conditions help us to finally "let go"[15] of the problematic assumptions which have haunted "evidence-based policy," and move toward an approach which allows us to more carefully attend to the complexity and politics of evidence in its situations of policy deliberation? In an urgent and evolving situation of need, rather than holding on to a "hopeless illusion," is it now possible to shift toward a conceptualization of evidence *in* policy—emergent, contingent, and co-affective—rather than perpetuate an imagined separation of one from the other? And if we did reimagine the relations between evidence and policy in this more performative way, by consequence, what would it mean to "follow the evidence"? How, then, is evidence constituted as a thing being followed?

We suggest the case of the COVID-19 pandemic as a prompt to new thinking not only about how evidence comes to bear on policy, but also about the nature of the thing we call "evidence" itself, reconfiguring it as a situated achievement, made in the here and now of policy deliberation events.[16] We propose a critical move toward a situated ontology of evidence *in* policy, disrupting how evidence has hitherto been characterized as being used *for* policy. The object of evidence itself has been relatively uncontested in the policy literature, generally deemed "capable of being translated into policy under the rubric of 'what works.'"[17] Where difficulties arise, and the relationship between evidence and policy has proven slippery or elusive, these difficulties have been treated as "technical" or "managerial" concerns[18] to be resolved, for example, through the generation of better evidence, better translation models, or better understandings of the policymaking process. Given the multiple evident complexities which have been thrown into sharper relief by the COVID-19 crisis, and calls for a different approach, we suggest that a reconceptualization of the thing we call evidence is warranted. To be clear, by rethinking the question of how evidence is constituted, we are not suggesting further discussion of the particular kind of research or information that may come to bear on policy, the criteria for judging its strength and validity, what kind of study designs might be suited to informing intervention implementation at scale, or how they might be translated better. Nor are we suggesting that scientific or other knowledge undergoes different interpretations as it moves from "objective" research environments and into "politicized" policy settings. These primarily epistemological questions about how the evidenced effects of interventions might be best known, evaluated, and translated have been well rehearsed as foundational concerns within the evidence-based policy literature, and generally hold on to an idea of "the evidence" as a natural, stable, and distinct knowledge object to be used and translated for policy. By proposing a critical move toward a situated ontology of evidence *in* policy we seek to destabilise this assumption, making an intervention to draw renewed critical attention to *a thing* that "might otherwise appear 'finished' or 'ready-made'" and allowing us to "scrutinise those entities that [...] analysis would often consider 'black-boxed.'"[19] The thing we call evidence is one such black-boxed entity. We suggest that redrawing the debate about the relations between evidence and policy on ontopolitical terms,[20] and re-configuring evidence as a situated achievement, helps us to appreciate how the thing we call evidence is enacted in events of policy deliberation, and not merely before it, with its translations constituting *transformations* into different things.[21] By shifting toward a situated ontology of evidence in policy, we can notice how the practices and relations of policy deliberation which enact the thing we call evidence perform the "following" of evidence, even as it is being made, performing "evidence-enough" for action.

Thus, drawing on thinking from science and technology studies, and especially what has been termed "the ontological turn"[22] we invite a shift away from thinking with evidence primarily as a matter of *epistemology* toward thinking with evidence as a matter of *ontology*. We put forward an "evidence-making intervention" approach[23] as a framework for conceptualizing how the thing we call evidence is done and relationally made, thus working with the politics, contingencies, and uncertainties of policy, and making visible how evidence is constituted, gathered, and made to matter within its particular sociomaterial conditions. In this approach, "what is messy is not defective"—a technical problem to erase or manage—but an ontopolitical concern that we can "learn to live with and think with."[24] We suggest that this approach might afford a more critical way of conceptualizing evidence in policy which does not simply ask "What *is* the evidence?," but also asks "*How* is evidence *made*?," "How is evidence *put-to-use*?," and "How is evidence *made-to-matter*?."

The Thing We Call Evidence

We use the turn of phrase "the thing we call evidence" deliberately to open-up the simple familiarity and presumed stability of "evidence" for policy, reimagining this "thing" as complex and entangled, relationally constituted, and transformed, in events of policy deliberation. Following Latour, "a thing is, in one sense, an object out there and, in another sense, an *issue* very much *in* there, at any rate, a *gathering*";[25] it is "both material object and site of dealing and dispute."[26] By emphasizing the thing-ness of evidence we attend to it as an entity in-the-making, not ready-made or preceding but rather continually emergent and enacted in the relations and practices of policy. The term enactment draws specific attention to how things "are brought into being [...] *realised* in the course of a certain practical activity," thus emphasizing "the generative power of the practices involved in the constitution of reality."[27] What is posited here is "a thoroughgoing relational materiality" in which things do not exist in and of themselves.[28] Our conceptualization of policy deliberation as *event* further promotes active attention to "how the coming together of various elements produces material *transformations*."[29] An event is a transformative situation, entailing an ontological reconfiguration, disrupting the entities it gathers, thus directing attention to the transformations of all the elements which enter into it.[30] Following Race,[31] we can say that "an event may be considered the creator of a difference between a before and an after," which here, we suggest, allows us to attend to how evidence is made and performed as a thing followed (i.e., enacted in the event of policy deliberation, and thus is the difference which comes after). Attending to things-in-the-making in transformative situations requires us to notice the practices of crafting that enactments depend upon, and to ask how these relations work,[32] giving attention not only to the processes by which things come to be, but also to "the manner in which their constituent elements come together, their contingencies and the differences these make to worlds and lives."[33]

Drawing on this thinking, we have proposed an "evidence-making intervention" approach,[34] which conceptualizes evidence as emerging through the situated relational dynamics involved in its accomplishment. An "evidence-making intervention" approach focuses on the practices and processes through which evidence and interventions come to be, troubling assumptions about how evidence and interventions are "transferred" into different sites, and instead paying attention to the transformations that occur in situated relations as they are put to use and made to work.[35] From this perspective, evidence and policy cannot be held as stable and separate from one another, but rather are co-affective with their milieu, and thus always multiple in their emergent realities. Taken together along with the conceptual tools available in science and technology studies, this approach provides a way to grapple with the complexity of evidence in policy, sharpening our attention to the transformative effects of the event of policy deliberation. It is an approach that allows us to attend to how practices and relations work to enact the thing we call evidence (a situated achievement), as well as how the making of evidence in the event of policy deliberation performs evidence *as a thing having been followed* (a difference made after).

While "what is the evidence?" is often posed as a question that seeks to remove politics and bias, assuming a technical and rational response within an "evidence-based" paradigm, acknowledging the *ontological politics* at work[36] allows for a different mode of questioning that seeks to notice the realities being negotiated and that it could be otherwise. It allows us to take evidence as a thing—a gathering, a reality—that is "historically, culturally and materially located."[37] Echoing Mol, then we might ask "located where?" To answer this question, we pay attention to the sites in which the thing we call evidence is made; to events of policy deliberation as the transformative situations where evidence is done. While much work in the field of science and technology studies has focused on the laboratory to rearticulate the making of a scientific fact,[38] the thing we call evidence is made not in laboratories or scientific practices alone. Evidence is made in the extending chains of relations in and through which science *travels into* and figures in sites of policy deliberation and public life. Too often, in the ordinary parlance of evidence-based policymaking, a scientific study or body of research is taken to be "the evidence," presumed to be stable and available to be picked-up and used. Rather than envisaging science as having made evidence which is then put into translation, we focus on tracing the *ontological* transformations in how science is made to perform as knowledge *in* a specific situated assemblage of policy practices and relations.[39] Thinking this way requires us to investigate the thing we call evidence together with (and never isolated from) the practices and relations which enact it, and as a political concern.

A Scene: The Whiteboard

Destabilizing assumptions about evidence as a fixed and stable thing capable of being translated for policy under the rubric of "what works" pushes us to rethink what counts as an evidence object. Suddenly, we find before us "a new curiosity"[40] about the thing we call evidence. If evidence becomes with its situations, in events of policy deliberation, a range of knowledge may be brought to bear and made as the thing we call evidence. Science, research or other knowledge is not necessarily consequential or authoritative for decision making until it is *made* as such. In Latour's[41] terms, our task is to move "*toward* the gathering" to attend to the contingency of how this thing we call evidence is made to exist. How do we do this? If practices are assemblages of relations, and those assemblages do realities, then, as Law suggests, we "need to proceed *empirically*."[42] That is, there is no way to examine the thing we call evidence except by paying attention to the performative effects of practices in specific sites. There is no "evidence" in general to analyze, except as it is enacted. Law[43] suggests that to study the performative effects of practices is

> to undertake the analytical and empirical task of exploring possible patterns of relations, and how it is that these get assembled in particular locations. It is to treat the real as whatever it is that is being assembled, materially and semiotically in a scene of analytical interest.

And here we introduce one scene, so as to attend to the material-semiotic relations assembled at "a particular place, moment, and occasion":[44] an otherwise mundane and dirty whiteboard, in an office, in London.

In May 2021, Dominic Cummings, former Chief Advisor to UK Prime Minister Boris Johnson, was called to testify before the House of Commons' Health and Social Care Committee and Science and Technology Committee. The committees were holding a joint inquiry into "lessons to be learned" from the response to the coronavirus pandemic so far. The remit of the joint inquiry was to examine the "impact and effectiveness of action taken by government and the advice it has received" across a range of issues including the deployment of nonpharmaceutical interventions such as lockdown and social distancing, the impact on the social care sector, the impact on at-risk groups, testing and contact tracing, modeling and the use of statistics, government communications and public health messaging, the United Kingdom's prior preparedness for the pandemic, and the development of treatments and vaccines.[45] The remit of the joint inquiry was, explicitly, about the relationship between evidence and policy in a time of public health crisis response, convened as the United Kingdom continued to be at the epicenter of international attention as the pandemic unfolded.

Cummings' appearance before the Committees was anticipated as a "box office moment."[46] Cummings' candid account attracted significant controversy as he described the "monumental incompetence" and dishonesty of government officials, a lack of planning, and a failure to "follow the science" and take advice.[47] In the days leading up to his testimony, Cummings posted a series of comments to Twitter, outlining his observations of the decision-making processes and events that had unfolded inside the highest echelons of the United Kingdom's government in the early months of 2020. Among the highly publicized tweets was this one:

> 65/ First sketch of Plan B, PM study, Fri 13/3 eve - shown PM Sat 14/ 4: NB. Plan A 'our plan' breaks NHS, >4k p/day dead min. Plan B: lockdown, suppress, crash programs (tests/treatments/vaccines etc), escape 1st AND 2nd wave (squiggly line instead of 1 or 2 peaks) … details later.[48]

Accompanying the tweeted text was an image, labeled "Prime Minister's Office, 10 Downing Street Westminster, England, United Kingdom. 13 March 2020 at 20:09:23" (see Figure 8.1). The image zooms in on a dirty whiteboard, with text scrawled in red pen, outlining in crude terms Cummings' rationale for dramatically shifting away from the United Kingdom's previously announced mitigation strategy to instead implement a 'full lockdown' suppression strategy to prevent the National Health Service (NHS) from collapsing. In the top right-hand corner of the whiteboard were the initials of key actors including the prime minister, the chief medical officer, and scientific advisors. Laid out on the rest of the board were six points, "1/ No vaccine in 2020," "2/ Must avoid NHS collapse," "3/ To stop NHS collapse, we will probably have to 'lockdown'," "4/ Lockdown = e/o stays home; pubs etc close," "5/ What's diff btw Plan A + B? […] full lockdown before collapse which means ~ 2 weeks (?) before we catch up with Italy." The sixth and final point at the bottom of the whiteboard posed, in stark terms, what was fast becoming far from a hypothetical question: "6/ Who do we <u>not</u> save?" To the side of the whiteboard were jotted notes, outlining what "full lockdown" might mean ("less contact; no contact; contact illegal") and other matters of concern ("ventilators; testing," "employee status; vulnerability status"). Among the text on the whiteboard were a series of hand-drawn curves; figures

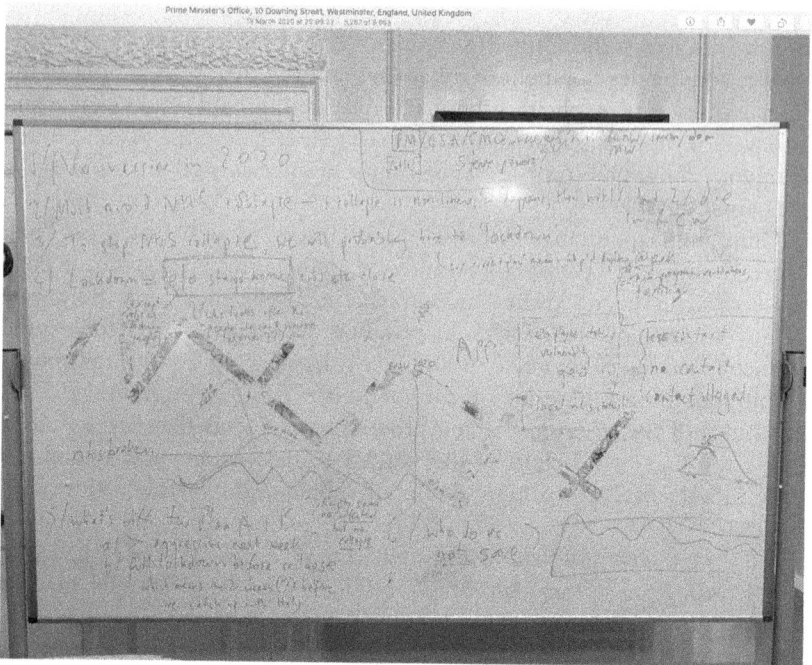

Figure 8.1 The whiteboard, Reproduced from Twitter: Dominic Cummings (@ Dominic2306), Twitter, 26 May 2021.

representing modeled projections of transmission dynamics and the possible effects of a shift toward a suppression strategy. Figures we have since become all too familiar with, as calls to "flatten the curve" reverberated globally.[49]

This extraordinary image garnered media attention, with much speculation as to what it revealed about the evidence and advice underpinning the government's deliberations and shifting priorities in March 2020 amid what was fast becoming a catastrophic public health crisis in the United Kingdom and around the world.[50] As well as being shared widely on Twitter and in the press, this image of the whiteboard and associated tweet also featured explicitly in Cummings' testimony before the inquiry, figuring as a material object in his storied account of events which had taken place at Downing Street:

> The evening of Friday the 13th, I am sitting with Ben Warner and the Prime Minister's Private Secretary in the Prime Minister's study. We were basically saying that we are going to have to sit down with the Prime Minister tomorrow and explain to him that we think that we are going to have to ditch the whole official plan, and we are heading for

the biggest disaster this country has seen since 1940. *This is the whiteboard—I put it on my Twitter account, and I have also sent it to some journalists, so hopefully you will be able to see on the internet, and you have got copies of it here.* This has, essentially, plan B sketched on it. The timestamp is nine minutes past 8 on the Friday night. Essentially, what is happening at this point is, we are thinking, "What do we do on this?" At this point, the second most powerful official in the country, Helen MacNamara— the Deputy Cabinet Secretary—walked into the office *while we are looking at this whiteboard.* She says, "I have just been talking to the official Mark Sweeney, who is in charge of coordinating with the Department for Health. He said 'I have been told for years that there is a whole plan for this. There is no plan. We are in huge trouble.' I have come through here to the Prime Minister's office to tell you all that I think we are absolutely fucked. I think this country is heading for a disaster. I think we are going to kill thousands of people. As soon as I have been told this, I have come through to see you. It seems from the conversation you are having that that is correct." I said, "I think you are right. I think it is a disaster. I am going to speak to the Prime Minister about it tomorrow. We are trying to sketch out here what plan B is."[51]

In many ways, this mundane and dirty whiteboard might seem like an odd choice of scene. Why have we selected this event of policy deliberation to attend to how evidence is in the process of being made? After all, this is not "evidence" as we might usually have it represented to us within "evidence-based policy" as conventionally construed. It comes in a very different shape and form, publicly shared not to provide reassurance of an authoritative technical decision-making process but to perform the chaos, panic, and dysfunction which had circulated behind closed doors. The squiggly lines are not carefully translated models of projected intervention effect. The plan is literally and figuratively a "sketch." The figures of the curve are free drawn, unenumerated, unreferenced. Erased are the complexities of method, data, parameters, R, and any careful translational efforts. This modeling, if we are to recognize it as such, has escaped its reasoned calculus.[52] But inside this intimate moment of deliberation and crisis, which we have been retrospectively allowed to glimpse through its being made public, these modeled projections have been not carefully translated but rather *transformed*. The squiggly line sketched on this dirty whiteboard is a powerful actor in this scene. For the moment, on this whiteboard, this *is* the evidence, enacted as *a thing which was being followed*. In the here and now of the policy event, this hand-drawn squiggly line constitutes "evidence-enough." We suggest that this rather unconventional scene offers a case for analysis that allows us to attend to how evidence is performed in fluid, flexible, and malleable ways, through invocation, implication, imagination, and relational connection. It allows us to attend to the situated ontology of evidence in policy, in all its politics and mess.

Our interest here is "to consider the practices that enable the mutual constitution of the properties of the entities involved and the relevance of the context in which they are situated."[53] It is to ask how the material-semiotic relations here work to assemble a putative reality.[54] How is the thing we call evidence enacted in this image and its associated story as a thing being followed, even in the chaos of crisis? This image is, after all, a product of practices (practices moreover, that we rarely have such privileged access to observe) and could well have been assembled differently (and might never have been made public; indeed, "going public" is part of the assemblage of relations to which we must attend). To begin, the image of the whiteboard and its story tells us something about the relevant characters and their relations. This image performs COVID-19 response evidence in a set of relations that are simultaneously technological (no vaccine in 2020; the development of an app), medical (NHS collapse), financial (pubs etc. close), social (everyone stays home, contact illegal), sociomaterial (employee status), geographical (local, geo), and ethico-political (who do we not save?). It also enacts futures (Plan A and Plan B) with governing effects in the present and authorizes who has the power to speak (the PM). This image and its story tell us who and what is important, in the room in Downing Street, in the inquiry process, and publicly, and about the hinterland of sociopolitical practices to which the thing we call evidence relates. The whiteboard is, after all, multiply located: in the prime minister's office on the 13 March 2020, privately snapped in the inner sanctum of government by a key advisor at 8 pm amid an uncertain and unfolding crisis, seen by few; and on Twitter on the 26 May 2021 (enduring online in multiple media formats since), shared publicly and strategically amid the controversial revelations of a highly politicized and much scrutinized inquiry. The whiteboard performs differently in these various sites of practice. Different realities are being done in different temporal locations, even as the futures imagined in this image and its squiggly lines continue to unfold. This case is useful for seeing how evidence is gathered as a thing altogether differently in its situation and across different assemblage relations, for instance, from the inside relations of a closed government policy deliberation event to its re-eventuation in assemblages of public relations. Evidence is here constituted multiply according to its assemblages (the performance of the whiteboard, the making known and public of the whiteboard event). Evidence does not settle or stabilize within these events, as if becoming located to or locked into a past, but is continually in-the-making. We caution then, from reading into the whiteboard scene as an after the event realization of how evidence was at the time, because evidence, in its every eventuation, is always in the process of being enacted anew. The whiteboard, its display and how this enacts the thing we call evidence cannot be detached from the assemblage relations of its eventuation, that is, Cummings' appearance in a public inquiry and our analyses of this for our purposes now.

Evidence Made in Relations

The whiteboard, and the complex assemblage of relations and practices of which it is a material artifact, is thus working to enact the thing we call evidence itself, as a thing which is being followed. How is it doing this? Let us notice *the curves*. The figure of the model, and the curve itself, has become ubiquitous in the context of COVID-19. Models play a critical role in producing evidence in response to novel viral outbreaks, generating evidence for rapid decision making. Within "evidence-based policy" as conventionally construed, models are imagined as a "bridge" to knowing, when empirical data are absent, or in an interim of uncertainty, thus closing down unknowns and making governance possible.[55] Modeled projections make "evidence of *potentials* and *not yet knowns* as a means to *intervene*"[56] and, in doing so, entangle in social life as public concerns.[57] By the time this image "goes public" on Twitter and during the inquiry in 2021, these curves are legible and familiar household figures. It can become as the thing we call evidence because it is enacted in relation to this hinterland of social, political, scientific, and cultural practices, in particular networks or systems of circulation.[58] Tracing the figure of the curve against this hinterland, we can begin to notice how the thing we call evidence is enacted, detaching from science and becoming remade as a distinct object, and thus, here, performed as not only recognizable but also consequential for policy-making. The figure of the curve, nestled in the heart of the six points on the whiteboard, is implicitly rendered as a central object around which decisions are taking place. These hand-drawn curves of projected futures—Plan A, with NHS collapse, and Plan B, the "squiggly line instead of 1 or 2 peaks," to avoid the potential of such collapse—are being done as evidence, as having the capacity to bring about the governance of possible futures, as having the attributes of a thing which makes policy action possible, even in crisis and uncertainty.

These hand-drawn curves might at first appear to offer a "window onto reality,"[59] in conventional "evidence-based" parlance "translating" the findings of mathematical modeling into a decision-making space. But these connections are part of the illusionary work being done as evidence is eventuated as a thing being followed. One version of the "evidence-based policy" story of March 2020 is that one particular modeling study dramatically shifted the UK government's previously announced mitigation strategy to one of suppression, through presentation of the projected effects of interventions, including home isolation of cases and quarantining of their household members, social distancing of the entire population, and lock-down measures, to drive down infections until a vaccine became available.[60] Given that the suppression strategy potentially needed to be maintained for months, the authors of this modeling study examined the impact of an "adaptive policy in which social distancing (plus school and university closure, if used) is only initiated after weekly confirmed case

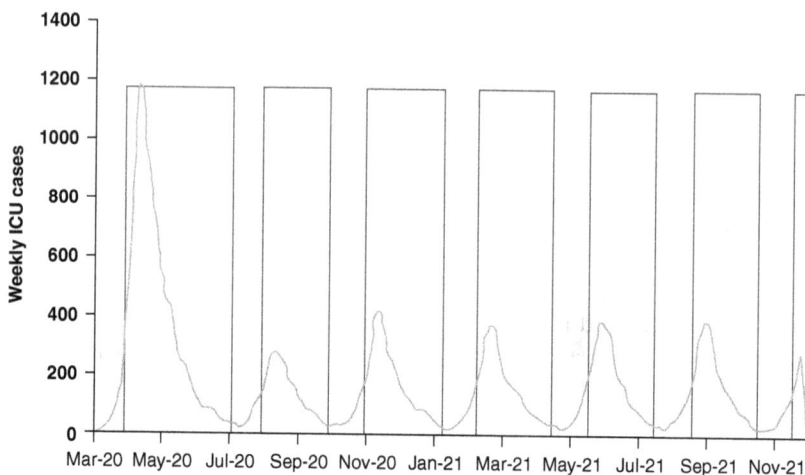

Figure 8.2 Suppression strategy, reproduced from Neil Ferguson et al., *Impact of non-pharmaceutical interventions (NPIs) to reduce COVID-19 mortality and healthcare demand*, Imperial College COVID-19 Response Team (2020), 12. https://www.imperial.ac.uk/media/imperial-college/medicine/sph/ide/gida-fellowships/Imperial-College-COVID19-NPI-modelling-16-03-2020.pdf

incidence in ICU patients (a group of patients highly likely to be tested) exceeds a certain 'on' threshold, and is relaxed when ICU case incidence falls below a certain 'off' threshold"[61] (see Figure 8.2).

This report, led by Professor Neil Ferguson at Imperial College, was released to the public on the 16 March 2020 and was used by government to explain and rationalize the policy change. The authors of the report concluded that "epidemic suppression is the only viable strategy at the current time" and that "the social and economic effects of the measures which are needed to achieve this policy goal will be profound."[62] While considering the feasibility of the strategy, the authors carefully emphasized in their report that they did "not consider the ethical or economic implications of either strategy here, except to note that there is no easy policy decision to be made,"[63] explicitly attempting to separate their science from the realm of policy and politics, while nonetheless hoping that the report might be consequential for decision making in a time of crisis.

We can perhaps see the traces of this particular model and its suppression strategy on the whiteboard and in Cummings' tweet: "Plan B: lockdown, suppress, crash programs (tests/treatments/vaccines etc), escape 1st AND 2nd wave (*squiggly line* instead of 1 or 2 peaks)" (emphasis added). We might assume that the "squiggly line" corresponds to that in the report (Figure 8.2), "brought in" via the translational device of the curve. One might see the "squiggly lines" as reference points to tacit evidence that is assumed to sit behind and before. But here on the whiteboard, the methods, the maths, the science, the careful equivocation, the distancing of the conclusion recommending a suppression strategy from matters of ethical and economic concern, and the delineation of the roles of the scientist and the policy decision maker, have all been erased. There is implicit work being done here; realities are being made "in the wake of explicit research findings."[64] It is this hand-drawn curve, with its erasure of method and scientific precision, resituated in its complex assembled relations, that is the thing being followed, that legitimates decisions to be made. This is not mere representation via translation but *transformation*. It is the hand-drawn curve, on the whiteboard, in the story, and not the model or the scientific report per se, that *is* the evidence.

Evidence Made Public

One might argue that it is not *this curve*, this hand-drawn squiggle, which informs decisions, but rather mathematical modeling generated by scientists (indeed, the work of these scientists has been highly profiled in the United Kingdom and around the world throughout the pandemic). One might suggest that this curve somehow offers a window onto some other stable reality, an evidence object "out there" being brought in and translated "in here." But what we seek to notice in the figure of the curve in this image and story of the whiteboard is how science, generated elsewhere by mathematical modelers with all kinds of precision and assumptions, is being enacted anew as the thing we call evidence, in particular and distinctive ways, in the context of Cummings' tweet on the 26 May 2021, and as it continued to circulate in media reports, into the joint inquiry and beyond. We attend to these curves not to trace the concrete preexistence of evidence as if it had come before, but rather to notice the qualities of evidence in-the-making and *how* it does its illusionary work. It is this image and the account of the lockdown deliberation event, released to the public and media in the context of an inquiry, that allows us to see the curves enacted as evidence, as the thing *being followed*. It is the "going public" of this moment, that allows for an account of action being taken on "the evidence," even if that action was mired in chaos, failure, and politics. It is in the public demand to *account* for the relation between evidence and policy, including in the explicit remit of the inquiry to uncover "lessons learned," that the thing we call evidence is made. Evidence is not outside of the political and the public but enacted within it. As Cummings shared in his testimony before the inquiry:

> I texted the group with the Prime Minister and the chief scientific adviser, and I said: "People can see the trajectory and how social distancing will be needed to flatten the curve. Very sensible people, including former CDC officials, etc. and doctors are saying the risks of delay are much, much higher than the risks of going too soon. If we're

not going to say tomorrow, starting social distancing today, we're waiting, and effectively just keep telling people to wash hands, there's going to be massive pushback saying, 'Why wait five days? Why not move now? Why not flatten earlier?' Proposing tomorrow that we delay action until next week will require extremely clear justification with supporting data, models, etc. We would have to make it public for global scrutiny." This was on the night of the 11th. [...] I said on the 11th, *"We're going to have to make all these models public and whatnot."*[65]

The thing we call evidence is being made in the putative reality being assembled in this account of failure, not because the evidence was prior, definite, "out there," "on hold," waiting to be taken up and used but rather because it has been performed publicly as a thing which had been followed *after,* and which made the difference. This is evidence made *in* its extending chains of relations, as both action and failure to act are justified publicly and politically; not evidence "for" or "before" policy. This accounting is an instance of not only evidence made in practice but evidencing *as practice,* as a political and relational doing, as it goes public. Gone is a linear process of a stable evidence object translated for policymakers and communicated to publics. Publics are bound up in and part of this deliberation event's effects and participate relationally in the process of evidence's eventuation as a thing followed.

Evidence Made Political

As evidence is enacted and transformed in its situations, it is inextricably entangled in a network of relations which is thoroughly and inescapably *political.* We find that the thing we call evidence is both an object and a site of dealing and dispute.[66] Here, on the whiteboard, situated in the Prime Minister's Office and on Twitter amid an ongoing crisis, it is not possible to separate science from the social, as the authors attempted to do in the report when they stated that they did "not consider the ethical or economic implications of either strategy."[67] Attending to evidence as enacted in policy events allows us to notice and work with its entanglement in these ethico-political concerns, without seeing this as defective or beyond the realm of what evidence can or should do. As the image of this deliberation event goes public, this is not a mere translation of scientific knowledge but a far more radical remaking. These hand-drawn squiggly curves can *count* as evidence, and have effects, not in spite of these ethico-political entanglements but precisely *because of* the complex relations in which they are enacted—made as evidence in an event of policy deliberation, and re-enacted again in the context of the inquiry, as evidence which was "followed," and which was made to matter socially and politically. This enactment of the thing we call evidence exposes and challenges the notion of a technical-rational decision-making process, performing it instead as ethically and politically charged, and sometimes messy and failing. What is performed as evidence, as a thing followed, forms part of the contestation of the event. Evidence is not outside of this politics; it is enacted *in* it. Here, we suggest, a "standard" conception of politics (based on power, ideology, etc.) is productively "augmented" by the "politics of ontological constitution."[68] Politics is frequently invoked as a barrier to rational decision making based on the best available evidence, within an evidence-based policy paradigm. But by thinking with evidence as a matter of ontology, we can see that something altogether different is at stake. This is evidence which can be made as *thoroughly* political and ethical in its concerns, not seeking sanitized separation from charges of bias and lack of objectivity. What we encounter is not a smooth translation or a fluid object, gently taking on a new shape as it moves from science to policy, but rather a *fiery* object, characterized by jumps and discontinuities.[69] As we attend differently to evidence in its situations and trace its extending chains of relations, the metaphor of translation fails us; evidence is not translated but *transformed* in ways that are made to matter socially, ethically, and politically. By attending to the practices and relations in which the thing we call evidence is done as a following, a difference made after, we find that there is an explicit politics of evidence-making at work.

Conclusion: Toward a Situated Ontology of Evidence in Policy

We have proposed an approach which makes a critical move toward a situated ontology of evidence in policy. Reconfiguring evidence as a situated achievement, enacted in policy deliberation events, allows us to attend to the thing we call evidence in new ways, noticing it as an "evidence-making intervention."[70] An evidence-making intervention approach does away with delimiting assumptions underpinning "evidence-based policy" as conventionally construed, which have hitherto conceived of evidence and policy as distinct and separate domains to be "bridged" through "translation," instead proposing an approach which attends to the transformations which occur in deliberation events. Noticing evidence as a situated achievement (a relational effect, a difference made in an event) shifts our focus to the practices and relations involved in its becoming, and opens up possibilities for alternative configurations of the worlds in which the realities of evidence are produced.[71] This approach helps appreciate how evidence is done and performed as a thing "followed," emerging as a thoroughly political and public concern, and as "evidence-enough" in the face of uncertainty. This approach, we suggest, allows us to encounter evidence in new ways.

It has been said that the fundamental contribution to science and technology studies of the ontological turn is "its power to draw renewed critical attention to objects that might otherwise appear 'finished' or 'ready-made,'" scrutinizing objects that analysis would often consider "'black-boxed' and no longer controversial."[72] Evidence is one such object, too often assumed to be fixed, stable, ready to be taken up and used in policy deliberation. Probing the ontology of the thing we call evidence interferes with assumptions about a singular, rational and ordered "evidence-based policymaking" paradigm, allowing us to more carefully attend to the thing we call evidence in the mess and complexity of policy practices, enacted as a thing which is relational and entangled, rendered political and public in its eventuations. We find that evidence-making events necessarily force a "letting go" of reasoned calculus, and of the locations in which scientific knowledge is made. But it is precisely this detachment which is the difference made in evidence-making events. It is through "letting go" that the thing we call evidence is enacted and transformed as a thing followed, and as a thoroughly social and political concern. Evidence does not sit behind or before policy deliberation, but is inside it, the enactment of it as a thing followed part of its eventuation and how it is made to matter. By thinking with this approach, what is messy is not defective but an ontopolitical concern that we might attend to more closely, noticing the productive effects which make a difference to worlds and lives.

This chapter has offered a proposal, "testing the value of engaging more openly and attentively"[73] to evidence in-the-making. This is not to claim "anything goes" but rather to "shift our understanding of the *sources* of the relative immutability and obduracy of the world: to move these from 'reality itself' into the choreographies of practice."[74] A move toward a situated ontology of evidence—from evidence *for*, to evidence *in* policy and its extending chains of relations—will have implications for the sociotechnical and sociopolitical arrangements of the "evidence-based policy" movement, and its place as a governing paradigm. However, in a time of public health emergency and global disease, we are more aware than ever that it is no longer desirable to proceed as if there is an imagined separation of science and politics which might be maintained, under the rubric of "what works." Science and the social are entangled and relational, and decisions are value-laden and political, with multiple unanticipated effects. Rather than seeing these political entanglements as a failure of the technical-rational ideal of "evidence-based policy," the approach we have proposed allows us to attend to how evidence is being done in its extended chains of relations. By moving toward a situated ontology of evidence in policy, we can attend to how evidence is made, how evidence is put to use, and how evidence is made to matter, without reverting to failing metaphors that have artificially sustained the separation of these two domains. Taking evidence as enacted in events of policy deliberation, and not as a thing separate or before it, is analytically and *politically* productive as it asks us to explore how evidence is constituted, what policy events *do*, and whether or not these effects are desirable.[75] It is an approach which leads us to ask if this thing we call evidence is done well, or could be made otherwise, thus promoting a mode of questioning that attends to other matters of concern and worldings. This approach, we suggest, allows for a mode of questioning that brings to the fore ethico-political matterings that might ordinarily be obscured.

Notes

1 Justin Parkhurst, *The Politics of Evidence: From Evidence-Based Policy to the Good Governance of Evidence* (London: Routledge, 2017), 4.
2 Annette Boaz et al., "Does Evidence-Based Policy Work? Learning from the UK Experience," *Evidence & Policy* 4, no. 2 (2008): 247.
3 Kari Lancaster and Tim Rhodes, "What Prevents Health Policy Being 'Evidence-Based'? New Ways to Think about Evidence, Policy and Interventions in Health," *British Medical Bulletin* 135, no. 1 (2020).
4 Katherine Smith, *Beyond Evidence-Based Policy in Public Health: The Interplay of Ideas*, Palgrave Studies in Science, Knowledge and Policy (London: Palgrave Macmillan, 2013); Kathryn Oliver et al., "A Systematic Review of Barriers to and Facilitators of the Use of Evidence by Policymakers," *BMC Health Services Research* 14, no. 1 (2014/01/03 2014).
5 Brian Head, "Reconsidering Evidence-Based Policy: Key Issues and Challenges," *Policy and Society* 29, no. 2 (2010/05/01 2010): 87.
6 Trisha Greenhalgh and Sietse Wieringa, "Is It Time to Drop the 'Knowledge Translation' Metaphor? A Critical Literature Review," *Journal of the Royal Society of Medicine* 104, no. 12 (2011).
7 Greenhalgh and Wieringa, "Is It Time to Drop the 'Knowledge Translation' Metaphor? A Critical Literature Review," 507–508.
8 Trisha Greenhalgh and Jill Russell, "Reframing Evidence Synthesis as Rhetorical Action in the Policy Making Drama," *Healthcare Policy* 1, no. 2 (2006): 34.
9 Greenhalgh and Russell, "Reframing Evidence Synthesis as Rhetorical Action in the Policy Making Drama," 34.
10 Judith Green, "Epistemology, Evidence and Experience: Evidence Based Health Care in the Work of Accident Alliances," *Sociology of Health & Illness* 22, no. 4 (2000).
11 Kari Lancaster, Tim Rhodes, and Marsha Rosengarten, "Making Evidence and Policy in Public Health Emergencies: Lessons from COVID-19 for Adaptive Evidence-Making and Intervention," *Evidence & Policy* 16, no. 3 (2020).
12 Nason Maani and Sandro Galea, "What Science Can and Cannot Do in a Time of Pandemic," *Scientific American* 2 February 2021, https://www.scientificamerican.com/article/what-science-can-and-cannot-do-in-a-time-of-pandemic/.
13 Tim Rhodes, Kari Lancaster, and Marsha Rosengarten, "A Model Society: Maths, Models and Expertise in Viral Outbreaks," *Critical Public Health* 30, no. 3 (2020).
14 Lancaster, Rhodes, and Rosengarten, "Making Evidence and Policy in Public Health Emergencies: Lessons from COVID-19 for Adaptive Evidence-Making and Intervention."
15 Smith, *Beyond Evidence-Based Policy in Public Health: The Interplay of Ideas*, 4.
16 Martin Savransky and Marsha Rosengarten, "What Is Nature Capable of? Evidence, Ontology and Speculative Medical Humanities," *Medical Humanities* 42 (24 May 2016); Tim Rhodes and Kari Lancaster, "Evidence-Making Interventions in Health: A Conceptual Framing," *Social Science and Medicine* 238, no. October (2019); Kari Lancaster, "Performing the Evidence-Based Drug Policy Paradigm," *Contemporary Drug Problems* 43, no. 2 (2016).

17 Carol Bacchi and Susan Goodwin, *Poststructural Policy Analysis* (New York: Palgrave, 2016), 10.
18 John Law and Vicky Singleton, "Object Lessons," *Organization* 12, no. 3 (2005/05/01 2005).
19 Steve Woolgar and Javier Lezaun, "The Wrong Bin Bag: A Turn to Ontology in Science and Technology Studies?" *Social Studies of Science* 43, no. 3 (2013/06/01 2013): 323.
20 Annemarie Mol, "Ontological Politics. A Word and Some Questions," *The Sociological Review* 47, no. S1 (1999); Annemarie Mol, *The Body Multiple: Ontology in Medical Practice* (Durham Duke University Press, 2002).
21 Rhodes and Lancaster, "Evidence-Making Interventions in Health: A Conceptual Framing."
22 Woolgar and Lezaun, "The Wrong Bin Bag: A Turn to Ontology in Science and Technology Studies?"
23 Rhodes and Lancaster, "Evidence-Making Interventions in Health: A Conceptual Framing."
24 Isabelle Stengers, *Another Science Is Possible: A Manifesto for Slow Science* (Cambridge: Polity Press, 2018), 120; Rhodes and Lancaster, "Evidence-Making Interventions in Health: A Conceptual Framing."
25 Bruno Latour, "Why Has Critique Run Out of Steam? From Matters of Fact to Matters of Concern," *Critical Inquiry* 30, no. 2 (2004): 233, emphasis original.
26 Suzanne Fraser and Kate Seear, *Making Disease, Making Citizens: The Politics of Hepatitis C* (London: Routledge, 2011), 9.
27 Woolgar and Lezaun, "The Wrong Bin Bag: A Turn to Ontology in Science and Technology Studies?" 323–324, emphasis original.
28 John Law, *After Method: Mess in Social Science Research* (Oxon: Routledge, 2004), 83.
29 Kane Race, "Reluctant Objects: Sexual Pleasure as a Problem for HIV Biomedical Prevention," *GLQ: A Journal of Lesbian and Gay Studies* 22, no. 1 (2015): 3, emphasis added.
30 Race, "Reluctant Objects: Sexual Pleasure as a Problem for HIV Biomedical Prevention"; Kane Race, *The Gay Science: Intimate Experiments with the Problem of HIV* (Oxon: Routledge, 2018); Mike Michael and Marsha Rosengarten, *Innovation and Biomedicine: Ethics, Evidence and Expectation in HIV* (Basingstoke: Palgrave Macmillan, 2013).
31 Race, "Reluctant Objects: Sexual Pleasure as a Problem for HIV Biomedical Prevention," 4.
32 Law, *After Method: Mess in Social Science Research*; John Law, "Collateral Realities," in *The Politics of Knowledge*, ed. Fernando Domingues Rubio and Patrick Baert (Oxon: Routledge, 2012).
33 Race, *The Gay Science: Intimate Experiments with the Problem of HIV*, 4.
34 Rhodes and Lancaster, "Evidence-Making Interventions in Health: A Conceptual Framing."
35 Rhodes and Lancaster, "Evidence-Making Interventions in Health: A Conceptual Framing."
36 Mol, *The Body Multiple: Ontology in Medical Practice*; Mol, "Ontological Politics. A Word and Some Questions"; Law, *After Method: Mess in Social Science Research*.
37 Mol, "Ontological Politics. A Word and Some Questions," 75.
38 Bruno Latour and Steve Woolgar, *Laboratory Life: The Construction of Scientific Facts* (Princeton: Princeton University Press, 2013).
39 Lancaster, "Performing the Evidence-Based Drug Policy Paradigm."
40 Woolgar and Lezaun, "The Wrong Bin Bag: A Turn to Ontology in Science and Technology Studies?" 323.

41 Latour, "Why Has Critique Run Out of Steam? From Matters of Fact to Matters of Concern," 246.

42 Law, "Collateral Realities," 157, emphasis original.

43 Law, "Collateral realities," 157.

44 Law, "Collateral realities," 157.

45 "Coronavirus: Lessons Learnt," UK Parliament, 2021, accessed 1 July 2021, https://committees.parliament.uk/work/657/coronavirus-lessons-learnt/.

46 Laura Kuenssberg, "Dominic Cummings Sketches Out Script as Grilling Approaches," *BBC News*, 25 May 2021, https://www.bbc.com/news/uk-politics-57246133.

47 Martin McKee, "Martin McKee: What Did We Learn from Dominic Cummings' Evidence to MPs on the Covid Crisis?" *The BMJ Opinion* 25 July 2021, https://blogs.bmj.com/bmj/2021/05/26/martin-mckee-what-did-we-learn-from-dominic-cummings-evidence-to-mps-on-the-covid-crisis/.

48 Cummings.

49 Rhodes, Lancaster, and Rosengarten, "A Model Society: Maths, Models and Expertise in Viral Outbreaks."

50 Ellie Cambridge, "DOMSDAY Dominic Cummings Shares '2020 Lockdown Plan' Pic That Says 'Who Do We NOT Save' Hours before Explosive Grilling," *The Sun*, 26 May 2021, https://www.thesun.co.uk/news/15070125/dominic-cummings-picture-lockdown-plan-before-grilling/; Stephen Matthews and Sam Blanchard, "Dominic Cummings Reveals How the Wheels Came Off Inside Number 10: Extraordinary Timeline of Chaos as Coronavirus Spread across Britain … before Civil Servant Screamed 'We're f******' and Aides Drew up 'Plan B' on a Whiteboard Asking 'Who Do We Not Save?'" *Daily Mail*, 26 May 2021, https://www.dailymail.co.uk/news/article-9620599/Coronavirus-Dominic-Cummings-shares-lockdown-brainstorm-question-not-save.html; Rachel Schraer, "Dominic Cummings: What Was on His Whiteboard?," *BBC News*, 26 May 2021, https://www.bbc.com/news/health-57254654; McKee Martin McKee: What did we learn from Dominic Cummings' evidence to MPs on the covid crisis?

51 Health and Social Care Committee and Science and Technology Committee, Oral evidence: Coronavirus: Lessons learnt, HC 95, 19 (London 2021), Q1003, emphasis added.

52 Tim Rhodes and Kari Lancaster, "Excitable Models: Projections, Targets, and the Making of Futures without Disease," *Sociology of Health & Illness* 43, no. 4 (2021).

53 Woolgar and Lezaun, "The Wrong Bin Bag: A Turn to Ontology in Science and Technology Studies?" 328.

54 Law, "Collateral Realities."

55 Tim Rhodes and Kari Lancaster, "Mathematical Models as Public Troubles in COVID-19 Infection Control: Following the Numbers," *Health Sociology Review* 29, no. 2 (2020/05/03 2020).

56 Rhodes, Lancaster, and Rosengarten, "A Model Society: Maths, Models and Expertise in Viral Outbreaks," 255, emphasis original.

57 Rhodes and Lancaster, "Mathematical Models as Public Troubles in COVID-19 Infection Control: Following the Numbers."

58 John Law, "Seeing Like a Survey," *Cultural Sociology* 3, no. 2 (2009/07/01 2009).

59 Law, "Collateral Realities," 166.

60 Neil Ferguson et al., *Impact of non-pharmaceutical interventions (NPIs) to reduce COVID-19 mortality and healthcare demand*, Imperial College COVID-19 Response Team (2020), https://www.imperial.ac.uk/media/imperial-college/medicine/sph/ide/gida-fellowships/Imperial-College-COVID19-NPI-modelling-16-03-2020.pdf; Sarah Boseley, "New Data, New Policy: Why UK's Coronavirus Strategy Changed," *The Guardian*,

17 March 2020, https://www.theguardian.com/world/2020/mar/16/new-data-new-policy-why-uks-coronavirus-strategy-has-changed?CMP=Share_iOSApp_Other.

61 Ferguson et al., *Impact of non-pharmaceutical interventions (NPIs) to reduce COVID-19 mortality and healthcare demand*, 11.

62 Ferguson et al., *Impact of non-pharmaceutical interventions (NPIs) to reduce COVID-19 mortality and healthcare demand*, 16.

63 Ferguson et al., *Impact of non-pharmaceutical interventions (NPIs) to reduce COVID-19 mortality and healthcare demand*, 4.

64 Law, "Collateral Realities," 167.

65 Health and Social Care Committee and Science and Technology Committee, Short Oral evidence: Coronavirus: Lessons learnt, HC 95, 15–16., Q998–Q1000, emphasis added.

66 Fraser and Seear, *Making Disease, Making Citizens: The Politics of Hepatitis C*; Latour, "Why Has Critique Run Out of Steam? From Matters of Fact to Matters of Concern."

67 Ferguson et al., *Impact of non-pharmaceutical interventions (NPIs) to reduce COVID-19 mortality and healthcare demand*, 4.

68 Woolgar and Lezaun, "The Wrong Bin Bag: A Turn to Ontology in Science and Technology Studies?" 335.

69 Law and Singleton, "Object Lessons."

70 Rhodes and Lancaster, "Evidence-Making Interventions in Health: A Conceptual Framing."

71 Savransky and Rosengarten, "What Is Nature Capable of? Evidence, Ontology and Speculative Medical Humanities"; Rhodes and Lancaster, "Evidence-Making Interventions in Health: A Conceptual Framing."

72 Woolgar and Lezaun, "The Wrong Bin Bag: A Turn to Ontology in Science and Technology Studies?" 323.

73 Race, "Reluctant Objects: Sexual Pleasure as a Problem for HIV Biomedical Prevention," 25.

74 Law, "Collateral Realities," 172, emphasis original.

75 Law, "Seeing Like a Survey."

9 "Drawing Thresholds That Make Sense": Diagrammatic Evidence and Urgency in Automatic Outbreak Detection

Steffen Krämer

During the current COVID-19 pandemic, epidemiological evidence practices are closely observed, and have received tremendous public interest. They have given way to a spectrum of public epidemiologies and contestations of who counts as an epidemiological expert and who does not. What counts as evidence in epidemiology—in a narrow understanding of the discipline—derives to a large extent from a "statistical style reasoning."[1] Historically there have been numerous substrands of epidemiology, but those at the forefront today are mainly studies of population-level health and disease that make use of statistical calculations and simulations rather than experimental studies in a laboratory.[2] Studying epidemiological evidence practices would appear to be, for the most part, a case study of contemporary statistics and mathematical modeling. However, the perspective adopted in this chapter is different: here, epidemiological evidence practices are situated in a particular institutional, political, and cultural juncture, where epidemiology, data visualization, and policy discourses impinge upon each other. The adopted perspective is influenced by research in the field of media studies and science and technology studies, and more particularly, by research on "knowledge infrastructures" and complementary theoretical developments sometimes subsumed under the idiom of "coproduction."[3] Even though this chapter focuses on practices, it also proceeds from a minimal working definition of the object *evidence*, defining evidence as that which is validated and that which is obvious or apparent. The first element of this operationalization of evidence means, in turn, that there is usually something yet to be validated, and which holds an *evidentiary potential*. The second element, by contrast, has a rather aesthetic meaning, as illustrated by the German translation of evident as *augenfällig*, to fall into one's eyes. Between both elements of this operationalization of evidence is an obvious tension, as one emphasizes the friction-laden and explicit process of validation, and the other the seemingly smooth and often implicit process of apperception. To stretch this polemic opposition further, this chapter will revolve around the conundrum of how objects of evidentiary potential are designed to be smoothly intelligible.

DOI: 10.4324/9781003188612-12

The objects of evidentiary potential discussed here are called *signals*, and they are the result of complex algorithmic operations. They make computationally derived suggestions of whether a relevant epidemiological event might be unfolding at a given moment based on extrapolations from historical data. Signals are a crucial digital entity in the contemporary alarm and alert systems of public health agencies around the world, though they are not necessarily accepted without reservations in day-to-day practice. Signals are not usually publicized outside the professional working environment of epidemiologists and public health officers. But nonetheless, the design and development of the systems that generate these signals involve multiple publics along the way. They are affected by tendencies and decisions at the policy level as well as by developments in the fields of data science, data visualization, and machine learning. These different stakeholders also express the signal's evidentiary potential in different affective degrees. For some, signals might hold an evidentiary promise to be timelier, and avoid missing out important critical events before it is too late. For others, a signal might express a form of urgency to act without being sufficiently grounded. Wherever the emphasis is put, how the signal's evidentiary potential is expressed is part of the affective structuring of epidemiological evidence practices.

In the following, I will first introduce the conceptual and technological frame of reference of this chapter's analysis by providing a brief and selective presentation of recent developments of epidemic surveillance and early warning systems. I will then move on to a more in-depth presentation of the design of signal reports in Germany's national agency for disease control, the Robert Koch Institute (RKI). Finally, I will discuss these findings along media-theoretical, sociological, and political dimensions of interest. Much research has been done in the last two decades on anticipatory biosecurity, preemption, and preparedness.[4] My own line of theorization, however, zooms in on the particular media affordances that determine what might be suggested as evidentiary material in the context of anticipatory work. For developing theories about these affordances, philosophy is as much of a companion as qualitative social research.[5] Over all, this chapter picks up one of the tensions described in the book's introduction: the tension between different demarcations of expertise, and their respective claims of how evidence practice should be construed.

Early Warning Systems and Automatic Outbreak Detection

The development of early warning systems for epidemiological security has been largely affected by international arrangements in global public health, starting with the Sanitary Conferences in the nineteenth century, the League of Nations Health Organization in the interwar period, or the World Health Organization and its regional bureaus in the second half of the twentieth century.[6] In addition, the design and planning of early warning and surveillance systems in public health was strongly influenced by the development of new information and communication technologies.[7] A rather recent trend in this lineage has been the turn toward so-called "syndromic surveillance" since the beginning of the 2000s, which complements routine surveillance by including prevalidated information with the intent to sense emerging epidemic events even quicker.[8] Information sources could include "preclinical" data such as over-the-counter sales at pharmacies, work absentee lists, or information from help lines; or it could be "clinical prediagnostic data sources" from ambulances and hospital visits not validated by a doctor's diagnosis;[9] and also social media data has been prompted as the next frontier of pre-validated data sources.[10] The preemptive rationale that undergirds syndromic surveillance posits data timeliness and accuracy as the two most important variables in judging for or against specific system designs. The promise to detect and warn against an epidemic event as early as possible has to be weighed against the possibility, that the alarm signal is inaccurate and that the warning system is not sensitive or specific enough. In other words, binary distinctions between unvalidated and validated do not fully grasp the trade-off decisions that must be made between timeliness and urgency. One does not just choose either validated or non-validated evidentiary entities, but rather moves across a continuum of degrees of potential validity in relation to promised timeliness.

Syndromic surveillance is only the most recent keyword that received attention, and it remains a vaguely defined and contested term, with differing understandings from one health agency to another.[11] It crystallizes, however, a socio-technical and affective tension of contemporary evidence practices in epidemic surveillance and early warning procedures. This characteristic tension can be also posited in a more phenomenological register as the tension between two different kinds of sensing; between the vigilant sensing of an unfolding event with various interconnected data sources at one's disposal, and the necessity to provide a robust enough infrastructure of validating the warning signal against a misleading sense of urgency. As the case study below will show, this tension is not specific to syndromic surveillance but it equally characterizes routine surveillance, where cases are reported by traditional sources such as hospitals or general practitioners. Single cases are aggregated and compared with historical values to check whether they have reached a critical limit. These processes are automatized today to make the early detection of clusters even quicker and support public health staff in the decision whether to initiate further steps of investigation. Again, these automatic computations and recommendations do not just generate warning signals but also a socio-technical milieu of urgency, where the promised capacity to act fast is met with the risk and fear of false positives. This puts increasing importance on the acceptance of the supporting computer algorithms, in order to win the trust of public health officers and epidemiologists, and to integrate these automatized detection procedures into the "epistemic machinery"[12] of public health surveillance.

Automatic outbreak detection mobilizes what one could call a relational form of "regulatory knowledge,"[13] where a threshold is defined not as an absolute but in relation to a historical baseline, and whose comparison becomes automatized. The cultural technique of comparison is key in this arrangement. On the one hand it is delegated to algorithmic background processes; on the other hand it still has to be made visible at times to test the algorithm's proper functioning by human operators. My interest in this turn to automatic outbreak detection, especially in the last thirty years, is the way in which this new infrastructure of detection generates a certain testing environment and regime of visibility for alarm signals, or for computational objects with evidentiary potential, to become trustworthy and reliable. Before turning to the case study, I will first provide brief background information about the development of outbreak detection algorithms and their wider popularization through collaborative online platforms with a particular focus on how algorithmic processes have been made visible. These developments have directly affected the automatic detection system that was implemented in Germany and whose analyses will follow thereafter.

Aberration Detection

Contemporary automatic outbreak detection algorithms are built on mathematical formulas and techniques for statistical aberration detection that have been used for much longer. Many of the institutional proposals and designs for using automatic aberration detection for disease surveillance developed in the 1990s, but they almost unanimously referenced statistical process control and quality control methods, which have been used since the 1920s.[14] In the cultural history of aberration detection in epidemiology this is an unexpected lineage, whereby the checking of success/failure-ratios and times of low performance in industrial manufacturing is translated into the seasonal variations of epidemic disease. In fact, the industrial logic of quality control, and its formalization in mathematical forms and graphical charts, was applied to two different semantic fields in the context of epidemiological surveillance; on a first level, the success/failure ratio and margin of optimal performance was mapped onto conspicuous case numbers; and on a second meta level, it was used to test the performance of alert systems and examine how many false alarms could be tolerated while remaining informative enough.

During the 1990s, a number of different methods for aberration detection came into use by local and national health institutions in the United States and in the United Kingdom,[15] such as the cumulative sum method (CUSUM), the moving average method, or the historical limit method, as well as a mix of different methods. Around 2000, an "Early Aberration

Reporting System" (EARS) was developed by the Center for Disease Control (CDC)[16] that sought to integrate and compare these different procedures. In a summary article from 2003, the developers of the EARS project described their technical understanding of aberration detection as "the change in the distribution or frequency of important health-related events when compared with historical ... or recent" data.[17] In their definition, historical data referred to data at least three years old, and recent data meant data published in the last nine days. Depending on which temporal depth one chooses as a baseline, different mathematical methods of analysis have more salience. The authors of EARS tested five different methods to calculate aberrations, and the outcome of these tests was the selection of three refined methods to be used in contexts where little historical data was available.

The algorithmic infrastructure of EARS and the moving average method of the CDC have received attention beyond the institutional context of the United States, becoming, for example, an important reference for the design of detection systems in Germany.[18] Their development also crystallized what had been an already implicit part of the methods that were developed earlier; a particular way of dealing with the computational contingency of early warning systems, including the contingency of the temporal frames of reference, and of the alarm signals that the systems generated. The development of EARS showed the necessity to test different algorithms and mathematical models—depending on the amount of historical data available—which may lead to very different kinds of outbreak signals, and the need to assess or control the working of these algorithms comparatively.

In addition to iteratively consolidating the algorithmic and mathematical infrastructure of aberration detection in epidemiological surveillance, the technical discourse of the 1990s and early 2000s also shaped the visualization regime of alarm signaling. Most authors cited above proposed to use quality control charts; for example the "p charts" or CUSUM charts that had begun to take hold in public health and medicine during the previous decade.[19] In 1990, the CDC's Weekly Report began using a bar chart design, based on the proposal by Stroup et al.,[20] which showed the current number of diseases and how this compared to a previous historical baseline. In other words, the search for the most adequate graphical method to visualize epidemiological aberrations from a historical reference set had become institutionally recognized, and would remain an explicit consideration for the development of automatic early warning infrastructures in years to come.

Exchanging Operative Scripts and Visual Styles

Whereas the implementation of automatic aberration detection algorithms and graphical methods to monitor them occurred on the level of institutionally backed discourses that would help to build the "knowledge infrastructure" of contemporary epidemic surveillance,[21] it is worth emphasizing a second transformation that affected such processes of infra-structuring and pushed toward the integration of a wider public. Since 1997, statisticians and mathematicians have begun sharing ideas and operations for how to implement some of the standard institutional procedures in computer applications online, for example, by exchanging scripts for the programming language "R" on a community platform called *The Comprehensive R Archive Network* (CRAN).[22] Among the programming packages that have been shared on the platform are some that specifically make proposals for how to implement aberration detection in R. For example, Michael Höhle has published an R application package called "surveillance," which offered the possibility to compare existing aberration detection algorithms and the mathematical methods they include, to see how they would respond differently to the same data set.[23] The same R package became a core part of the early warning system that was later put into use by the RKI.[24]

Similar to the results of the EARS paper, the comparison made in R showed that different mathematical models and algorithms triggered alarms differently. Moreover, it made visible that the whole computational context in which an alarm comes into existence is different; not only because one uses a different temporal depth of historical baseline data, but because each of the algorithms generates a different threshold above which an excess is recognized and an alarm is flagged. In order to visualize this difference in the computational context, the original presentation of the R package from 2007, and the one explaining the implemented version of aberration detection ten years later, used a "p chart" in the tradition of the previously mentioned quality control methods.[25] The chart offered a visual testing environment, bringing together in one image the alarm signal and its computational context—the relative distribution of case numbers and of the expected threshold line (Figure 9.1).

The reason for emphasizing the contribution of R programming packages to the development of outbreak detection algorithms is not simply because they equally suggested quality controlling different detection algorithms, or because they emphasized the contingency of the detection system and its output, but because they helped to disseminate an infrastructure of aberration detection, including techniques of making visible and operational scripts that can be quickly implemented elsewhere. Because such platforms of knowledge exchange have begun to play an increasingly important role in the last two decades, the development of the contemporary practice of designing, controlling, and visualizing outbreak detection algorithms cannot be sufficiently described nor explained by turning to semantic shifts on the level of policy discourse—such as the shift toward syndromic surveillance—nor to the discursive records of scientific journals. It is of equal importance to recognize the development of "infrastructural publics" beyond public health institutions,[26] which may additionally stabilize or destabilize how the practice of surveillance is being technically realized.

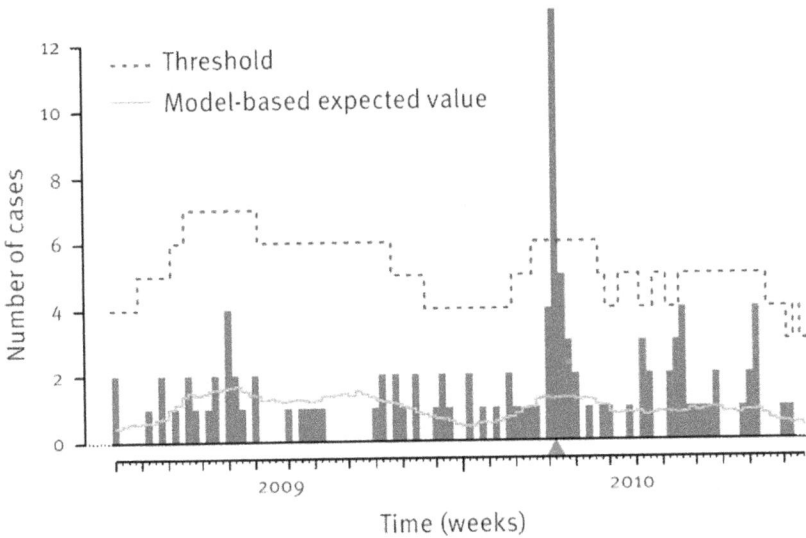

Figure 9.1 Control chart explaining the detection algorithm used by the RKI, from Maëlle Salmon, Dirk Schumacher, and Michael Höhle, "Monitoring Count Time Series in R: Aberration Detection in Public Health Surveillance," *Journal of Statistical Software* 70, no. 10 (2016). The chart is almost identical to a chart published in Michael Höhle, "Surveillance: An R Package for the Monitoring of Infectious Diseases," *Computational Statistics* 22, no. 4 (2007) about the R programming package "surveillance." The triangle on the bottom of the x-axis points to the position of an alarm signal.

Making Alarm Signals Visible

In Germany, epidemiological surveillance is distributed among local health departments on the one hand and the national agency for disease control, the Robert Koch Institute (RKI), on the other. The reporting, sharing, and assessing of epidemiologically relevant data is facilitated through software called "SurvNet@RKI."[27] While the reporting of data to the RKI is mandatory, the use of the software is not. Over the years, the RKI has continuously developed new data analysis and visualization applications, either in addition to the SurvNet software or as part of a new reporting framework called DEMIS that is scheduled to be fully implemented by 2021.[28]

Among these technological offerings from the RKI has been a so-called *signal report*, via which subscribers receive a weekly disease-specific report that indicates where current case numbers have exceeded a historical baseline, based on aberration detection algorithms of the kind previously mentioned. In 2016 and again in 2021, I conducted interviews with developers working on the redesign of these signal reports. Signal reports have been contested at the beginning for potentially creating false alarms and burdening epidemiological investigations with more extra work. Against the background of this skepticism, the developers of the second-generation signal report at the RKI voiced their interest in providing "thresholds that make sense." Three versions of the signal report and its extensions can be distinguished: an initial version that was in use from 2013 until 2017; a second and redesigned HTML version of the report, in use from April 2017 until today; and a web interface that is currently under development and which has not yet been released.

At the time of my first interviews, I could not obtain permission to observe the use of signal reports in situ and I had to focus on the appearance of the reports and the explanations of the design by its developers. However, comparing the different versions of the reports is already revealing because it shows how design expectations change over time, including aspects of visibility and affordances for interactive control. Moreover, as I tracked the changes in the design of these signal reports, I found a gap between what was initially proposed and what was eventually realized, which points to the quarrels and boundaries of how the evidentiary potential of a signal should be made tangible. Because some of the developers of such reports have a background in data science, mathematics, or web development, the changes in the medium of the signal report points to transformations that are institutionally situated on the one hand, yet embedded in trends in data visualization and data science more generally on the other. The signal report indirectly serves as record of the boundary negotiations between epidemiology and data science; crystallizing in the relationship between tacit epidemiological expertise and algorithmic recommendations.

The first digital signal reports from the RKI were table-based, presenting case numbers and highlighting those in red, whose value represented an aberration from an expected historical dataset. Epidemiologists at the RKI or subscribers from health departments could then "move into" or "open up" the signal by leaving the report's interface and navigating into a linked database in order to retrieve more detailed information about the specific clustering of the disease in question at a particular place and time. However, at the time of my interviews in 2016, a redesign of the interface of the signal report had just begun. The mockup for the possible new design moved away from the table view to include different kinds of data representations in parallel on one screen (Figure 9.2). This proposal iterated design development that had become common in contemporary data visualizations more generally; the multichart tableaus or "dashboards," with a number of different visual displays of either the same dataset or different aspects of it.[29] The visual comparison across different kinds of plots—be it maps or charts—is an epistemic practice well founded in the history of epidemiology, but in the contemporary context of dashboards it has received a new technical shaping.[30]

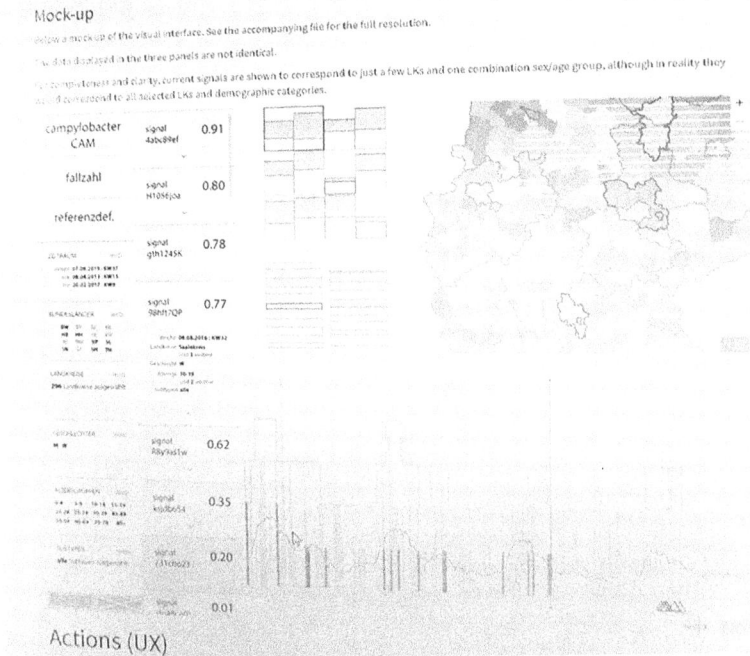

Figure 9.2 Unrealized mockup for the design of the new signal report of the RKI in November 2016. Photograph of the author.

At the bottom right of Figure 9.2, one can see that among the different diagrams for the proposed visual interface of the new signal report was also a control chart similar to the one presented above in the background section, alongside a map and bar charts. It appears as if the control chart had become a common interface for visualizing the proper functioning of the automatic detection system at the time. However the control chart never made it into publication—as I learned during my follow-up interview in early 2021. The idea was not entirely abandoned, and control charts might still be integrated into a new design at a later point in the future, but for the time being, the developers had chosen to concentrate on "easier" display types. What remains in question is whether "easy" referred to the ease of usability and intuitiveness, or to easier implementation. The currently used signal report uses a combination of a table view, a map view, and a bar chart on one screen (Figure 9.3).

Innovation narratives tend to focus on what has succeeded over time rather than ideas discarded along the way. However, I believe one can draw some, albeit speculative, interpretations from the omission of this control chart: while the chart was of particular relevance for those who are testing and controlling the algorithms—who aimed for some form of algorithmic

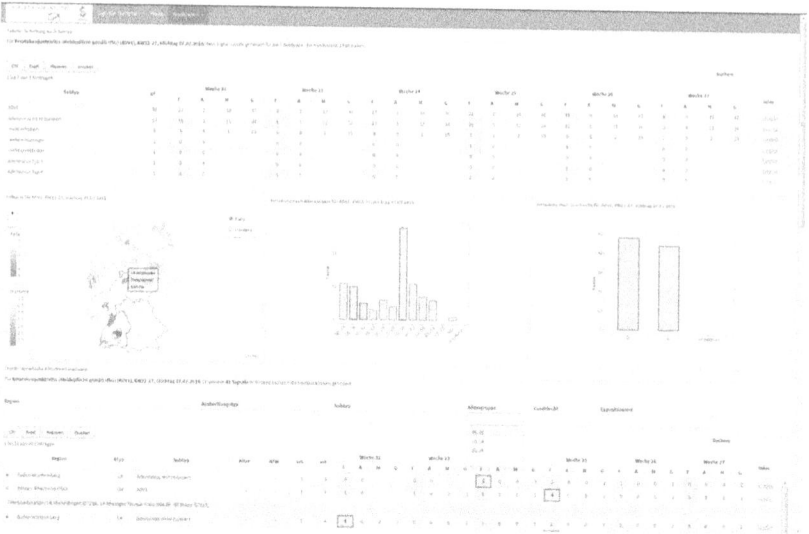

Figure 9.3 Current design of the signal report at the RKI since April 2017, reprinted with permission from: https://www.rki.de/DE/Content/Infekt/IfSG/Signale/Projekte/Signale_Projekte_node.html

intelligibility—the same intelligibility was ultimately perceived as not necessary for, and maybe even by, epidemiologists. The data scientists and programmers who redesigned the interface of the signal report had observed and interviewed epidemiologists at the RKI about the requirements that any design should meet. It might have been an outcome of these exchanges that traditional plot types such as tables, bar charts, and maps simply ranked higher than the control chart. This line of interpretation fits with a remark made during my initial interviews in 2016, when I was told that there is little benefit in opening the black box of an algorithm, not because of secrecy, but because it is complicated. The assumption was that knowledge work is targeted and focused, and should not be clouded by additional complications but rather cut straight to the relevant point. The self-presented role of the programmer and designer of the system was to facilitate this kind of targeted assessment. This remark represented a form of professional boundary drawing, where the expertise of the programmer and system's engineer was distinguished from the knowledge work of the epidemiologist.

The affirmation of the black box by the programmers and developers, and the presumably indifference toward the visual control of what's inside the black box on the side of public health officers and epidemiologists, points to the different epistemic engagements with the virtual technical milieu of an instrument by different users. For example, the technical milieu may be made visible for monitoring and description, or its internal structuring may be ignored and rather engaged with its productive capacity through acts of trial and error. This does not mean that public health officers and epidemiologists would not be wary of the contingency of a signal and how it depends on the apparatus and methodical decisions by which it is produced. But approaching this contingency by quality controlling the detection algorithm in the manner pursued by developers was perhaps not pragmatically useful or relevant for epidemiologists. Different forms of engagement may be distributed unequally across the members of an organization, between developers and users of a technology, and they characterize part of the socio-technical negotiation between these groups.

An extensive amount of scholars in the field of science and technology studies have developed self-reflexive conceptual devices to study the black-boxing of technological and sociological entities in general,[31] and more recently, sociologists have turned to the specific black-boxing of algorithms in particular.[32] Moreover, authors from the first generation of actor-network theory have identified different moments of "translation" that are taking place in the process of socio-technical stabilization and black-boxing.[33] Taking inspiration from these works, the omission of the control chart points us to the productivity of ignorance as part of technological stabilization, and to the practice of interface development as a material and aesthetic register of the negotiation that leads to this stability. Firstly, the stabilization of a technology depends not only on the dissensus or consensus about what one can do with the technology, but also on how some of the operative ingredients are shadowed to make other functions effectively available for appropriation and habituation. Secondly, the testing, getting acquainted and potentially even trusting the possibilities of these black boxes may involve different modalities and ways of engagement. The negotiations between these modalities and forms of engagement are not made explicit as such, but they are articulated in the descriptive register of interface development.

As already indicated, the omission of the control chart in further versions of the visual interface of the signal report, should by no means imply that the alarm signal itself was not a contested object. I was told that since their first introduction outbreak detection algorithms were contested by epidemiologists and other public health officers for potentially creating too many "false alarms," making epidemiological work inefficient. Such contestations also largely depend upon which diseases epidemiologists are focusing on. For example, those working on salmonella surveillance—and therefore with large numbers—are generally more sympathetic to signal reports than those who focus on smaller data sets, such as measles surveillance.

Knowing that signals reports are potentially contested, the general interest in drawing relevant thresholds has remained throughout all phases of development, but the ways of achieving this seem to have changed. The control chart once promised to visualize the computational context of a signal and assess its relative emergence within the diagrammatic testing environment of a chart. However, this graphical method has not evolved into a relevant procedure for all involved parties and the medium of the control chart has not become the primary means for stabilizing the issue of the contingency of alarm signals. Instead, for the upcoming planned interface of the signal report, the developers of the interface have decided to give epidemiologists the possibility to "adjust filters," and thereby tinker with the value of the signal that has been computed.

Moreover, rather than finding a visual way of marking the difference between alarm signals and actual occurring cases, it was agreed upon to use language and terminology to differentiate between what the algorithm suggests and what has been reported and confirmed. The new design of the signal report will distinguish between "AI clusters" on the one hand, which are the outcome from automatic aberration detection and machine learning, and "outbreaks" on the other hand, which include the confirmed cases reported to the national agency by regional and local health departments.

Discussion: Diagrammatic and Affective Boundary Work

The development, alteration, and contestation of signal reporting at the RKI in the last five to ten years offers material for the discussion of evidence practices along three conceptual lines. It exemplifies the different activities and operational chains that revolved around (signal) reports as diagrammatic evidence practices, as affective evidence practices, and as sociopolitical boundary work.

Diagrammatic Evidence Practices

Visualization infrastructures are complex technological assemblages that go beyond the mere output of an image for a human onlooker. Yet, the very specificity of the output format should not be neglected either; it makes a difference whether information is plotted as text, image, or as a hybrid of text and image such as maps and charts, and whether they can be manipulated interactively on a screen, or need to be handed around, annotated manually, and redrawn. The signal visualization infrastructure at the RKI uses a variety of output formats from color coded tables to graphs, charts, and maps. A more recent strand of media theory has proposed to view these different formats through the unifying category of diagrams. In this understanding, diagram refers to a particular class of media that combines image and text, is inscribed upon a flat surface, engages thought, vision, and learning through schematic processes, and is based on the recognition of structural similarity.[34]

As far as diagrammatic evidencing is concerned, diagrams have been described as hybrids between image and text, with a particular kind of visual propositionality.[35] Behind this interest for the propositional status of diagrams stands a long discussion about the nonpropositionality of images as being supposedly unable to negate.[36] Particularly informative in this context is the description of image evidence by Heßler and Mersch, who point to the evidentiary action of images in their capacity to show and suggest, while continuing the non-propositional interpretation of images as being unable to negate and unable to "differentiate degrees of certainty."[37] In other words, even though images can show this or that is the case, according to Heßler and Mersch, they cannot lessen this demonstrative evidence by articulating grey zones of uncertainty in a way similar to language. Here diagrams differ. Because they often involve image and text, diagrammatic evidencing is equipped with the possibility to further contextualize what is being shown by the image, and therefore also allows for further degrees of articulating certainty.

From the perspective of a viewer, diagrammatic evidencing is enacted at a basic level through different kinds of comparison; between visual and textual information, between image and annotation, and between the different graphical elements in one image. Diagrams also make the strategic comparison between various classes of information possible. For example, a diagram such as the control chart presented above made different processes visually comparable within the image space, by placing curves of epidemiologically confirmed data points and outbreaks alongside simulated trends and alarm thresholds. The outbreak line represented the trusted standard of reported cases by the health department, while the alarm line was the automatically extrapolated and questionable alternative. The comparison between the lines (between structurally similar forms) affords a second grey zone of uncertainty that characterizes diagrammatic propositionality, beyond the one offered by combining image and text. Their comparison involves judgment by similarity based on degree; some parts of the lines are similar, others are not, but where the analogy is just good enough is not defined. At the same time, recognizing the breaching of the threshold enacts a binary comparison, where the threshold line is either hit or not. It is a digital event that has been illustrated as such in the visual logic of the chart through the crossing of two lines. In other words, diagrams shape evidence practices by centering acts of iconic structural comparison and visual propositionality, but also by staging an event structure in which an entity like the signal appears plausibly eventful against the background of other parts of the image.

While diagrammatic theories in media studies have to a large extent focused on the media specificity of the diagram, the discussion of diagrams in the social studies of science have integrated the coordinative action that diagrams facilitate among a group of scientists, engineers, or more broadly, within the scope of an epistemic culture. While Susan Leigh Star, James Griesemer, and Kathryn Henderson, for example, have emphasized the role of diagrammatic media for boundary work that literally draws together the epistemic action of different parties while remaining plastic enough to keep

them apart,[38] Hans-Jörg Rheinberger has emphasized the objectual referential work of diagrams, that generates "epistemic things" whose reality
guides the actions of the participants but which is at the same under constant revision.[39]

   This applies to the context of the signal report insofar as the signal itself is
an object, whose evidentiary potential has yet to be realized through validation, but which holds an evidentiary promise. The diagram of the
control chart provided a "graphematic space"[40] for the object of the signal
to be concretized for the human viewer; as a precarious and dynamic object
relative to the case and threshold lines in the same chart. However, with the
omission of the control chart the diagram apparently failed to coordinate
between developers and users of the report. But this failure of graphical
coordination did not rid the signal of its status as an epistemic object that
requires other kinds of coordination. Instead, the coordinating register was
shifted toward modes of collective engagement with the signal beyond
iconic demonstration and diagrammatic comparison.

Affective Evidence Practices

Coupling the specificity of a practice to an epistemic object—that has yet to
be realized in one way or another—links the diagrammatic practices described by Rheinberger to the affective aspects of epistemic practices described by Karin Knorr-Cetina. According to Knorr-Cetina, the power of
epistemic objects resides in their incomplete nature and partiality, and that
they "provide pointers to possible further explorations."[41] Moreover, she
suggests that this lack of completeness and the "unfolding character of
knowledge objects uniquely matches the 'structure of wanting,'" and
thereby connects this line of theorization to a Lacanian theory about drive
and affect.[42] Her emphasis on the epistemic object's partiality and
incompleteness—as a key lure, motivator, or drive for epistemic work—can
be generalized beyond the affective charge of epistemic practices as a crucial
ingredient of a variety of affect theories more generally, which situate affective power not in realized emotional states but in the potentiality of
affective tendencies. For example, Deleuzian affect theory has traditionally
based affect in the ability to affect and be affected,[43] and phenomenological
theories of affect have regularly emphasized the action potentials of certain
situations or environments.[44] However, rather than remaining on the level
of describing the affective dynamics of practices through a generic potentiality, this line of theorization also calls upon the researcher to explicate
the modes in which this potentiality is affectively expressed.

   What, then, is the mode in which the affective lure of the evidentiary
potential of signals is expressed? Alarm systems, it can be argued, affectively
realize a specific form of evidentiary urging. This mode combines the lure
of the epistemic object that still has to be validated with the practical
constraints and potentials of the wider context of production, where the

anticipatory norm of biosecurity or demands from public opinion might discredit an urgent call as a form of alarmism. Moreover, whether an evidentiary potential is endorsed as an alarm that calls for urgent follow-up action depends not only upon the trust in the alarm's context of production (the reliability of algorithms, for example), or the skepticism or affirmation of policy decisions, but also on how feelings of control are secured and subject positions are distributed.

The latest versions of the signal report were designed in a way that promised to put officers in control by being able to "customize filters" rather than being urged by an automatic signal, and rather than simply trusting a static visual presentation that is generated by rules set elsewhere. Such a distribution of agency and control is not particular to this case study, but it largely mirrors industry rationalizations for software and interface design more generally. To put the user in control—while at the same making services easier—is a contemporary mantra of many everyday applications. It would be a surprise if this did not also manifest in the language and design promises of developers at the RKI. It also reminds of the more general insight that a technology is not simply brought into the world by its designers, but that it is realized through various layers of "co-construction" between designers and users. In the present case, these layers of co-construction involved interviewing with the prospective users during the development of the signal reports and the "configuration" of user control through popular scripts such as filter customization.[45]

Whereas a diagrammatic coordination centers actions of showing and comparing, affective coordination centers techniques of subjective control vis-à-vis feelings of alertness and urgency. Both of these kinds of coordination, and the activities that support them, cannot, however, be restricted to an otherwise homogenous professional context. Instead, they stand at the crossroads of different working cultures and traditions; data science and programming on the one hand, and epidemiology on the other. In addition, both these strands may respond to different publics. The diagrammatic and affective evidence practices described are therefore also part of boundary negotiations in the "trading zone" between these work cultures.[46]

Sociopolitical Boundary Work

Given the ascribed action potential of the initial alarm signal, and the different takes on algorithmic knowledge within a public health institution, the drawing of alarm thresholds has become a political matter; both in the sense of micropolitical negotiations between different epistemic cultures within one institution, and in the sense of the interaction between institutional actors, policy frameworks, and the wider public. Moreover, in designing outbreak detection algorithms, programmers do not only mobilize statistical models that are common in epidemiology but are also affected by technological standards beyond epidemiology when implementing these algorithms in visualization systems. They may be influenced by current trends in the fields of data visualization and interface design, and by aspirations for open data frameworks.

For example, the engagement of a wider public through access to data or software code was emphasized by developers at the RKI both in 2016 and again in 2021. At the same time, data protection regulation and the protection of national sovereignty for health data puts restrictions on the way the data is publicized. It needs to be noted, however, that the RKI does already provide a public data interface, so it is rather a matter of the extent to which data, software code, or interface solutions are made public. Moreover, I witnessed the interest in current trends in data visualization during my interviews in 2016, when my interviewees had handbooks for information visualization piled on their desk, and when they had published new visualization scripts as repositories to the collaborative online platform GitHub which is commonly used for open-source software development. In the more recent development phases of the signal report, the effects of wider trends in data visualization became apparent through the developers' interest for dashboards. All of these examples testify again that not only the co-construction of developers and users matters during the development of new technological expert systems, but also the interaction between experts, policy frameworks and wider technical publics.

In addition, there is another kind of political interpretation of the signal report at RKI, which follows the affective politics of urgency claims and public pressure, where the right to decide when and what to act upon is contested. The implementation of automatic detection algorithms has introduced the computerized recommendation system as yet another actor to this field of urgency claims. The political question in these cases revolves around the right of the human expert's self-determination to decide when to act upon evidentiary promises. In this light, the process of designing and redesigning the signal report crystallized an affective micropolitics, which is based on the distribution of the capacities to follow-up or ignore certain action potentials, and which negotiates feelings and understandings of alert and urgency.

Altogether, the micropolitical conflict that the use of signal reports—and automatic aberration detection—generated has not been stabilized through diagrammatic coordination. Instead, the conflict was resolved by designing action possibilities for user's intervention and control; the possibility to filter, and by distinguishing between cluster and outbreak. In other words, micropolitical boundary work meant in this case redistributing agency between machinic recommendation and epistemic expertise, and it registered materially in the naming of data events and design templates for user interaction.

Conclusion

Evidence practices in the context of epidemiological surveillance might refer to a variety of routines: from the diagnostic recognition of notifiable diseases by general practitioners and their reporting to a central authority, through the tracing and analysis of those cases by local health departments and the laboratory confirmation of certain virus strains, to the work that is being done by statisticians or data scientists on the aggregate level, before and after cases have been confirmed. This chapter focused on the latter context, and on the institutional procedures of making automatic outbreak suggestions before the confirmation of an epidemic event. These activities belong to a set of anticipatory surveillance practices, and they do not generate evidence in a sense of after-the-fact, but rather they suggest what might become a candidate for routine evidence procedures to be activated. In other words, they work on the virtual milieu of evidence practices, of what and how evidentiary potentials are expressed.

To make such evidentiary potential visible or otherwise tangible may involve multiple points of negotiation between different epistemic traditions and work cultures. It is against the background of such negotiations that the "drawing of thresholds that make sense" is a particularly interesting practice. This practice has not been realized as expected. It has remained under constant revision in a similar way to how epistemic objects have been described in the context of experimental systems. Analogously, the drawing of thresholds and production of meaningful alarm signals represent the evidentiary lure of automatized surveillance systems, whose objects are always partial and incomplete, and it affectively structures the evidentiary boundary work between data scientists, mathematicians, and epidemiologists.

Notes

1 Ian Hacking, "Statistical Language, Statistical Truth and Statistical Reason: The Self-Authentification of a Style of Reasoning," in *The Social Dimensions of Science*, ed. Ernan McMullin (Indiana: Notre Dame University Press, 1992), 130–157.

2 Different from, for example, the 1920s, where a laboratory based "experimental epidemiology" was still an avenue worth pursuing (Simon Flexner, "Experimental Epidemiology—Introductory," *The Journal of Experimental Medicine* 36, no. 1 (1922): 9–14). For the multiple strands of epidemiology that mingled during the inter-war period, see J. Andrew Mendelsohn, "Von der Ausrottung zum Gleichgewicht: Wie Epidemien nach dem Ersten Weltkrieg komplex wurden," in *Bakteriologie und Moderne. Studien zur Biopolitik des Unsichtbaren 1870–1920*, eds. Philipp Sarasin, Silvia Berger, Marianne Hänseler, and Myriam Spörri (Frankfurt a.M.: Suhrkamp, 2007), 239–284.

3 Paul Edwards, *A Vast Machine: Computer Models, Climate Data, and the Politics of Global Warming* (Cambridge: MIT Press, 2010), 17–20; Sheila Jasanoff, "The Idiom of Co-production," in *States of Knowledge. The Co-production of Science and Social Order*, ed. Sheila Jasanoff (London: Routledge, 2004), 2–6.

4 Andrew Lakoff, "Preparing for the Next Emergency," *Public Culture* 19, no. 2 (2007): 247–271; Peter Adey and Ben Anderson, "Anticipating Emergencies: Technologies of Preparedness and the Matter of Security," *Security Dialogue* 43, no. 2 (2012): 99–117.

5 The initial research for this chapter was based on the document analysis of technical publications about the design of outbreak detection algorithms since the 2000s, as well as primary and secondary resources about early warning systems in public health more generally. In addition, I conducted two expert interviews and one group discussion with employees of the Robert Koch Institute in Germany in 2016, and

one follow-up interview with another employee in 2021. The interviewees had a background in either epidemiology and public health or data science.

6 cf. Huber, "The Unification of the Globe By Disease? The International Sanitary Conferences on Cholera, 1851–1894." *Historical Journal* 49, no. 2 (2006): 453–476; Lenore Manderson, "Wireless Wars in the Eastern Arena: Epidemiologic Surveillance, Disease Prevention and the Work of the Eastern Bureau of the League of Nationshealth Organisation, 1925–1940," in *International Health Organisations and Movements, 1918–1939*, ed. Paul Weindling (New York: Cambridge University Press, 1995), 109–133; Lorna Weir and Eric Mykhalovskiy. *Global Public Health Vigilance: Creating a World on Alert* (New York: Routledge, 2012).

7 For example, Jack Woodall, "Outbreak Meets the Internet: Global Epidemic Monitoring by Promed-Mail." *SIM Quarterly. The Newsletter of the Society for Internet in Medicine* 1 (1997); Louisa R. Beck, Bradley M. Lobitz, and Byron L. Wood, "Remote Sensing and Human Health: New Sensors and New Opportunities," *Emerging Infectious Diseases* 6, no. 3 (2000): 217–227.

8 Berger et al., "Review of Syndromic Surveillance: Implications for Waterborne Disease Detection," *Journal of Epidemiology and Community Health* 60, no. 6 (2005): 543–550; cf. Lyle Fearnley, "Signals Come and Go: Syndromic Surveillance and Styles of Biosecurity," *Environment and Planning A* 40 (2008): 1615–1632.

9 Berger et al., "Review of Syndromic Surveillance," 543–544.

10 cf. Annika Richterich, "Infodemiologie – Von 'Supply' zu 'Demand'. Google Flu Trends und transaktionale Big Data in der epidemiologischen Surveillance," in *Big Data. Analysen zum Digitalen Wandel von Wissen, Macht und Ökonomie*, ed. Ramón Reichert (Bielefeld: Transcript, 2014), 333–363; Lindsay Thomas, "Pandemics of the Future: Disease Surveillance in Real Time," *Surveillance & Society* 12, no. 2 (2014): 287–300.

11 Personal communication with employees of the RKI in Germany; for the term's original ambiguity in the US-American context, see Fearnley, "Signals Come and Go," 1619.

12 Karin Cetina Knorr, *Epistemic Cultures: How the Sciences Make Knowledge* (Cambridge: Harvard University Press, 1999), 3.

13 Reinhardt, Carsten, "Regulierungswissen und Regulierungskonzepte," *Berichte zur Wissenschaftsgeschichte* 33 (2010): 352–353.

14 Stroup et al., "Detection of Aberrations in the Occurence of Notifiable Diseases Surveillance Data," *Statistics in Medicine* 8 (1989): 323–329; Hutwagner et al., "Using Laboratory-Based Surveillance Data for Prevention: An Algorithm for Detecting *Salmonella* Oubreaks," *Emerging Infectious Diseases* 3, no. 3 (1997): 395–400; cf. Sheila M. Williams et al., "Quality Control: An Application of the Cusum," *BMJ* 304 (1992): 1359–1361; Denis Bayart, "How to Make Chance Manageable: Statistical Thinking and Cognitive Devices in Manufacturing Control," in *Cultures of Control*, ed. Miriam R. Levin (Amsterdam: Harwood, 2000), 151–174.

15 For the United States, see Stroup et al., "Detection of Aberrations"; Simonson et al. 1997; Hutwagner et al., "Laboratory-Based Surveillance"; for the United Kingdom, see C.P. Farrington et al., "A Statistical Algorithm for the Early Detection of Outbreaks of Infectious Disease," *Journal of the Royal Statistical Society. Series A (Statistics in Society)* 159, no. 3 (1996): 574–563.

16 Hutwagner et al., "Bioterrorism Preparedness."

17 Hutwagner et al., "Bioterrorism Preparedness," 90.

18 EARS and moving average were mentioned as an important reference during the author's first interviews with experts at the Robert-Koch-Institute in Germany in 2016.

19 cf. Williams, Parry, and Schlup, "Quality Control: An Application of the Cusum."

20 Stroup et al., "Detection of Aberrations."

21 Paul Edwards, *A Vast Machine,* 17–20; Karin Cetina Knorr, *Epistemic Cultures: How the Sciences Make Knowledge* (Cambridge: Harvard University Press, 1999), 3.

22 CRAN is accessible at https://cran.r-project.org/.

23 Michael Höhle, "Surveillance: An R Package for the Monitoring of Infectious Diseases," *Computational Statistics* 22, no. 4 (2007): 571–582.

24 Maëlle Salmon, Dirk Schumacher, and Michael Höhle, "Monitoring Count Time Series in R: Aberration Detection in Public Health Surveillance," *Journal of Statistical Software* 70, no. 10 (2016): 2; Maëlle Salmon et al., "A System for Automated Outbreak Detection of Communicable Diseases in Germany," *Euro Surveill* 21, no. 13 (2016): 1–7.

25 Salmon et al., "System," 3.

26 Matthias Korn et al., eds., *Infrastructuring Publics* (Wiesbaden: Springer VS, 2019), 11–47.

27 Sabine Ritter, *Survnet@rki – Das Meldesystem Zum IfSG. Anwenderhandbuch. Arbeitsversion 1.7* (Berlin: Robert-Koch-Institut, 2016).

28 Michaela Diercke, "Deutsches Elektronisches Melde- Und Informationssystem Für Den Infektionsschutz (Demis) – Entwicklungsschritte," *Epidemiologisches Bulletin* 25 (2018): 227.

29 cf. Shannon Mattern, "Mission Control: A History of the Urban Dashboard," *Places Journal* (online), Published March 2015. https://placesjournal.org/article/mission-control-a-history-of-the-urban-dashboard/?cn-reloaded=1.

30 Practices of comparing two maps or tables and maps side by side date back to the traditional atlas works of medical geographers and epidemiologists of the nineteenth century (Tom Koch, *Disease Maps. Epidemics on the Ground*) and visual comparison has been equally recalled as an essential epistemic technique of medical geographers after World War Two (cf. Steffen Krämer, *Diagrams of Epidemiological Knowledge in Medical Geography and Public Health Surveillance*, PhD Thesis (Hamburg: University of Hamburg, 2020)). In the 1990s, researchers at the US National Center for Health Statistics conducted design tests with statisticians to find out what would be the most efficient design for future disease atlases and, again, the practice of comparing between different maps was emphasized (Linda Pickle and Douglas Herrmann, "Cognitive Research for the Design of Statistical Rate Maps").

31 See, for example, the contributions by Trevor J. Pinch and Wiebe Bijker and by Michel Callon in Bijker, Wiebe E., Thomas Parke Hughes, and Trevor J. Pinch, eds. *The Social Construction of Technological Systems. New Directions in the Sociology and History of Technology* (Cambridge: The MIT Press, 1987).

32 Jonathan Roberge and Robert Seyfert, "Was Sind Algorithmuskulturen," in *Algorithmuskulturen: Über Die Rechnerische Konstruktion Der Wirklichkeit*, ed. Robert Seyfert and Jonathan Roberge (Bielefeld: transcript, 2017), 7–40.

33 cf. Michel Callon, "Some Elements of a Sociology of Translation: Domestification of the Scallops and the Fishermen of St Brieuc Bay." In *Power, Action and Belief: A New Sociology of Knowledge?*, ed. John Law (London: Routledge, 1986), 196–223; Bruno Latour, "Technology Is Society Made Durable." In *A Sociology of Monsters: Essays on Power, Technology and Domination*, ed. John Law (London: Routledge, 1991), 103–131.

34 cf. Sybille Krämer, *Figuration, Anschauung, Erkenntnis. Grundlinien Einer Diagrammatologie* (Frankfurt a.M.: Suhrkamp, 2016), 59–86; Matthias Bauer and Christoph Ernst, *Diagrammatik. Einführung in Ein Kultur- Und Medienwissenschaftliches Forschungsfeld* (Bielefeld: transcript, 2010), 40–64.

35 Krämer, *Figuration*, 61–62, 80.

36 cf. Emmanuel Alloa, "Ikonische Negation. Unter Welchen Umständen Können Bilder Verneinen," in *Bild Und Negativität*, ed. Lars Nowak (Würzburg: Königshausen und Neumann, 2019).

37 Martina Heßler and Dieter Mersch, "Bildlogik Oder Was Heißt Visuelles Denken?" in *Logik Des Bildlichen. Zur Kritik Der Ikonischen Vernunft*, eds. Martina Heßler and Dieter Mersch (Bielefeld: transcript, 2009), 29.

38 Susan Leigh Star and James R. Griesemer, "Institutional Ecology, 'Translations' and Boundary Objects: Amateurs and Professionals in Berkeley's Museum of Vertebrate Zoology, 1907–39," *Social Studies of Science* 19, no. 3 (1989): 410–411; Kathryn Henderson, *On Line and on Paper* (Cambridge: The MIT Press, 1998), 25–57.

39 Hans-Jörg Rheinberger, "Experimental Systems, Graphematic Spaces," in *Inscribing Science. Scientific Texts and the Materiality of Communication*, ed. Timothy Lenoir (Stanford: Stanford University Press, 1998), 295–302; cf. Karin Knorr Cetina, "Objectual Practice," in *The Practice Turn in Contemporary Theory*, ed. Theodore R. Schatzki, Karin Knorr Cetina, and Eike von Savigny (London: Routledge, 2001), 190.

40 Rheinberger, "Experimental Systems," 295–302.

41 Knorr Cetina, "Objectual Practice," 192.

42 Knorr Cetina, 194.

43 Brian Massumi, "Notes on the Translation and Acknowledgments," in *A Thousand Plateaus—Capitalism and Schizophrenia*. Written by Gilles Deleuze and Félix Guattari (Minneapolis: University of Minnesota Press, 1987), 16.

44 Jan Slaby, "Affektive Intentionalität—Hintergrundgefühle, Möglichkeitsräume, Handlungsorientierung," in *Affektive Intentionalität. Beiträge Zur Welterschließenden Funktion Der Menschlichen Gefühle*, ed. Jan Slaby et al. (Paderborn: mentis, 2011), 23–48.

45 Nelly Oudshoorn and Trevor Pinch, "Introduction: How Users and Non-Users Matter," in *How Users Matter. The Co-construction of Users and Technology* (Cambridge: The MIT Press, 2003), 7–11.

46 Peter Galison has reconstructed how in postwar physics and with the use of electronic images two different "instrument subcultures," an image subculture and a logic subculture, mingled in a "trading zone" that mutually affected both sides; Peter Galison, *Image and Logic: A material culture of microphysics* (Chicago: University of Chicago Press, 1997), 806–807.

10 Producing Migration Knowledge: From Big Data to Evidence-Based Policy?

Laura Stielike

Evidence and Governance in Big-Data-Based Migration Research and Policymaking

This chapter explores the relationship between evidence and governance in the field of big-data-based migration research and policymaking. Recently, migration has become an object of study for data scientists who analyze social media, search engine, satellite, and mobile phone positioning data.[1] At the same time, and in light of a postulated lack of accurate and reliable migration data, states and international organizations have discovered that big data can be used as a source of "evidence" in evidence-based policymaking.[2]

Evidence-based policy has been defined as "public policies, programs, and practices that are grounded in empirical evidence."[3] The concept can be linked to debates about better evaluations of policy outcomes in the 1970s, to discussions of "evidence-based medicine" in the 1990s, and to trends in organization and management studies.[4] It gained prominence with the New Labour Government in the United Kingdom in the late 1990s and early 2000s.[5] In 2000, the then Education Secretary David Blunkett stressed the new role of social science research for policy development and evaluation: "We need to be able to rely on social science and social scientists to tell us what works and why and what types of policy initiatives are likely to be most effective."[6]

The approach of evidence-based policymaking has been criticized for several reasons. First, it has been identified as an approach that aims at enhancing the techniques of managing strategic policymaking, and thereby depoliticizes the production and utilization of knowledge for governance purposes.[7] Second, it has been pointed out that evidence-based policymaking is based on the assumption that the relation between evidence and the resulting policy is linear and direct, instead of stressing that policy problems and their solutions are constructed through negotiation and deliberation.[8] A third critique of evidence-based policymaking is that it assumes that ethical, moral, and political questions can be answered by policymakers simply on the basis of "best evidence."[9] And fourth, the approach is based on the assumption that evidence is "value-free and context-neutral."[10]

DOI: 10.4324/9781003188612-13

In the field of migration research, Christina Boswell has been among the first to study the relationship between "evidence" and migration policy. Drawing on qualitative field studies in the United Kingdom and Germany, she points out that policymakers do not only use expert knowledge to improve migration policy but also to lend authority to their already existing preferences or to show their capacity to make sound decisions in controversial policy areas.[11] In a comparative study on the "research–policy nexus," Martin Bak Jørgensen has shown that in Sweden social scientists influenced the agenda-setting for a reform of immigrant integration policies, while in Denmark social science research was used very selectively to legitimize government policies.[12] In the same vein, Simon Warren argues that in the British education system "evidence" was politically mobilized to legitimize existing forms of racial and social discrimination against migrants.[13] Focusing on the so-called European "migration crisis" of 2015, Baldwin-Edwards, Blitz, and Crawley identify a significant "gap between evidence and policy," and carve out three factors that explain this gap: First, due to competing approaches in social science research, what counts as evidence is always contested and challenged. Second, not only research results but also policymakers' individual expertise and judgment, institutional capacity, and dominant policy narratives serve as evidence. Third, policymaking is an inherently political process.[14] Andrew Geddes has explored the relationship between research and policy at the level of EU migration governance, with a focus on the knowledge production surrounding "temporary and circular migration." Building on Boswell's work, he observes a "general tension between instrumental approaches to the use of information and knowledge ('evidence-based policymaking'), approaches that substantiate existing policy choices ('policy-based evidence-making') and approaches that legitimate institutional roles ('policy-based institution-building')."[15]

In this chapter, I will show that all three relationships between evidence and governance identified by Geddes can also be found in the field of big-data-based migration research and policymaking; an approach so far not applied to this field.[16] More generally, I will argue that the use of big data in migration research and governance perpetuates the belief that migration can be governed for the benefit of all if only there is enough "evidence" available. In such a view, lack of information is constructed as the main challenge for migration governance, ignoring that the often conflicting relationship between migratory practices and the attempts to govern them is a highly political question.

In the first section, I will introduce the field of big-data-based migration research and policymaking, show how actors in this field frame their work as a contribution to evidence-based policymaking, and argue that instead of translating big data into evidence-based policy the field is coproducing migration knowledge. In the second section, following Geddes' three identified strands of evidence-governance relationships, I will present data challenges as a possible example of evidence-based policymaking, data hubs as an example of policy-based institution-building, and feasibility studies on the use of big data for migration management as an example of policy-based evidence-making. I will argue that all three ways of using big data in the context of migration policy reinforce the assumption that lack of "evidence" is the central challenge for migration governance; thereby framing the ways in which societies should best deal with migration as a technical instead of a political question. Finally, I show the need to investigate more carefully how unequal power relations between migrants, and those who study and govern them, influence the coproduction process of migration knowledge and its outcomes.

Coproducing Big-Data-Based Migration Knowledge

In December 2018, 164 states signed the Global Compact for Safe, Orderly and Regular Migration. The compact's first objective is to "collect and utilize accurate and disaggregated data as a basis for evidence-based policies."[17] Among the proposed actions is the use of big data for the governance of international migration. Such a nexus between data and migration governance is not new. European colonial powers already tried to collect precise data on mobile populations in their colonies.[18] In the more recent past, the 2005 report of the Global Commission on International Migration pointed out that "effective migration policies" rely on "access to more timely, accurate and detailed migration data."[19] Ten years later—after the adoption of the United Nations Sustainable Development Goals (SDGs) in 2015—the calls for better data on migration grew significantly louder. Governments and international organizations identified large gaps in national and international migration statistics which made it impossible to measure the progress toward the achievement of the migration-related targets of the SDGs. Moreover, the guiding principle of the UN Sustainable Development Agenda to "leave no one behind" required making marginalized groups more visible in statistics, which increased the demand for migration-related data even more.[20] The so-called refugee crisis in Europe (2015–2016) created another discourse of urgency related to the need for better migration data to control migration flows, improve humanitarian action, enhance integration policies, and deliver more objective information on migration.[21] Against this backdrop, governments and international organizations increasingly support the use of alternative data sources, especially big data. The promise of big data for migration governance is the access to migration-related data that is virtually real-time—or can be updated frequently—that covers geographic areas with no or limited official migration statistics, and that has much larger sample sizes and more flexible definitions of migration than traditional surveys.[22]

In the last few years a transnational network has evolved; consisting of international organizations' data hubs, data scientists at universities, Internet and technology companies, and nonprofit organizations, all involved in the big-data-based production of knowledge on migration. Most of these actors frame their work as a contribution to evidence-based policymaking. For example, in June 2018, the European Commission's Knowledge Centre on Migration and Demography and the International Organization for Migration's Global Migration Data Analysis Centre launched the Big Data for Migration Alliance, to "advance discussions on how to harness the potential of big data sources for the analysis of migration and its relevance for policymaking."[23] At the same time, data scientists and computational social scientists consider their big-data-based research on migration as an "input for policymaking" and envisage a "systematic use of non-traditional data for policy support and migration governance."[24] The lines between big-data-based research and policymaking are thus blurred. Data scientists give policy advice,[25] and staff members of international organizations co-author big-data-based research papers.[26] Instead of a simple translation from big data to evidence-based policy, the relationship is one of a coproduction of migration knowledge.

Although most actors involved in the production of big-data-based knowledge on migration frame their work as a contribution to a more evidence-based migration policymaking, at least three dominant relation-ships between evidence and governance can be identified in the field.

Evidence-Based Policymaking: Data Challenges

Evidence-based policymaking is based on the assumption that more detailed and quantifiable knowledge leads—somewhat automatically—to better and more effective policies. In the context of big data, so-called data challenges are an example of this approach. The main idea of a data challenge is that international groups of researchers are invited to use data provided by states, international organizations, or private corporations to produce research that can help find solutions to societal challenges. For example the "Data for Development" (D4D) challenge organized by the French tele-communications company Orange in 2012 was probably the first data challenge with a strong focus on mobility.[27] Orange provided anonymized call detail records (CDR) of mobile phone calls and SMS exchanges of five million customers in Ivory Coast covering the time period December 2011 to April 2012.[28] In 2014, Orange and Sonatel—the main tele-communications provider in Senegal—organized a second "Data for Development" challenge providing the call detail records of more than nine million customers in Senegal.[29] Another data challenge with a focus on migration-related topics was organized in 2017 by the European Commission's Knowledge Centre on Migration and Demography. However, the "Data for Integration" (D4I) challenge did not provide big data sources but rather the spatially disaggregated 2011 census data collected by national statistical offices showing the concentration of migrants in cells of 100 by 100 meters in the cities of eight European countries.[30] Also in 2017, the "Data for Refugees Turkey" challenge was launched. The largest Turkish telecommunications company Türk Telekom provided anonymized mobile phone data of almost a million refugee and non-refugee users to international groups of researchers. In cooperation with Turkish research institutions, refugee communities, ministries, and international organizations, research teams were invited to use the provided data sources to produce research that could help to improve the situation of Syrian refugees in Turkey in the fields of security, health, education, work, and integration.[31]

The provision of better evidence for better policymaking is the central promise of data challenges. In the final chapter of the book collecting the studies carried out in the framework of the "Data for Refugees Turkey" challenge, the authors state their aim to "translate the knowledge gained from the challenge into policy recommendations."[32] The term "translate into" implies that "the knowledge gained" and the "policy recommendations" have a linear one-way relationship. However, the authors also argue that international governmental organizations or nongovernmental organizations representing refugee interests should be involved in all stages of the data challenge (design, implementation, dissemination, policymaking).[33] This would mean that political actors and interests actually influence the research process—starting from the development of research questions. This would imply the coproduction of knowledge rather than the simple, one-directional translation of big-data-based research results into policy.

Albert Salah and his colleagues who wrote the aforementioned chapter "Policy Implications of the D4R Challenge" touch on these important questions regarding the relationship between evidence and governance in the context of big-data-based knowledge production. They point to the fact that sometimes the opportunity for researchers to develop and test new methods in the framework of data challenges is much more important than policymakers' actual interest in the research findings.[34] They also stress that it is often unclear exactly how "results might be exploited for policymaking," and that privacy risks related to big data prevent policymakers from drawing on this kind of research.[35] Another "limitation" that "should temper policy-makers' broad reliance on big data for policy making" is bias in big data sources. For example, social media users and mobile phone users are two groups with quite different demographic characteristics; the penetration rate of mobile operators differs from region to region, and certain groups—such as refugee children and the larger percentage of refugee women than men who do not own a mobile phone—are not represented in call detail record data. Another source of bias are bots or bot-assisted social media accounts strategically utilized to influence online discussions.[36]

The researchers involved in the "Data for Refugees Turkey" challenge, as well as Linnet Taylor—who critically analyzed the first "Data for Development" challenge in Ivory Coast—draw attention to the risk of "function creep" in the use of mobile phone positioning data: "A gradual expansion of the functionality of some system or technology beyond what it was originally created for."[37] Salah et al. see the risk that a mechanism designed to serve a population can take on an unintended functionality that damages the interest of this population. Taylor identifies a potential move from care to control when call detail record data is not only used by humanitarian organizations but also by governments:

> In contrast to humanitarian organisations oriented towards care—responding to mobility in cases of natural disasters or conflict—governments have an interest in controlling mobility, and specifically in predicting, tracking and preventing unauthorized migration flows towards their borders. Mobile phone data used either as real-time surveillance data or in agent-based models clearly have the potential to help governments pre-empt undocumented migration, thus moving along the spectrum from care to control.[38]

To minimize the risk of function creep, Salah et al. propose to involve international governmental organizations and nongovernmental organizations who represent the interests of refugees in all stages of the data challenge.[39] Taylor argues that a better data protection system than self-regulation needs to be established for both corporations and researchers when geo-coded mobile data is shared and used for policy-oriented research.

A more general risk of the data challenge format is, in my view, that this framing of *better evidence for better policymaking* implies that finding answers to migration-related questions is primarily a question of better data. Such a perspective ignores conflicting interests and unequal power relations between migrants and government agencies, between newly arrived and long-established local populations, and between differently labeled groups of migrants. Incoherence, inefficiency, and ineffectiveness in the field of migration policy could be interpreted as a result of the permanent struggles between different groups and interests rather than stemming from a lack of evidence.

Policy-Based Institution-Building: Data Hubs

According to Geddes, the production and provision of scientific expertise can provide "legitimacy for institutional actors seeking to expand their role."[40] The recent launch of data hubs by international organizations is then an example of "policy-based institution-building," as international organizations try to position themselves as relevant advisors for a more evidence-based migration policy.

The International Organization for Migration's Global Migration Data Analysis Centre (GMDAC) was established in Berlin in 2015. Its central aims are to strengthen "the role of data in global migration governance," to support "IOM Member States' capacities to collect, analyse and use migration data" and to promote "evidence-based policies by compiling, sharing and analysing IOM and other sources of data."[41] In 2017, the GMDAC launched the Migration Data Portal which is presented as a "one-stop shop" for global migration data—still in the making.[42] The Migration Data Portal is structured according to six sections: (1) "Data" which provides an interactive world map that visualizes "international, publicly-available and internationally comparable" data on migration; (2) "Themes," which contextualize the data; (3) "Tools" which provide guidelines on data collection, interpretation, and dissemination; (4) "SDGs and GCMs"[43] which provide information on migration data needs for these international political initiatives; (5) "Overviews" which help to find migration statistics by country and region; and (6) "Blog" which offers the "migration data community" a space of exchange on new developments.[44] The Migration Data Portal positions itself in line with the discourse on evidence-based migration policy. The portal is not intended as a source for migration research but rather:

> To help policy makers, national statistics officers, journalists and the general public interested in the field of migration to navigate the increasingly complex landscape of international migration data, currently scattered across different organisations and agencies. Especially in critical times, such as those faced today, it is essential to ensure that responses to migration are based on sound facts and accurate analysis. By making the evidence about migration issues accessible and easy to understand, the Portal aims to contribute to a more informed public debate.[45]

The Migration Data Portal also offers a "Data Innovation Directory," which provides a database of innovative migration data projects searchable by keyword, data source, region, and topic.[46] Here, short presentations of big-data-based projects carried out by private companies, universities, NGOs, and international organizations drawing on mobile phone data, social media data, satellite data, and machine learning can be accessed, and further links to those projects are provided.

The European Commission's Knowledge Centre on Migration and Demography (KCMD)—established in 2016—can be seen as another data hub in the field of evidence-based migration policymaking. The KCMD "aims to provide independent scientific evidence for strengthening the Commission's response to the opportunities and challenges related to migration, demography and related policies."[47] Besides its publications on issues related to migration data and its participation in important events of the international migration data community, the KCMD offers a "Knowledge Portal" which offers "a single point of entry to knowledge relevant to EU policies on migration and related fields."[48] The KCMD Knowledge Portal consists of four elements: (1) The "Data Catalogue" which provides a large collection of datasets on various migration-related topics;[49] (2) the "Information Catalogue" which gathers websites, actors, networks, projects, and activities relevant to EU policies on migration and related fields;[50] (3) the "Dynamic Data Hub" which provides access to migration datasets through interactive maps;[51] and (4) the "Atlas of Migration" which makes accessible national migration data and indicators of 198 countries and territories via country-specific dashboards.[52] The four main elements of the KCMD Knowledge Portal do not include big data sources. However, staff of the KCMD have authored studies on the use of big data in the production of knowledge on migration and mobility, and the KCMD has commissioned a study on this topic.[53]

GMDAC's and KCMD's main activity in the field of big data takes place within the Big Data for Migration Alliance (BD4M) founded in 2018 by both organizations.[54] The BD4M Alliance presents itself in line with the evidence-based policymaking discourse as "the first-ever dedicated network of stakeholders seeking to facilitate responsible data innovation and colla-boration to improve the evidence base on migration and human mobility and its use for policy making."[55] One major aim of the alliance—which has meanwhile partnered with the GovLab of New York Universities' Tandon School of Engineering—is to promote and facilitate "data collaboratives" as a new form of public-private collaboration "in which participants from different sectors—in particular companies—exchange their data to create public value."[56] The GovLab provides a "data collaboratives explorer" which lists more than 200 data collaboratives in all kinds of sectors (health, environment, agriculture, and so on) of which several work on tracking mobility in crisis situations but none on more general migration issues.[57]

Another data hub working on migration-related topics is UN Global Pulse, established in 2009 as a UN Secretary-General's "initiative on big data and artificial intelligence for development, humanitarian action, and peace."[58] The initiative works through a network of so-called Pulse Labs in Jakarta, Kampala, and New York. UN Global Pulse has—mostly in co-operation with other UN agencies—conducted quite a few big-data-based studies on migration; for example with the UN Population Fund on the use of online search data to estimate migration flows, and with the UNHCR Innovation Service on the use of big data analytics and machine learning in the context of social media and forced displacement.[59]

Finally, the World Bank and the UNHCR inaugurated the Joint Data Center on Forced Displacement (JDC) in Copenhagen in 2019. The Center's mission statement "to enhance the ability of stakeholders to make timely and evidence-informed decisions that can improve the lives of af-fected people" also speaks the language of evidence-based policymaking.[60] The JDC "focuses on the collection, analysis, and dissemination of primary microdata. This refers to demographic and socioeconomic data, which includes information on income, poverty, skills, health, and economic activity, among others."[61] So far the Center has no big-data-based projects. However, it might only be a matter of time until big data sources are used.

Such recent mushrooming of migration-related data hubs established by large and prominent international organizations is an example of "policy-based institution-building." It becomes clear that international organizations are positioning themselves strategically within the discourse of evidence-based policymaking, to expand their roles to act as indispensable advisors and knowledge providers for a more evidence-based migration policy. By addressing migration first and foremost as a problem of sufficient, reliable, and timely data such data hubs reinforce a discourse on migration that does not focus on conflicting interests, diverging political convictions, or unequal power relations but rather on the almost technical question how migration can best be managed.

Policy-Based Evidence-Making: Feasibility Studies on Big Data for Migration Management

Drawing on the work of Boswell, Geddes argues that expert knowledge is also mobilized and used to confirm or legitimize already existing policy choices.[62] Example of this in the context of big data are three feasibility studies on the use of big data for migration management conducted between 2017 and 2019. All three studies were cofinanced by the Business Applications Programme of the European Space Agency (ESA), aimed at "helping companies to integrate space data and technology into commercial services."[63] The program offers funding opportunities and consultancy to entrepreneurs who have a business idea that involves space data and technology.

The first feasibility study, "Migration Radar 2.0," was conducted by the IT and business consultancy firm CGI; based in the Netherlands with support from the Dutch national statistical office. The company aimed to develop a service that combines population statistics, social media analytics, and earth observation analytics to "improve the monitoring of migratory movements that may lead to migration flows into Europe," and "the forecasting of migratory flows towards Europe as a possible destination," and thereby create an "Early Warning System."[64]

The second feasibility study "Big Data Applications for Migration Management" was conducted by the Italian based business consultancy firm BIP with support from the large satellite manufacturer Thales Alenia Space. The main objective of the study was to develop three "services" to "improve migration emergency management": (1) "Migration Flow Mapping" through internet, mobile phone, and satellite data; (2) "Trends and Forecasts elaboration" based on social media text analysis and diverse Internet sources such as Google trends, news, and weather forecasts; and (3) "Advanced Surveillance & Rescue" through the use of drones for the monitoring of migration flows. The potential users of these newly developed services are—according to the study—policymakers and decision makers at EU level, international organizations engaged in migration management, and reception centers and NGOs.[65]

The third feasibility study titled "Big Data for Migration" was conducted by the Italian based technology company E-GEOS, and was supported by the University of Sheffield, the University of Salzburg, the Fraunhofer Institute for Intelligent Analysis and Information Systems, the Belgian consulting agency Evenflow, and the geo-data analytics company Spatial Services, based in Austria. The study explored mobile phone data, social media data, satellite data, intelligence data, administrative sources, and border statistics to develop "services" that focus on four migration-related topics: (1) "Mitigation," understood as the reduction of root causes of migration; (2) "Preparedness," understood as the establishment of an early warning system concerning "migration onsets"; (3) "Response," understood as a mix of "ascertaining human rights, registration and tracking, humanitarian assistance, monitoring hot spots and safe houses, border control, monitoring seas, integration at destination countries"; and finally (4) "Repatriation," understood as the support of "sustainable reintegration to avoid renewed mitigation."[66]

The "language of emergency" used in the feasibility studies and their focus on migration control show that the "services" developed by technology companies and consultancy firms react to a dominant political discourse in Europe that frames migration as a risk not so much to migrants themselves as to possible European transit and receiving countries and their populations. In this context, Linnet Taylor and Fran Meissner speak of a "process of mutual shaping between technology and policy" and argue that "an existing policy vision creates a market for technologies which then shape the world to fit that policy vision and make its enforcement possible."[67]

All three feasibility studies claim that it is possible to better monitor and forecast so-called migration flows by combining diverse big data sources with traditional migration statistics. Thereby they do not only corroborate the belief that migration is something that needs to be managed but also that it can be managed. As a consequence, the studies provide "evidence" of the governability of migration—in the sense that something that can be monitored and forecasted appears to be manageable. Thus, the "evidence" produced by the three "feasibility studies" serves to substantiate the dominant political concept of "migration management."[68]

Evidence for and against Migration Policy

The discourse on evidence-based migration policymaking proliferates in the field of big-data-based migration research and policymaking. In this chapter, drawing on Boswell and Geddes, I have argued that although most actors involved in the production of big-data-based knowledge on migration frame their work as a one-directional contribution to a more evidence-based policymaking, three dominant relationships between evidence and governance are at play in the field: evidence-based policymaking, policy-based institution-building, and policy-based evidence-making. Data challenges such as the "Data for Refugees Turkey" challenge present themselves as an approach to evidence-based policymaking. However, policymakers' lack of interest in research results, privacy risks for data subjects, bias in big data sources and the risk of function creep complicate the exploitation of "evidence" resulting from data challenges for policymaking. The recent mushrooming of migration-related data hubs established by international organizations can be interpreted as policy-based institution-building. Most data hubs provide big-data-based knowledge on migration, which could actually be seen as a strategic move to underline the innovative and pioneering qualities of international organizations as consultants and knowledge providers and to legitimize the expansion of their institutional roles. Feasibility studies on the use of big data for migration management like those cofinanced by the Business Applications Programme of the European Space Agency can be understood as policy-based evidence-making. Their focus on monitoring, forecasting, and managing so-called migration flows produces "evidence" of the governability of migration which substantiates the dominant political concept of "migration management."

All three ways to use big data in the context of migration policy reinforce the assumption that lack of "evidence" is the central challenge for migration governance, and migration could be managed for the benefit of all if there was only enough reliable and timely data available. This perspective ignores the conflicting interests, diverging political convictions, and unequal power relations between various actors involved in migration governance; and the conflictual relationship between migratory practices and the attempts to govern them. What good migration governance is, what reliable "evidence" on migration is, and which methods should be used to produce this "evidence" is highly contested. In light of different evidence cultures within the field of migration studies the recent emergence of big-data-based migration research is especially welcomed and promoted by the migration policy community as it promises to deliver "hard evidence" that can seemingly be translated more easily into migration policy compared to the results of qualitative migration research.

However, meanwhile big-data-based "evidence" on migration is also used to critique migration policy and its often negative consequences for migrants. The project Forensic Oceanography employs forensic methods to find "evidence" of practices of nonassistance and human rights violations against migrants by state and nonstate actors in the Mediterranean Sea. Drawing on various data sources, but especially on satellite images which are usually used to control migration, the initiators of the project, Charles

Heller and Lorenzo Pezzani, call this reading "against the grain" of data a "disobedient gaze."[69] In dialogue with NGOs, they have combined satellite data and testimonies of survivors to reconstructed the death of 63 migrants whose boats were drifting for 14 days in NATO's maritime surveillance area without any assistance. This "evidence" was then used in several law suits.[70]

Big-data-based migration knowledge is not simply translated into policy. Instead, assumptions about what migration is, which problems it poses, and how it should be dealt with are already at the core of every research question, data visualization, or study design. If we assume that knowledge on migration is always the outcome of a coproduction process we should, in my view, focus more closely on the diversity of producers, their unequal access to the means of production and dissemination, and on the ways in which the coproduction of knowledge on migration could become more inclusive, power-sensitive and self-reflexive.

Notes

1 For example, Emilio Zagheni et al., "Inferring International and Internal Migration Patterns from Twitter Data," in *WWW '14: Proceedings of the 23rd International Conference on World Wide Web*, ed. Chin-Wan Chung et al. (New York: The Association for Computing Machinery, 2014); Marcus H. Böhme, André Gröger, and Tobias Stöhr, "Searching for a Better Life: Predicting International Migration with Online Search Keywords," *Journal of Development Economics* 142 (January 2020): 102347; Linus Bengtsson et al., "Improved Response to Disasters and Outbreaks by Tracking Population Movements with Mobile Phone Network Data: A Post-earthquake Geospatial Study in Haiti," *PLoS Medicine* 8, no. 8 (20 August 2011): e1001083.

2 Marzia Rango and Michele Vespe, "Big Data and Alternative Data Sources on Migration: From Case-Studies to Policy Support," Summary Report of the European Commission—Joint Research Centre Conference, Ispra, Italy, 30 November 2017, https://knowledge4policy.ec.europa.eu/sites/default/files/BD4M-workshop-2017-summary-report.pdf; United Nations General Assembly, Resolution 73/195, Global Compact for Safe, Orderly and Regular Migration, A/RES/73/195 (19 December 2018), https://undocs.org/en/A/RES/73/195.

3 Dvora Yanow, "Evidence-Based Policy," in *Encyclopedia Britannica Online*. Article published 29 March, 2013, last modified 7 December 2018, https://www.britannica.com/topic/evidence-based-policy.

4 Yanow, "Evidence-Based Policy."

5 Wayne Parsons, "From Muddling through to Muddling up—Evidence Based Policy Making and the Modernisation of British Government," *Public Policy and Administration* 17, no. 3 (July 2002): 43–60; Peter Wells, "New Labour and Evidence Based Policy Making: 1997–2007," *People, Place and Policy Online*, 2007, 22–29.

6 David Blunkett, "Influence or Irrelevance: Can Social Science Improve Government?" (speech), ESRC and Department for Education and Employment, Swindon, 25 February 2000, cited in Ken Young et al., "Social Science and the Evidence-Based Policy Movement," *Social Policy and Society* 1, no. 3 (July 2002): 215.

7 Parsons, "From Muddling Through," 44.

8 Trisha Greenhalgh and Jill Russell, "Evidence-Based Policymaking: A Critique," *Perspectives in Biology and Medicine* 52, no. 2 (2009): 315.

9 Greenhalgh and Russell, "Evidence-Based Policymaking," 307.

10 Greenhalgh and Russell, 308.

11 Christina Boswell, *The Political Uses of Expert Knowledge: Immigration Policy and Social Research* (Cambridge: Cambridge University Press, 2009).

12 Martin Bak Jørgensen, "Understanding the Research—Policy Nexus in Denmark and Sweden: The Field of Migration and Integration," *The British Journal of Politics and International Relations* 13, no. 1 (February 2011): 93–109.

13 Simon Warren, "Migration, Race and Education: Evidence-Based Policy or Institutional Racism?" *Race Ethnicity and Education* 10, no. 4 (December 2007): 367–385.

14 Martin Baldwin-Edwards, Brad K. Blitz, and Heaven Crawley, "The Politics of Evidence-Based Policy in Europe's 'Migration Crisis,'" *Journal of Ethnic and Migration Studies* 45, no. 12 (10 September 2019): 2149.

15 Andrew Geddes, "Temporary and Circular Migration in the Construction of European Migration Governance," *Cambridge Review of International Affairs* 28, no. 4 (2 October 2015): 572.

16 Stephan Scheel and Funda Ustek-Spilda have argued that the introduction of big-data-based methodologies into migration statistics will not prevent the "politics of numbers" by migration policy actors who count migrants in particular ways to produce "evidence" in support of specific political agendas. These "politics of numbers" could be described as "policy-based evidence-making." However, their argument is of a theoretical nature and not based on empirical research in the field of big-data-based migration research or policymaking. "Big Data, Big Promises: Revisiting Migration Statistics in Context of the Datafication of Everything," *Border Criminologies* (blog), *University of Oxford Faculty of Law*, 1 June 2018, https://www.law.ox.ac.uk/research-subject-groups/centre-criminology/centreborder-criminologies/blog/2018/06/big-data-big.

17 United Nations General Assembly, A/RES/73/195, 6.

18 Samuël Coghe, "Tensions of Colonial Demography. Depopulation Anxieties and Population Statistics in Interwar Angola," *Contemporanea. Rivista Di Storia Dell'1800 e Del '1900* 18, no. 3 (2015): 472–78; John M. Cinnamon, "Counting and Recounting: Dislocation, Colonial Demography, and Historical Memory in Northern Gabon," in *The Demographics of Empire: The Colonial Order and the Creation of Knowledge*, ed. Karl Ittmann, Dennis D. Cordell, and Gregory Maddox (Athens: Ohio University Press, 2010), 130–156.

19 Global Commission on International Migration, *Migration in an Interconnected World: New Directions for Action. Report of the Global Commission on International Migration* (GCIM, 2005), 2, https://www.refworld.org/publisher,GCIM,,,435f81814,0.html.

20 UN Statistics Division, Expert Group Meeting, Improving Migration Data in the Context of the 2030 Agenda, ESA/STAT/AC.339/1 (20–22 June 2017), https://unstats.un.org/unsd/demographic-social/meetings/2017/new-york-egm-migration-data/Background%20paper.pdf; and from the same source, ESA/STAT/AC.339/1, "Recommendations," https://unstats.un.org/unsd/demographic-social/meetings/2017/new-york--egm-migration-data/EGM%20Recommendations_FINAL.pdf. United Nations Expert Group Meeting, "Improving Migration Data in the Context of the 2030 Agenda. Recommendations, New York Headquarters, 20–22 June 2017," 2017, https://www.unescap.org/sites/default/files/UNSD_EGM_Recommendations_20-22Jun2017.pdf.

21 Laura Stielike, "Migration Multiple? Big Data, Knowledge Practices and the Governability of Migration," in *Research Methodologies and Ethical Challenges in Digital Migration Studies. Caring for (Big) Data?*, ed. Marie Sandberg et al. (Basingstoke: Palgrave Macmillan, 2022), 113–138.

22 Rango and Vespe, "Big Data and Alternative Data," 6.

23 Knowledge Centre on Migration and Demography, and IOMs Global Migration Data Analysis Centre, "BD4M Launch Event," *Big Data and Alternative Data Sources on Migration* (webpage), 25 June 2018, https://knowledge4policy.ec.europa.eu/migration-demography/big-data-alternative-data-sources-migration_en#launch.
24 Spyridon Spyratos et al., "Quantifying International Human Mobility Patterns Using Facebook Network Data," *PLOS ONE* 14, no. 10 (24 October 2019): 19.
25 Christina Hughes et al., *Inferring Migrations, Traditional Methods and New Approaches Based on Mobile Phone, Social Media, and Other Big Data […]* (Luxembourg: Publications Office, 2016); Spyridon Spyratos et al., *Migration Data Using Social Media: A European Perspective* (Luxembourg: Publications Office of the European Union, 2018).
26 See, for example, Spyratos et al., "Quantifying International Human Mobility"; Joao Palotti et al., "Monitoring of the Venezuelan Exodus through Facebook's Advertising Platform," *PLOS ONE* 15, no. 2 (21 February 2020): e0229175; Böhme, Gröger, and Stöhr, "Searching for a Better Life."
27 Two thirds of the 74 papers presented at the final conference of the data challenge dealt with tracking mobility. Linnet Taylor, "No Place to Hide? The Ethics and Analytics of Tracking Mobility Using Mobile Phone Data," *Environment and Planning D: Society and Space* 34, no. 2 (April 2016): 324.
28 Vincent D. Blondel et al., "Data for Development: The D4D Challenge on Mobile Phone Data," arXiv.org open-access archive, Computers and Society, 28 January 2013, http://arxiv.org/abs/1210.0137.
29 Yves-Alexandre de Montjoye et al., "D4D-Senegal: The Second Mobile Phone Data for Development Challenge," arXiv.org open-access archive, Computers and Society, 30 July 2014, http://arxiv.org/abs/1407.4885.
30 European Commission, *Data for Integration (D4I)* (website), accessed 13 April 2021. https://knowledge4policy.ec.europa.eu/migration-demography/data-integration-d4i_en; Alfredo Alessandrini et al., *High Resolution Map of Migrants in the EU* (Luxembourg: Publications Office of the European Union, 2017), https://data.europa.eu/doi/10.2760/0199.
31 Albert Ali Salah et al., "Data for Refugees: The D4R Challenge on Mobility of Syrian Refugees in Turkey," arXiv.org open-access archive, Computers and Society, 14 October 2018, http://arxiv.org/abs/1807.00523; Albert Ali Salah et al., *Guide to Mobile Data Analytics in Refugee Scenarios: The "Data for Refugees Challenge" Study* (Cham: Springer, 2019); "Data 4 Refugees: Turkey," Data-Pop Alliance, accessed 14 April 2021, https://datapopalliance.org/d4r/.
32 Salah et al., *Guide to Mobile Data Analytics*, vi.
33 Albert Ali Salah et al., "Policy Implications of the D4R Challenge," in *Guide to Mobile Data Analytics in Refugee Scenarios*, ed. Albert Ali Salah et al. (Cham: Springer International Publishing, 2019), 490.
34 Salah et al., "Policy Implications," 492.
35 Salah et al., "Policy Implications," 490.
36 Salah et al., "Policy Implications," 489.
37 Bert-Jaap Koops, "The Concept of Function Creep," *Law, Innovation and Technology* 13, no. 1 (16 March 2021), 2.
38 Taylor, "No Place to Hide?," 330.
39 Salah et al., "Policy Implications," 490.
40 Geddes, "Temporary and Circular Migration," 586.
41 "About GMDAC," Global Migration Data Analysis Centre (website), International Organization for Migration, accessed 16 April 2021, https://gmdac.iom.int/about-gmdac.
42 "About the Migration Data Portal," Migration Data Portal (website), accessed 19 April 2021, https://migrationdataportal.org/about.

43 The abbreviations SDGs and GCM stand for Sustainable Development Goals and Global Compact for Safe, Orderly and Regular Migration.

44 Migration Data Portal, "About the Migration Data Portal."

45 Migration Data Portal, "About the Migration Data Portal."

46 "Data Innovation Directory," Migration Data Portal (website), accessed 19 April 2021, https://migrationdataportal.org/data-innovation.

47 "About the Knowledge Centre for Migration and Demography," Knowledge Centre on Migration and Demography, European Commission, accessed 19 April 2021, https://knowledge4policy.ec.europa.eu/migration-demography/about_en.

48 "Knowledge Centre on Migration and Demography (KCMD) Data Portal," Knowledge Centre on Migration and Demography, European Commission, accessed 19 April 2021, https://bluehub.jrc.ec.europa.eu/.

49 "Data+ Catalogue," Knowledge Centre on Migration and Demography Data Portal, European Commission, accessed 19 April 2021, https://bluehub.jrc.ec. europa.eu/catalogues/data/.

50 Knowledge Centre on Migration and Demography, "KCMD Information Catalogue," accessed 19 April 2021, https://bluehub.jrc.ec.europa.eu/catalogues/info/.

51 "Dynamic Data Hub," Knowledge Centre on Migration and Demography, European Commission, accessed 19 April 2021, https://bluehub.jrc.ec.europa.eu/ migration/app/index.html.

52 "Atlas of Migration," Knowledge Centre on Migration and Demography, European Commission, accessed 19 April 2021, https://bluehub.jrc.ec.europa.eu/migration/ app/atlas.html.

53 Hughes et al., *Inferring Migrations*; Spyratos et al., *Migration Data*; Stefano Iacus et al., *How Human Mobility Explains the Initial Spread of COVID-19: A European Regional Analysis* (Luxembourg: Publications Office of the European Union, 2020), https:// op.europa.eu/publication/manifestation_identifier/PUB_KJNA30292ENN.

54 Knowledge Centre on Migration and Demography, and IOMs Global Migration Data Analysis Centre, "BD4M Launch Event."

55 "The Challenge," BD4M—Big Data for Migration Alliance (website), accessed 8 April 2021, https://data4migration.org/.

56 "How We Work," BD4M—Big Data for Migration Alliance, accessed 19 April 2021, https://data4migration.org/.

57 "Data Collaboratives Explorer," GovLab, accessed 19 April 2021, https:// datacollaboratives.org/explorer.html.

58 "About UN Global Pulse," UN Global Pulse, United Nations, accessed 19 April 2021, https://www.unglobalpulse.org/about/.

59 UN Global Pulse, "Estimating Migration Flows Using Online Search Data," (PDF) Global Pulse Project Series, 2014, https://www.unglobalpulse.org/wp-content/ uploads/2014/04/UNGP_ProjectSeries_Search_Migration_2014_0.pdf; UN Global Pulse, UNHCR Innovation Service, "Social Media and Forced Displacement: Big Data Analytics & Machine-Learning," (PDF), 2017, https://www.unhcr.org/ innovation/wp-content/uploads/2017/09/FINAL-White-Paper.pdf.

60 "Who We Are," Joint Data Center on Forced Displacement (website), World Bank and UNHCR, accessed 19 April 2021, https://www.jointdatacenter.org/who-we-are/#mission.

61 "Who We Are," Joint Data Center on Forced Displacement.

62 Boswell, *Political Uses of Expert Knowledge*; Geddes, "Temporary and Circular Migration," 572.

63 ESA Space Solutions, "About ESA Business Applications," European Space Agency, accessed 26 April 2021, https://business.esa.int/.

64 "Migration Radar 2.0," ESA Space Solutions, European Space Agency, updated 29 May 2018, https://business.esa.int/projects/migration-radar-20.

65 "Big Data Applications to Boost Preparedness and Response to Migration," ESA Space Solutions, European Space Agency, updated 25 May 2018, https://business.esa.int/projects/big-data-applications-to-boost-preparedness-and-response-to-migration.

66 "Big Data for Migration Study—Big Data Applications to Boost Mitigation Preparedness and Response to Migration Feasibility Study," ESA Space Solutions, European Space Agency, updated 13 September 2019, https://business.esa.int/projects/big-data-for-migration-study.

67 Linnet Taylor and Fran Meissner, "A Crisis of Opportunity: Market-Making, Big Data, and the Consolidation of Migration as Risk," *Antipode* 52, no. 1 (January 2020): 285.

68 Martin Geiger and Antoine Pécoud, "The Politics of International Migration Management," in *The Politics of International Migration Management*, ed. Martin Geiger and Antoine Pécoud (Basingstoke: Palgrave Macmillan, 2010), 1–20.

69 Sophie Hinger, "Transformative Trajectories—the Shifting Mediterranean Border Regime and the Challenges of Critical Knowledge Production. An Interview with Charles Heller and Lorenzo Pezzani," *Movements. Journal für kritische Migrations- und Grenzregimeforschung* 4, no. 1 (2018): 196; Lorenzo Pezzani and Charles Heller, "A Disobedient Gaze: Strategic Interventions in the Knowledge(s) of Maritime Borders," *Postcolonial Studies* 16, no. 3 (2013): 289–298.

70 Hinger, "Transformative Trajectories" 195; Charles Heller et al., "Report on The »Left-To-Die Boat«," 11 April 2012. https://www.fidh.org/IMG/pdf/fo-report.pdf.

Part IV

Contesting Evidence:
The Politics of Heterodox
Evidence

11 Fearful Narratives: Evidence Production in the Visual Rhetoric of the Historic Anti-vaccine Movement in the German States

Christiane Arndt

The Historic Anti-vaccine Movement and Historic Media Development

The nineteenth century, and particularly the second half of the nineteenth century, featured rapid developments in medicine, visual technology, and the dissemination of information through printed material. In medicine, it was bacteriology that had its heyday. During the 20 years from 1873 to 1893, around 30 pathogens were first described, including the tuberculosis bacillus, and the cholera and typhus bacilli. Beginning prominently with the development of the smallpox vaccine by Edward Jenner in the 1790s, more immunizations were subsequently developed during the nineteenth century, for example, the cholera, anthrax, and rabies vaccines. With the introduction of vaccines in Europe and the United States came also legislation introduced to prescribe public health measures to achieve a reduction in fatalities and ultimately mass immunity. Thus, through legislation and popularization, medical progress directly affected people's lives.

Such rapid development coincided with the similar advance in contemporaneous media. Photography, as one of the most culturally relevant technological inventions of the nineteenth century, precipitously progressed from a complicated procedure that involved physical strain and showed meager results right after its invention in 1824 by Joseph Nicéphore Niépce, to a widely used and increasingly convenient technology toward the end of the century, with George Eastman famously coining the catchphrase "you press the button, we do the rest."[1] In view of this, the nineteenth century has been deemed the century of the visual and the image.[2] This trend took hold decidedly in the sciences and medicine, too, and here specifically when it came to communicating to the public. In medical photography, the new medium augmented the observation of the body by documenting symptoms in medical publications, archiving visible renditions of patients in patient files (with hysteria being one prominent example), including microscopical images in scientific and popular scientific publications through the technique of microphotography,[3] supporting

DOI: 10.4324/9781003188612-15

diagnostics with x-ray images, and documenting bodily movement in Étienne-Jules Marey's photo sequences.

Photography was not the only technological advancement that had an impact on the popularization of health-related topics. Starting in the mid-nineteenth century, Germany, like other Western cultures, saw a rapid rise in print media. The new technologies of rotational printing (from 1846 onward)[4] and xylographic image printing enabled the journals, newspapers, and magazines that disseminated information—both in textual and visual form—to develop into modern mass media, stressing the importance of the visual in these media.[5] The trend for reading circles and a growing public library system further supported the popular press in establishing itself as a common, accessible, and influential source of information for the wider population.[6] At the same time, the various magazines, often called "family magazines," also decidedly shaped and accelerated the dissemination of specifically hygiene-related information. With respect to vaccine related information in the realm of popular science, family magazines and other venues such as hygiene exhibitions usually featured visuals that displayed the context of vaccination, such as bucolic scenes of country vaccine clinics,[7] or arms exhibiting vaccination marks.

This chapter examines the intersection of bacteriology, photography, and print culture, and addresses the ways in which perceived evidence is constructed through images and text. Anti-vaccine publications of the nineteenth century were reactions to medical developments, and their use of images—often photos—and texts exemplifies the wider media framework of nineteenth century culture. Scientific developments of the time also included reflections on the methodology of science, which Claude Bernard prominently exemplified in his *An Introduction to the Study of Experimental Medicine* from 1865, and scientists made invested attempts to communicate the advances through evolving media. Nonetheless, the anti-vaccine league was successful in promoting their material as evidence to large parts of the population. Prominent German language publications of the anti-vaccine league were, for example, *Der Impfgegner: Monatsschrift für praktische Volkswohlfahrt und naturgemäße Gesundheitspflege* (The anti-vaccinist: monthly publication to support practical preventative health and hygiene for the people; from 1876 to 1926) or *Der Impfzwanggegner: Monatsschrift der Reichs-Impfgegner-Zentrale* (The anti-compulsory-vaccinist. Monthly publication of the national center of anti-vaccinists; from 1928 to 1932). Indeed, some publications were specifically presented to resemble contemporary periodicals, including characteristics such as advertisements and title pages.[8] The situation was comparable among all contemporaneous Western countries, especially as photographs and information material were shared between anti-vaccine organizations.

Around the middle of the nineteenth century, opposition against vaccines and vaccine legislation shifted from being centered on individual figures such as the physician Heinrich Oidtmann to a form of protest that was organized in groups of mostly lay people.[9] This shift resulted in changed publication practices from the single-author books that had dominated the anti-vaccine material dissemination until then[10] to discussions around vaccines featured in daily papers and family journals.[11] Essentially, this form of organization fit with the general movement toward *Vereine* (associations) in Germany during the latter half of the nineteenth century, namely associations that gathered people with similar interests in an organized fashion.[12] Enabling a network of anti-vaccine campaigners, these associations staged public speaking events, and sent out information material, including the aforementioned magazines, as well as pamphlets, brochures, and other print material like postcards and photographs. Already in 1880, the international appeal of anti-vaccinist propaganda consolidated the foundation of the International Organization of Anti-Vaccinists in Paris, followed by similar national and international umbrella organizations around the turn of the twentieth century.[13] In the United States, various anti-vaccine leagues were also founded around that time: the Anti-vaccination Society of America in 1879, the New England Anti-compulsory Vaccination League in 1882, then the Anti-vaccination League of New York City in 1885.[14] While publication practices and associations indicate that the topic concerned the society at large, membership statistics of anti-vaccinist organizations were not recorded and thus an exact number of anti-vaccinists cannot be provided. As further indication for the mass appeal of the movement, a highly frequented mass protest that took place in Leicester in March 1885, with an estimated 100,000 protesters is often referred to.[15] The rising number of organizations, a similarly growing number of publications[16] and the size of events such as the Leicester protest suggest that the movement broadened its base by including people who might not have been opposed to vaccines in general, but who specifically rejected mandatory vaccines, in the United Kingdom as well as in the German States.[17] Nadja Durbach summarizes the significance of the anti-vaccine sentiment when she states that the movement "is particularly historically significant not only because it was arguably the largest medical-resistant campaign ever mounted in Europe, but because it articulated these anxieties around the safety of the body and the role of the modern state."[18]

The anti-vaccine movement's base comprised a diverse spectrum of the general public. Particular to the German States, the *Lebensreform* (life reform) movement, which accumulated various alternative living philosophies such as the nudist, vegetarian, garden community, or reformed clothing creeds fed into anti-vaccinist organizations. The followers of these principles were, as Officer of Health Prof. Martin Kirchner evaluated at the time, joined by other social groups:

> The circles, from which anti-vaccinists are recruited, are very diverse. The majority are friends of natural living, supporters of naturopathy, anti-vivisectionists, homeopaths and supporters of similar movements.

Many are opposed to physicians and scientific medicine in general. For others, political motivations are part of their opposition to vaccines, they refuse any kind of political paternalism, maintaining personal freedom and the inviolability of body and soul. Others refer to religious concerns and regard the vaccine as an interference with God's world order. There are also physicians amongst the anti-vaccine proponents; these even claim that the majority of physicians is in agreement with the anti-vaccinists, and that they merely due to convenience, profitability or oppression proceed with vaccinations against smallpox. The anti-vaccine proponents amongst physicians are by majority naturopaths who believe that they will find the salvation of humankind outside of scientific medicine.[19]

Malte Thießen summarizes accordingly that the life reform movement, social hygiene and political liberalism were the "three roots of vaccine criticism."[20]

Opposition against vaccines had been present since the introduction of the smallpox vaccine at the end of the eighteenth century.[21] Its growth, however, was directly related to vaccine legislation, which was introduced for example in Germany and the United Kingdom during the first half of the nineteenth century. The German states established the Bavarian Vaccination law as early as 1807, followed by the *Reichsimpfgesetz* in 1874, and the United Kingdom introduced the British Vaccination Act in 1840.[22] The *Reichsimpfgesetz* consolidated a variety of localized laws in the German States, particularly centered on the question of mandatory vaccination based on the 1835 Prussian legislation for mandatory vaccines in case of an epidemic. While the fines that could theoretically be issued were substantial,[23] records show that legally possible penalties were rarely issued in cases of actual or assumed vaccine evasion.[24] If a case was actually seen through by juridical authorities, the ensuing bureaucratic burden was much more problematic than the usually rather low fines.[25] With regard to the legislation, the main criticism expressed by the anti-vaccinists was that it presented a violation of personal integrity. Anti-vaccinist Paul Förster, for example, stated, "The issue is the right to a free disposal of one's own body, the questioned parental right, freedom of thought and conscience; how should we, in this respect, let everything be enacted upon us, against our knowledge and conscience, reminiscent of an enforcement of times that we transcended long ago?"[26]

Anti-vaccine Rhetoric and Public Education Media

As anti-vaccine campaigns used both photography and print media, they had at hand a powerful apparatus that was geared toward influencing the public chiefly by producing the desired reality and disseminating information efficiently. The volume of publications issued by anti-vaccine organizations increased significantly in the second half of the nineteenth century, which provides one indicator for the rising popularity of anti-vaccine ideas.[27] One of these publications was anti-vaccinist Hugo Wegener's 1912 *Der Impf-Friedhof* (The vaccine cemetery). Wegener was one of the main anti-vaccine activists, who published three anti-vaccination monographs between 1911 and 1912: *Segen der Impfung* and *Unerhört!* in 1911, *Der Impf-Friedhof* in 1912.[28] Patrick Mayr describes the engineer from Frankfurt as a "dazzling character" due to his proliferous publication output and his sharp, acidic rebukes of pro-vaccine advocates.[29] As a contributor to the periodical *Der Impfgegner*[30] Wegener displayed a high level of in-dustriousness and continuity; he served as collector of vaccine accident reports for the journal 1912 and 1914[31] and copublished the periodical *Die Impffrage* with teacher and fellow anti-vaccinist Paul Mirus after 1912.[32] With a tendency to focus on details that were irrelevant for the central argument, Wegener targeted officials such as prominently the aforemen-tioned Prof. Kirchner in his publication *Unerhört! Verteidigung und Angriff eines Staatsbürgers. Gegen Kirchner* (1911) with a tirade of insults that avoided engagement in actual medical arguments.[33] In line with the arguably re-latively rare court proceedings against vaccine opponents, there seem to have been no charges against him.[34] Regarding Wegener's popularity, Mayr explains that with a lack of sources proving how widely his writings were received, or documented print runs for his publications, any as-sumptions are limited to Wegener's own statements, which cannot be verified and might very well have been exaggerated to foster self-advertisement. Wegener's motivation is unclear; it is possible that he was altruistically invested in protecting children from the vaccines he deemed unsafe. It is, however, equally likely that he gained financially from the revenues of his publications, printed by the "Luise Wegener Verlag," a name which indicates that the publishing house was run by his wife.[35] The shock factor of the publications must have certainly boosted the sales figures.[36] *Der Impf-Friedhof* together with Wegener's *Segen der Impfung* from 1911 displayed a collection of vaccine victim narratives procured from different sources (such as submitted letters from individuals which detailed supposed vaccine accidents, contributions from anti-vaccine associations, etc.) and presented these accident reports in the form of numbered lists.[37] *Der Impf-Friedhof* proclaimed on its title page that the reader would be introduced to 36,000 cases, including 139 images.[38] The cases were mostly from German-speaking countries, but included several international vaccine "victims." Parents, family members, or friends of supposed vaccine victims who witnessed disease or death allegedly reported these cases, either to magazines, to anti-vaccine organizations, or their members.[39] While Wegener's publication was not itself a periodical, it reused cases and images—of which some had previously been published in periodicals, for example in *Der Impfgegner*.

The image-text combinations Wegener employs in his publications were similar to medical case narratives, and so the anti-vaccine material imposed an evidential gesture by referring to the scientific, medical genre. The presentation of vaccine victims in Wegener's collections simulated medical case studies, including the typical use of patient photographs as part of the patient file. This similarity in turn provided a shortcut to a provision of proof that induced the process of scientific evidence without having to (re)establish it. This reference to science was merely a performative gesture associated with the genre of the case study and the characteristics of medical image use. The material disseminated by the anti-vaccine campaigners aimed at convincing its audience through establishing a kind of evidence that merely fit the *notion* of practical evidence, mainly since it superficially *looked* like evidence. In this instance, evidence was not based on the communication of results that are the product of a methodology which is imbedded in, follows the methodology of, and is vetted by a discipline—as Ludwik Fleck pivotally describes part of the epistemological structure of evidence.[40] Rather, the *evidential character* had its origin in the medial implications of the material, its context, and the way in which the material was presented.[41] The end result, however, resembled a compiled file of cases presented as witness statements that in addition to medical practices also alluded to juridical evidence,[42] and thus simulated a range of established performative gestures. In addition to insufficient medical reasoning to explain the symptoms depicted in the photographs,[43] there was furthermore no methodically sound reference given for the presented images beyond the declared sources.[44] Wegener's presentation was based on the supposed validity of the photographs as scientific observations; while they superficially mimicked practices of medical photography, the effect of the image considerably depended on immediacy as an effect of photography.

First of all, Wegener's publications captured the attention of readers due to the shock value of the images (Figure 11.1).[45] This effect certainly originated in the drastic physical marks that are visible on the displayed bodies, and the fact that the pictured victims were mostly children added to the distress.[46] The notable impact of the collection of anti-vaccination material compiled by Wegener for *Impf-Friedhof* is documented in the German imperial parliament (*Reichstag*) protocols, and Thießen quotes parliamentarian Maximilian Pfeiffer attributing a "veritable shudder" to the "images that have been put again on the table of the house."[47] However, in addition to shock due to distress and empathy, the rhetoric potential of the photographs lay in their ability to instigate a biographical story.[48] The display of the image piqued curiosity; the readers of the anti-vaccine journals, pamphlets, and flyers, and the audience at anti-vaccine talks were prompted to relate to the lives of the displayed individuals. The images first created, and then partially filled an informational void: they showed that the numbers in hygiene-related statistics were tied to individuals, and they prompted the reader to augment the fragments of the presented individual medical cases into a relatable story. The rhetoric device of

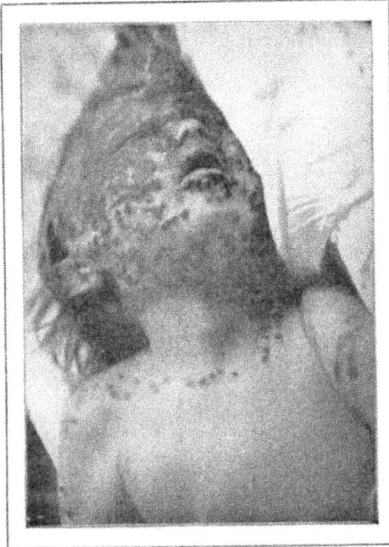

Figure 11.1 Title page of Hugo Wegener, *Segen der Impfung. Wenig von Vielem* (Frankfurt a. M.: Verlag Luise Wegener, 1911).

providing an example played on the readers' curiosity toward an individual image, and the active involvement of the reader enforced the photograph's message. With a vaccine-induced crisis at the center of each of the supplied biographical case narratives, the narrative formed a dramatic arc/plot in a nutshell, created empathy toward a fellow human being, and fostered acceptance of shared human vulnerability. "Narrative formats are regarded as a special kind of communication because they are deeply rooted in every-day life. [...] In addition, they make emotions accessible and produce empathy."[49] The common, typical characteristics of the biographical sketches thus underlined the potential for identification with the individuals in the images: this person is like me, what happened to them could happen to anyone—it could happen to me, to my child![50]

In addition to the shock value and to the performative gesture mimicking scientific evidence, the portraits of individual vaccine "victims" made explicit use of the intrinsic narrative potential of images. Two cases from the *Impf-Friedhof* exemplify this employment of narrative techniques. Case 65 from 1908 was originally published in the journal *Erziehung und Unterricht* (Education and instruction) of the same year.[51] Wegener's republication showed a montage of an image within an image (Figure 11.2).

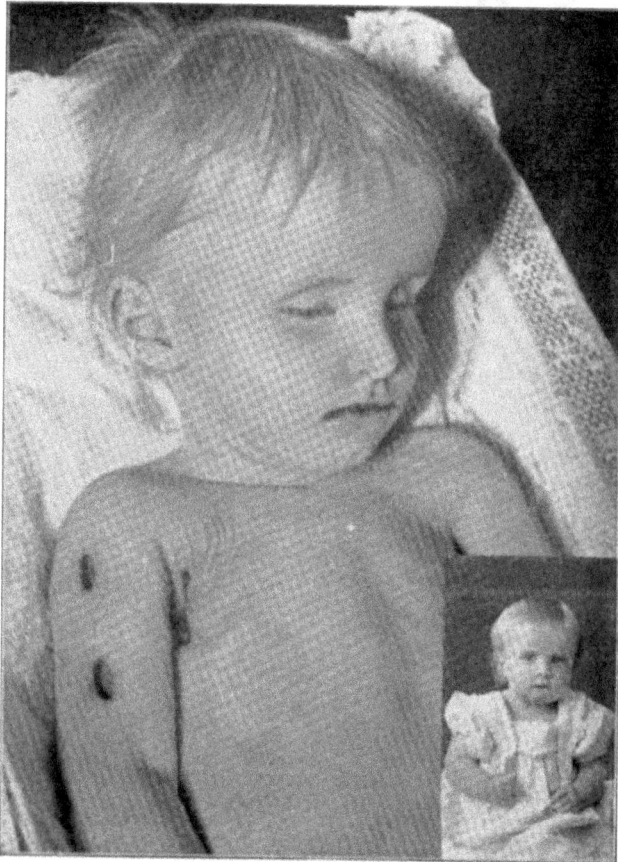

Figure 11.2 Hugo Wegener, *Der Impf-Friedhof. Was das Volk, die Sachverständigen und die Regierungen vom "Segen der Impfung" wissen* (Frankfurt a. M.: Verlag Luise Wegener, 1912), 21.

The larger image displayed a dead child, with an inset of a smaller image of the same child from before the supposed vaccine accident. The montage fueled the curiosity of the readership as it supplied additional visual information about the individual life of the vaccine victim. In many other cases, supplementary information was given in the accompanying text. But when, as seen in the example of Figure 11.2, the image included information through its visual and spatial presentation, it both stressed and furthered its narrative potential compared to a single image, or mere text. This relationship between narration and photography has been prominently described by Susan Sontag, who summarizes, "Only that which narrates can make us understand."[52] The depicted individual was not only displayed at one stage of their life, but two instances were given, each representative of a phase in life; namely healthy, and diseased or deceased. As is characteristic for a visual argument,[53] the reader is prompted to literally fill in the blanks—supported by the accompanying biographical sketch—and create a narrative.

The active involvement of the reader not only increased the impact of the images through the potential of encouraging a relatable narrative, it also supported the perception initiated by the photograph of having been a direct witness to this child's fate. Roland Barthes terms this effect of photography the awareness of "having-been-there."[54] For Barthes, this evidential effect the photograph invokes does not prove a specific factuality of an event (namely a vaccine accident). Rather, the use of a photograph results in a recognition of a past reality that forgoes any semiotic referential act aimed at producing irreducible scientific evidence, as Barthes describes the photograph: "[…] in Photography I can never deny that *the thing has been there*. There is a superimposition here. Of reality and of the past."[55] The photograph has the effect of inducing a tangible, physical connection to something that is perceived as having-been-there *and* as at the same time "contingent (and thereby outside of meaning)."[56] This describes the referential gesture that mimics scientific observation, a gesture that remained ultimately without a concrete referee. Peter Geimer amplifies Barthes' argument and, by referring to Helmut Lethen, terms this tangible immediateness of photography as "shock": "This tangency—'shock, irritation and fascination'—proves an access to the images, which has little in common with the processes of evidence. Therefore, it can also not be disproven by referring to the incontrovertibleness of semiology."[57] The anti-vaccine league utilized the shock potential which resulted from the described tangible immediacy of photography. The direct visual communication through the image triggered an emotionally amplified response for which there was no specific, verifiable reference.

Another example of the use of narrative techniques is seen in paired images, which claim to display the patient's condition before and after the vaccine; again, these images add visual information to the accompanying text and draw out the potential to turn the medical case into a biographical narrative. In case 770 of *Impf-Friedhof*, Rosina Sandall reported the fate of her coworker who underwent compulsory vaccination (see Figure 11.3).[58]

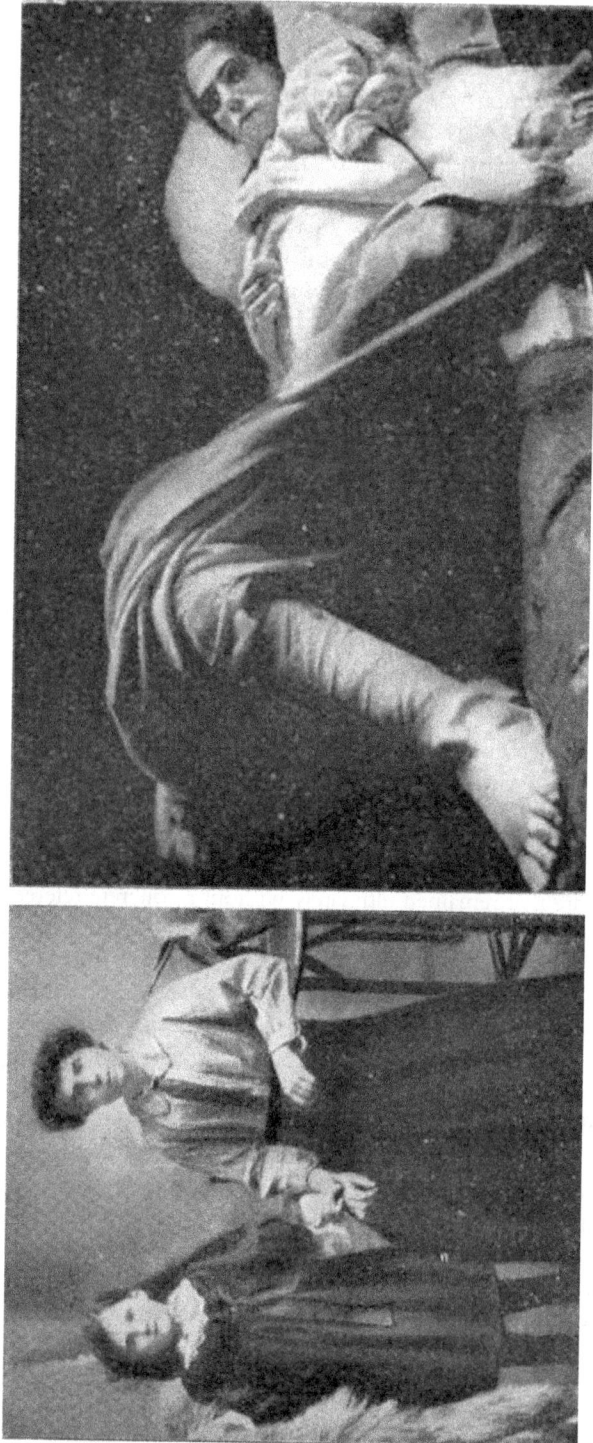

Figure 11.3 Wegener, *Impf-Friedhof* (1912), p. 281.

In this case, the images present two distinct points in time, which the reader needs to connect through a narrative. In addition, the impact of the images lies in their potential to refer closely to someone's individual life. The images are subtitled *"vor der Impfung"* (before the vaccination) and *"kurz vor dem Tode"* (immediately before death). This same double image technique displaying a change in appearance due to the impact of a disease was used in medical photography at the time, as for example in the portrayal of the Gloucestershire smallpox epidemic in 1896.[59]

Such a suggestion of biography, achieved through presenting different stages of the patients' life, increased the readers' feeling of being directly acquainted with the patient. The effect was even further reinforced, as *Der Impfgegner* (The anti-vaccinist) ostensibly requested case narratives and images from its readers, as specified on the title page.[60] These case reports thus came from the group that the typical readers identified and empathized with, and the implication that photos came directly from like-minded readers contributed to the effect that the images were taken at face value. Even though the images actually lacked verifiable credentials, readers were not likely going to question shocking documents that illustrated the pitiable fate of fellow anti-vaccinists, especially when these reports amplified their own fear that they might suffer the same fate. Wegener remarked in his *Impf-Friedhof* that he had summarized the solicited vaccine victim cases with respect to their main type of disease in order to reduce the amount of information, which suggests that the request to send in testimonies had been taken up in considerable volume, again indicating a sizable number of vaccine opponents.[61]

The witness potential attributed to photography per se, the (actual or perceived) acquaintance with an individual's fate, the construction of a biographical account, and the openly solicited material that included the reporting sources, all contribute to the production of a narrative. Furthermore, as the photographs are republished, they become, to use Barthes' term, coded as objective evidence. Coding—as Barthes elucidates by way of images in advertisement—describes the connotated content of an image. It includes cultural knowledge, but also comprises other kinds of "knowledge on which the sign depends,"[62] such as publication context and framing of the image. Barthes demonstrates that the denoted pictorial information of paintings—those elements that convey a message without any code[63]—is compounded by connotation in form of the artist's influence, any choice the artist makes (arrangement of scene, framing, color, etc.).[64] For photography, this connotation at first seems to be prevented—hence the phrase that photography presents itself as a "message without a code."[65] This would make photography the perfect medium for science, it seems, as Bruno Latour phrased it, *"acheiropoeiete,* not made by human hand."[66] However, as Barthes' analysis shows, photographic images actually draw heavily on connotational coding. Especially during the nineteenth century, when photography was still a new or fairly new medium, the connotational aspect has been masked by the supposed immediacy of photography: "This is without doubt an important historical paradox: the more technology develops the diffusion of information (and notably of images), the more it provides the means of masking the constructed meaning under the appearance of the given meaning."[67] This effect of supposed mechanical and thus objective reproduction through photography[68] included the anti-vaccine images, which drew on the supposed immediacy to mask the coding. The message of their images relied on the context, and increasingly on the recognition of the code. The images used by the anti-vaccine league over time became culturally associated with a certain message, and this message became the fixed general message of any similar image. As Mary S. Morgan's reflects, the evidential potential of exemplary narratives (such as case studies) within science, and the assessment of their implementation during the late nineteenth century, reveals that the exemplary power of the genre remained vague and unregulated. For the medical discipline, case studies had not yet gained the "autonomy to function more broadly as instruments of inquiry."[69] While case narratives can be presented and perceived as a kind of soft evidence, their use in the nineteenth century was not regulated by a scientific community. Such vagueness and lack of regulation amplified their potential for agents such as the anti-vaccine campaigners.

Circumstantial Evidence

The visual coding that the anti-vaccine league successfully established for their photographs had its precedent in the use of visuals in academically supported health promotion. Both relied on displaying the associative circumstances of vaccination instead of its actual physiological process. In general, the legitimation for the use of vaccines at the end of the nineteenth century was still largely based on the observation of their success: while the general concept of their mechanism was known, the complex details were still to be examined. This situation set the stage for further uses of contextual imagery in medical culture that largely conveyed what can be observed with the naked eye, even if the displayed aspects were secondary features of vaccines.[70] Karen Walloch describes how photos of the wider circumstances of production, such as the cows in the field, were used in the promotion of commercial lymph production instead of data obtained through microscopic analysis.[71] Examples for the circumstantial use of visuals that is more closely associated with the administration of vaccines and its effect on the vaccinated person are the well-known images of the milkmaid's hand, initially published by Edward Jenner in 1798.[72] This approach provided vaccine sceptics with a manual for employing visual rhetoric for their purpose and against hygiene education, and so the development of mass media played into the hands of the anti-vaccine league. Jenner's image of the milkmaid's hand for example, was referenced by a compositionally similar image used as part of the anti-vaccine propaganda

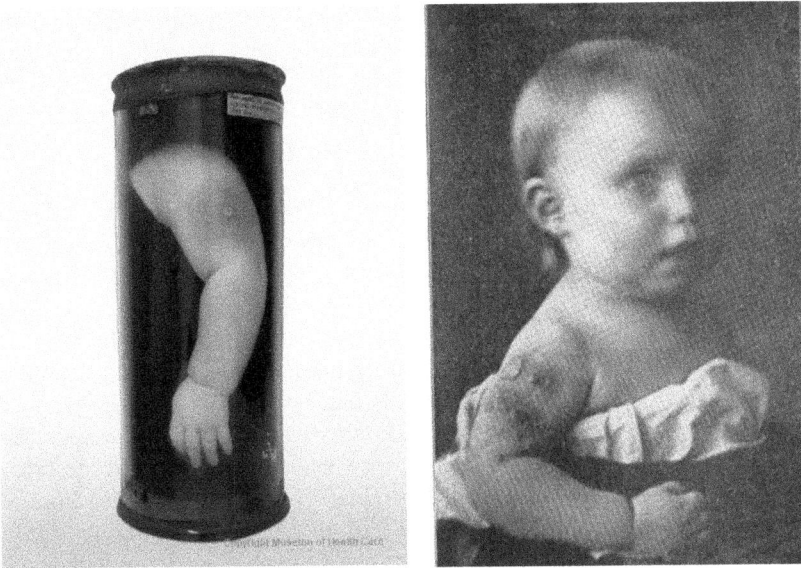

Figure 11.4 Appropriation of medical imagery (left) by the anti-vaccinist movement (right). Queen's University Collection of the Museum of Health Care, Wax Model, Joseph Towne 1850, Kingston, item #997002031; Wegener, *Impf-Friedhof* (1912), case number 222, p. 92.

(see Figure 11.4). The 1912 publication *Der Impf-Friedhof* turned the reference established by the original image on its head. This utilization was possible because Jenner's original image merely displayed an observation, at least for the lay public. It was not presented to readily include or replace an explanation that would become common knowledge; rather the image establishes a reference to scientific objectivity due to its generic recognition value which points to the medical context.

By using a photograph of an arm that displays a supposed vaccine accident (images of arms with wounds or scars were frequently used by the anti-vaccine campaigners), the anti-vaccine movement employed the established practice of images in health education, resorting to an established system of coded images, and expanding it to its advantage. In repeatedly performing this metonymical gesture, the images ultimately established a mental connection which was over time perceived as evidence. The evidential potential was thus connected to both the characteristics of the medium and the established use of images in hegemonial, academically supported health promotion. As readers additionally saw the same images in multiple publications, this repetition strengthened the evidential potential through recognition value, as its symbolic, generic, and ultimately iconographic values were further established.[73] Relying on what is visible on the outside both supports and disrupts the public promotion of vaccines. Rather than developing a visual archive of the invisible function of vaccines—which might indeed have been an option, given the understanding of its broad mechanism and the discovery and photographic representation of bacteria throughout the nineteenth century—medical popularization went with the intuitively more accessible idea of presenting visuals of patients, or visual renditions of the practice of vaccines.

In addition to the imagery displaying arms with vaccine side-effects, *Die Gartenlaube,* for example, used genre images of the "vaccine parlor" (see Figure 11.5), in order to display the down-to-earth normality of vaccines, even in the countryside. The specific promotion of vaccinations to the rural population was likely an attempt to counteract the discrepancy between city and country with regard to the development of scientific medical practices.[74] At the same time the use of such circumstantial imagery furthered the general acceptance of contextual images to represent the process of vaccination. However, this acceptance in turn boosted the cause of the anti-vaccine campaigners. Genre images, such as the displayed image from the *Gartenlaube,* were typical illustrations for mainstream media at the time. As they became typical, the images also became less dependent on an accompanying text.[75] Ultimately, using the circumstances of vaccination to stand in for the medical process itself became part of the information that

Figure 11.5 Die Gartenlaube 38 (1867), image b 605. The image is titled *Impfstube*; it is notably displaying a style resembling the 1858 painting *Impfstube* by R.S. Zimmermann.

both promoted *and* opposed vaccines, and established a metonymical use of visual medical information. The blemished arm stood for vaccine accidents in general. Furthermore, the decision to use images which focus on aspects of vaccines that were readily observable and thus easily grasped could insinuate a feeling of expertise in the audience. As they related to the accessible images, viewers considered themselves enabled to form their own opinion, which included rejecting the available (but not visually presented) scientific explanations as manipulative and conspiratorial.

Accidents and Uncertainties

The rhetoric of the image is unique to its historic situation both with respect to the advances in contemporaneous medical science, and to photography. The specific context of vaccine development during the nineteenth century and the aforementioned false sense of expertise in light of the complexity of the immunization process contributed to the success of the anti-vaccine propaganda. While medical investigations gradually explained the physiological process of immunization, the medical layperson could not necessarily understand the function of vaccination intuitively, and thus the immunological processes accordingly remained a "black box." Nadja Durbach describes the situation:

> Indeed, while vaccination has come to represent the triumph of germ theory, it predates it by almost a century and as a practice was based entirely on empirical evidence rather than on scientific theory or controlled experimental practices. Because no verifiable explanation for its success could be found, it appeared, charged an anti-vaccinator in 1876, that vaccination worked by "some mysterious cause."[76]

Despite the assertion of scientists and the government that vaccines were safe,[77] and the gradual but noticeable positive changes in everyday life in Europe due to the developments in medicine, advances in public health, especially regarding hygiene, contagion, and vaccination were often met with suspicion and fear, summarized in *Die Gartenlaube* as a "general suspicion against medicine."[78] Parts of the general public distrusted and took a stance against the procedure, compelled by incidents caused by substandard hygiene in the production and administration of vaccines.[79] A hygiene context that did not reach today's standard frequently lead to minor diseases following the administration of vaccines, mostly due to the quality of the lymph, contamination, substandard antiseptic practices and negligence in the aftercare.[80] These incidents were efficiently recorded and broadcasted by the anti-vaccine league, as Sharma summarizes: "Thousands of children suffered from more or less severe reactions to the procedure, from high-grade fever and ulcerated abscesses, to blindness and even death, and these stories consistently found their way into anti-vaccine pamphlets, the popular press, and parliamentary hearings."[81] While the major vaccine accidents are reasonably well-documented, it is considerably more difficult to assess the reports concerning small-scale and minor vaccine incidents and accidents, and this uncertainty about authenticity was also an issue at the time. Censorship of vaccine accident reports and public health ambitions to label accidents as minor[82] on one, and the opposite desire to amplify the message through exaggeration and dramatization like in Wegener's publications on the other side prevented and continue to prevent a clear assessment of the facts with regard to anti-vaccine publications. It is, however, palpable that historic anti-vaccine campaigners were concerned about people's health and that the campaigns were successful because this concern was shared by large parts of the wider population. The resulting fear of significant vaccine damages and uncertainty about vaccine effects pushed people toward alternative therapeutic methods. Hygiene issues surrounding the administration of vaccines affected the lower classes of society disproportionally, since wealthier people were able to access private care for the administration of vaccines and could avoid the public vaccine clinics. Sharma comments that the 1874 law, which made vaccines mandatory, was implemented without taking into account the social realities of vaccination practice.[83] These significant social issues majorly factored in the aforementioned public stance against *compulsory* vaccination in particular, since the mandate to invest time that had to be taken off work in travel to the site of the vaccine and the necessary aftercare (the dressings needed to be changed regularly) disproportionally affected people with lower income and a higher number of children, as well as those who lived in a rural area, who might not have been against vaccination per se. "This helps us to understand why so many parents worked so hard to evade the vaccine mandate," Sharma comments.[84]

The social issue concerning mandatory vaccination, however, also had a more fundamental angle to it. In the debate around the introduction of mandatory vaccines, social democratic and catholic religious parties (namely the SPD and the Zentrumspartei) both, despite very different political agendas concerning other issues, argued against mandatory vaccines.[85] The debate became part of the wider question whether children were to be consigned to the authority of the state as a group for the greater good of the wider population, or if parental authority for the individual child superseded this. The dispute around mandatory vaccines was thus incorporated in the wider context of the *Kulturkampf* (culture war), in which, broadly speaking and with respect to mandatory vaccination, each party sought to limit the power of the state in favor of the rights of citizens and parents, albeit in order to further secularization or to maintain a religious-oriented educational agenda respectively. Politics, Thießen concludes, instrumentalized the vaccine debate to push other issues, and the discussion exemplified the growing tensions between medicine and society, in which medicinal research claimed superior knowledge, but its power was limited by democratic institutions.[86] Ultimately, the vaccine discourse pivotally exhibited the growing tension between a rising awareness and readiness to defend individual agency on the one, and policies that pushed a scientific agenda often with little consideration of the opponents' arguments (or emotions) on the other hand. This conflict determined public discourse during the second half of the nineteenth century. An indicator for the difficulties anti-vaccinists faced is for example the blunt dismissal of the information material they issued to parliament together with their petitions as "routine" in parliamentary documents.[87] Parliamentarian Gustav Reiniger, who pleaded for taking the case of the anti-vaccine league seriously, urged his fellow parliamentarians to at least care to look at photographs provided by the anti-vaccine league, asking "Shall we rather bring the mutilated bodies or the corpses of the sacrificed children in person?"[88] Acknowledging the palpable desperation in the effort to make alleged vaccine victims visible reveals that the anti-vaccine photographs Reiniger refers to were (and at least partially continue to be) perceived differently than what was typical for medical photography within the scientific community. In this latter mode of medical photographic presentation and perception "focus falls on the technical and scientific challenges of visualization, concern for how pictures contributed historically to knowing a disease medically, how pictures have enabled identification and diagnosis and how they have structured bio-medical practices,"[89] as Lukas Engelmann categorizes medical photography. Anti-vaccinists, however, framed the photographs they used not to document diseases and make them medically identifiable, but rather to raise awareness of the general danger of vaccines they had identified. Engelmann describes this particular framing of disease photography, which differs from the intention that is more geared toward scientific categorization: "Seeing neglected diseases, previously unknown conditions and symptoms whose definition lacked specificity, promises recognition of patients' suffering."[90] Rather than providing scientific documentation, the anti-vaccine visuals displayed suffering and threat with the intention to initiate "an ethical process." The use of photographs by the anti-vaccine league thus aimed at "seeing disease as a metaphor for re-presentation"[91]—they wanted their cause to be heard, or rather *seen*. The routine omission of a discussion puts the extreme and blunt alternative medial discourse of the anti-vaccine campaigners somewhat in perspective.[92]

Public health communication at the time was not only challenged by a legitimately growing sense of agency and a lack of legal transparency, it was furthermore complicated because the epidemiological findings themselves were at the time not without challenges. Vaccine opponent Paul Mirus utilizes the deficient vaccine information: "Denn das Impfdogma kann nur deshalb sein trauriges Dasein fristen, weil das Volk über den wahren Sachverhalt noch in Unkenntnis schwebt." (The vaccine dogma can only continue its sad existence, because the people are kept in ignorance concerning the actual situation.)[93] The promotion of a health practice which was in its essence and results unobservable for the layperson and not in-tuitively understood was (and remains) challenging for public health au-thorities. Latour has described the communicative issues at hand, noting, "An idea, even an ingenious one, that will save millions of lives, is not going to move forward on its own."[94] While disease and death as ob-servable and very likely consequences of epidemics were frightening, the new reality of invisible and potentially ubiquitous microbes added to the horror. With respect of this fear factor, increasing government force, especially making vaccines mandatory, was decidedly counterproductive. As public health authorities apparently lacked the ability to comprehensively explain the function of vaccine immunity, the process became a "black box" of a reality, a process that cannot be seen with the "naked eye."[95] As such, the "black box" of vaccination made room for imagination and invited hypotyposis—namely the supplement of imagery to a described scene. This imagery was often either already provided by or instigated through the text and the use of visuals. The different components contributed to an augmented reality, promoting ideas about medical and scientific developments—a process that Martina King und Thomas Rütten have pointedly termed "mythoscientific."[96] The dizzying visualization con-glomerated bacilli, microbes, news items, and ghosts in a confusing manner, fueled by the increased speed with which information spread with the use of the rotation printing press since 1846, as the accelerated "spread" of news heightened the perception of print media as somewhat uncanny. Anti-vaccine activists cashed in on the virulent doubts when promoting their argument that vaccines were harmful for the body, and that they acted like poison. Instead of receiving vaccines, the movement recommended strengthening the body naturally, underlining the argument with the in-tuitive allegory of the body as a castle that is fortified against an outside enemy.[97] This more straightforward yet deceiving explanation added to the sense of expertise that is likewise an effect of the simplifying contextual images. In comparison, the paradoxical phenomenology of an intruder which has the ability to change the body for the benefit of the body is much more complex to convey. The theoretical vacuum of both available and readily explicable theories of what is unknown, unseen, and theoretically complex caused an "epidemiological vision of fear."[98] Philosopher Roberto Esposito explains in *Immunitas* how immunity serves as "the general para-digm of modernity," based on a nonmetaphorical reading of society as a system of communication.[99]

In a period where photography was considered to reveal what cannot be seen with the "naked eye," the implementation of spirit photography as an experimental application of the new medium insinuated that this unseen world included (at least for some) the worlds of ghosts. Thus, the medium itself comprised the characteristics that undermined an emerging relation-ship of hegemonic scientific knowledge and photography. Walter Benjamin's notion of the "optical unconscious"[100] speaks to the theoretical implications of this aspect of photography at the time, and given the scope of the theoretical concept, it is not entirely surprising that ghosts crept in. Ultimately, spirit photographs reveal how the medium of photography included in its technological setup and epistemological framework an element of fear that materializes in turn in the use of the medium.[101] The use of images in anti-vaccine rhetoric utilized this potential of the medial uncanny with precise intuition, and produced a rhetoric that exploited both the lack of an easily communicable explanation, and the narrative void of the photographs.

In summary, the historic anti-vaccine movement made use of the fear associated with the practice of vaccines and the uncanny potential the media of photography and mass printing include. This fear hypotypotically filled the void of a scientific visual explanation, utilizing the fear factor associated with photography at the time, and the achieved congruency of content and form furthered the impact of the anti-vaccine material. In addition, the anti-vaccine rhetoric employed images to instigate an individual narrative as a way to persuade their readers—through empathy and by means of an accessible argument that started with observable symptoms as part of an individual's fate. The evidential potential of this strategy lay in the corresponding aspects of the media used (photography and mass media), and in the similarities to established epistemological frameworks in medicine and hygiene education, especially the focus on visible evidence generated by the hygiene movement. These factors allowed for the visual anti-vaccine material to establish itself as a practice of evidence.

Notes

1 John Swinnen, "History: 7. 1880s" in *Encyclopedia of Nineteenth-Century Photography. Volume I, A-I*, ed. John Hannavy (New York: Routledge, 2008), 700.

2 Theoretical writings on this topic are broad; the most notable theoreticians of culture who focus on the visual culture of the nineteenth century are, e.g. Walter Benjamin, Michel Foucault, Susan Sontag, Alan Sekula, W. J. T. Mitchell, Jonathan Crary, and Tom Gunning.

3 See Christiane Arndt, "Fantastische Bazillen. Fotografie in der Gesundheitserziehung der Familienzeitschriften des 19. Jahrhunderts," in *Fotogeschichte. Beiträge zur Geschichte und Ästhetik der Fotografie*. Heft 97/2015 *Fotografie und Medizin. Von der Glasplatte zur Simulation*, ed. Anna Lammers.

4 For more on this topic, see Friedrich Kittler, *Grammophon, Film, Typewriter* (Berlin: Brinkmann und Bose, 1986), 281; Hedwig Pompe, "Die Zeitung—Eine Büchse der Pandora?" in *Die Kommunikation der Medien*, ed. Jürgen Fohrmann and Erhard Schüttpelz (Tübingen: Max Niemeyer Verlag, 2004), 196; Helmut Müller-Sievers, "Experiment im Einsatz. Bewegungszwingung und Erzähltechnik im 19. Jahrhundert," in *"Wir sind Experimente, wir wollen es auch sein!" Experiment und Literatur II 1790–1890*, ed. Michael Gamper, Martina Wernli, and Jörg Zimmer, 278–299 (Göttingen: Wallstein Verlag, 2010), 291–292.

5 The *Illustrirte Zeitung* says of the use of images: "Die Anschaulichkeit, mit welcher die Dinge gegeben wurden, rückte dieselben dem Verständniß näher und regte die Lust allgemeiner an, sich darüber zu unterrichten." *Illustrirte Zeitung*, no. 1279

(1868), 6. This can be translated as: The clarity with which the items are presented makes these more accessible and fuels the general desire to learn more about them. See Frank Heidtmann, *Wie das Photo ins Buch kam. Der Weg zum photographisch illustrierten Buch anhand einer bibliographischen Skizze der frühen deutschen Publikationen* (Berlin: Berlin Verlag Arno Spitz, 1984), 475–476 and 507–509.

6 See Heidtmann, *Wie das Photo*, 514.

7 See, for example, "Ländliche Kinderimpfung. Nach der Natur aufgenommen von Rudolph Geißler," in *Die Gartenlaube* 38 (1867), 605.

8 See Patrick Mayr, *Die Impfgegnerschaft in Hessen. Motivation und Netzwerk (1874–1914)* (Berlin: Peter Lang, 2020), 72. The magazine with the highest print run among German-speaking readers at the time was *Die Gartenlaube*, with a print run of ca. 400,000 issues at the peak of its popularity, see Rudolf Helmstetter, "'Kunst nur für Künstler' und Literatur fürs Familienblatt: Nietzsche und die Poetischen Realisten (Storm, Raabe, Fontane)," in *Kunstautonomie und literarischer Markt*, ed. Heinrich Detering and Gerd Eversberg (Berlin: Erich Schmidt, 2003), 54. The title page and general design of *Die Gartenlaube* is one example of how the anti-vaccine publications resembled the typical features of the popular press.

9 See Eberhard Wolff, "Medizinkritik der Impfgegner im Spannungsfeld zwischen Lebenswelt- und Wissenschaftsorientierung," in *Medizinkritische Bewegungen im Deutschen Reich (ca. 1870–ca. 1933)*, ed. Martin Dinges (Stuttgart: Franz Steiner, 1996), 83. This issue is recognized in the contemporaneous debate, see *Reichstagsprotokolle* 06.06.1883, 2859.

10 See Malte Thießen, *Immunisierte Gesellschaft. Impfen in Deutschland im 19. und 20. Jahrhundert* (Göttingen: Vandenhoeck und Ruprecht, 2017), 35–36; Robert M. Wolfe and Lisa K. Sharp, "Anti-vaccinationists Past and Present," in *British Medical Journal* 325, no. 7361 (2002): 430–432; Nadja Durbach, *Bodily Matters. The Anti-vaccination Movement in England, 1853–1907* (Durham, NC: Duke University Press, 2005), 7.

11 See Wolff, "Lebenswelt," 84.

12 See Otto Dann (ed.), *Vereinswesen und bürgerliche Gesellschaft in Deutschland*. Beiheft/Special Issue, *Historische Zeitschrift*, 9 (1984).

13 See Thießen, *Immunisierte Gesellschaft*, 35–37.

14 "History of Anti-vaccination Movements," The College of Physicians of Philadelphia (Updated 10 January 2018), https://wwwhistoryofvaccines.org/index.php/content/articles/history-anti-vaccination-movements.

15 The procession crowd at the Leicester march in 1885 was estimated to be "from 80,000 to 100,000." Christopher Charlton, "The Fight against Vaccination: The Leicester Demonstration of 1885," *Local Population Studies* 30 (Spring 1983): 63.

16 See Axel Helmstädter, "Post hoc—ergo propter hoc? Zur Geschichte der deutschen Impfgegnerbewegung." *Geschichte der Pharmazie* 42, no. 2 (1990): 21. The graph Helmstädter presents is based on data from Heinrich Molenaar, *Impftod. Bibliographie der internationalen medizinischen Literatur über Impfschäden, Nutzlosigkeit der Impfung und Verwandtes (Jahrbuch des internationalen Impfgegnerbundes Antivaccinator)* (Leipzig: Winkler, 1912) in its entirety.

17 See Thießen, *Immunisierte Gesellschaft*, 35.

18 Durbach, *Bodily Matters*, 5–6.

19 Martin Kirchner, *Schutzpockenimpfung und Impfgesetz. Unter Benutzung amtlicher Quellen* (Berlin: Schoetz Verlag, 1911), 100 (my translation).

20 Thießen, *Immunisierte Gesellschaft*, 32 (my translation).

21 Wolff, "Lebenswelt," 83.

22 The *Reichsimpfgesetz* of 1874 made vaccination mandatory for every child during its birth year, and for students at a public or private institution during their 12th year of life (see C.R.E. Jacobi, *Das Reichs-Impf-Gesetz vom 8. April 1874 nebst*

Ausführungs-Bestimmungen des Bundesraths und den in Geltung gebliebenen Landes-Gesetzen über Zwangs-Impfungen bei Pocken-Epidemien. [...]. (Berlin: Kortkampf, 1875), 24). In Germany, mandatory vaccination was based on this law until 1975. For the situation in England—where the first of a series of vaccine acts was established in 1840—see Durbach, *Bodily Matters*, 2.

23 The punishment for refusing vaccination was stated in 1874 as follows: "§. 14. Parents, foster parents and guardians, who omit to keep record of the document issued with respect to §. 12 will be issued a fine of up to twenty Marks. Parents, foster parents and guardians, whose children and fosterlings have eluded vaccination or the requirement to present themselves (§. 5), will be issued a monetary penalty of up to fifty Mark or imprisoned for up to three days." *Gesetz-Sammlung für die Königlichen Preußischen Staaten* (1835), 255–257. The particular enforcements are described on page 256, §54. *Deutsches Reichsgesetzblatt* Band 1874, Nr. 11, 33–34 (my translation).

24 See Thießen*, Immunisierte Gesellschaft*, 190.

25 For an example, see Avi Sharma, "Anti-vaccine Agitation, Parliamentary Politics, and the State in Germany, 1874–1914," in *We Lived for the Body* (Ithaca: Cornell UP, 2014), 129–130.

26 Paul Förster, ed., *Pocken und Schutzimpfung: Im Namen des Deutschen Bundes der Impfgegner* (Berlin: Berlin Deutsch. Bund d. Impfgegner Berlin Möller, 1900), iv (my translation).

27 See Helmstädter, "Post hoc," 21.

28 Mayr, *Die Impfgegnerschaft*, 142–143. See also Dietrich Krauß, "Impfgegner: Wir können alles außer impfen," in *Kontext*: Wochenzeitung. 8. Juli 2020.

29 Mayr. *Die Impfgegnerschaft*, 129 (my translation).

30 See Mayr, *Die Impfgegnerschaft*, 139. Wegener's *Der Impf-Friedhof* displays a collection of cases in a monograph-like fashion.

31 See Mayr, *Die Impfgegnerschaft*, 69, 139.

32 See Mayr, *Die Impfgegnerschaft*, 139.

33 See Mayr, *Die Impfgegnerschaft*, 40, 130–135, 141.

34 See Mayr, *Die Impfgegnerschaft*, 204.

35 See Mayr, *Die Impfgegnerschaft*, 130–136, 141, 227.

36 See Mayr, *Die Impfgegnerschaft*, 142.

37 Mayr, *Die Impfgegnerschaft*, 140.

38 Anthony Blair gives an account of image rhetoric, see Anthony J. Blair, "The Rhetoric of Visual Arguments," in *Defining Visual Rhetorics*, eds. Charles A. Hill and Marguerite Helmers (Mahwah, NJ: 2004), 49.

39 It is not possible to decide whether the cases in which Wegener claims "it has been reported to me" or "it has been brought to my attention" are direct submissions by family members or other first-hand witnesses, or if these are cases that have been communicated by third parties.

40 See, Ludwik Fleck, *Genesis and Development of a Scientific Fact* (Chicago: University of Chicago Press, 1979), 39.

41 "Rather, the evidentiary character of any picture – its accepted capacity to represent not just itself but an external object of knowledge – is endowed by the specific historical and geographical context in which pictures are produced, circulated and perceived." Lukas Engelmann, *Mapping AIDS. Visual Histories of an Enduring Epidemic* (Cambridge: Cambridge UP, 2018), 29.

42 Albeit for the North American context, Karen Walloch describes the use of anti-vaccine images in court: Karen Walloch, *The Antivaccine Heresy. Jacobson V. Massachusetts and the Troubled History of Compulsory Vaccination in the United States* (Rochester, N.Y.: University of Rochester Press, 2015), 86.

43 See Mayr, *Die Impfgegnerschaft*, 141.

44 See Mayr, *Die Impfgegnerschaft*, 47, 130–135.
45 Roland Barthes, *Camera Lucida. Reflections on Photography* (New York: Hill and Wang, 1981), 32. See also Peter Geimer, "Vom Schein, der übrig bleibt. Bild-Evidenz und ihre Kritik," in *Auf die Wirklichkeit zeigen*, ed. Helmut Lethen, Ludwig Jäger, and Albrecht Koschorke (Frankfurt a.m.: Campus, 2016), 206.
46 See Blair, "Rhetoric of Visual Arguments," 54.
47 *Reichstagsprotokolle*, 4274. The quote from the Reichstagsprotokolle reads: "Mit wahrhaftem Schauder [schaue man] auf die entsetzlichen Bilder, die auf dem Tische des Hauses wieder niedergelegt sind." The member of parliament, however, attributes the problematic outcome to accidents, not to the "characteristic of vaccines" per se. See also Thießen, *Immunisierte Gesellschaft*, 33.
48 Kinnebrock, Bilandzic and Klingler identify the narrative mode as integral to evidential practices in mass media. See Susanne Kinnebrock, Helena Bilandzic, and Magdalena Klingler, "Erzählen und Analysieren: Narrativierungen in der Wissenschaftsberichterstattung," in *Wissen und Begründen. Evidenz als umkämpfte Ressource in der Wissensgesellschaft*, ed. Karin Zachmann and Sarah Ehlers (Baden–Baden: Nomos, 2019), 137.
49 Kinnebrock, Bilandzic and Klingler, "Erzählen und Analysieren," 141.
50 Many aspects of the anti-vaccine material are in place to date, see for example the use of biographical narratives in M.S. Jus, ed., "Impfschäden." Special issue *Similia. Die Zeitschrift für klassische Homöopathie* 17, no. 36 (1996).
51 Hugo Wegener, *Der Impf-Friedhof. Was das Volk, die Sachverständigen und die Regierungen vom "Segen der Impfung" wissen* (Frankfurt a.m.: Verlag Luise Wegener, 1912), 20–22.
52 Susan Sontag, *On Photography* (New York: Picador, 2011), 23.
53 See Blair, "Rhetoric of Visual Arguments," 50–52.
54 Roland Barthes, *Image. Music. Text*, trans. Stephen Heath (London: Fontana Press, 1977), 44; "What makes visual arguments distinctive is how much greater is their potential for rhetorical power than that of purely verbal arguments." (Blair, "Rhetoric of Visual Arguments," 52.)
55 Barthes, *Camera Lucida*, 76.
56 Barthes, *Camera Lucida*, 34.
57 Geimer, "Vom Schein," 206 (my translation). Geimer refers to Helmut Lethen, *Der Schatten des Fotografen. Bilder und ihre Wirklichkeit* (Berlin: Rowohlt, 2014), 173–174.
58 Wegener, *Der Impf-Friedhof*, 281–282. The example is case 770.
59 For example, the photo described as: Ethel Cromwell, aged about 14 years, as a smallpox patient. Photograph by H.C.F., glass 10.6 × 8 cm, 1896 negatives of the Gloucestershire smallpox epidemic No. 8 and No. 19, 1896 Wellcome Library, London.
60 See, for example, *Der Impfgegner* 1883 (1. Jahrgang), no. 1 (Januar), title page. The paragraph soliciting readers to send in news was later changed to specifically include photographs.
61 See Wegener, *Der Impf-Friedhof*, 336. The quote reads "Da bei vielen Meldungen über Impfschäden verschiedene und verschiedenartige Krankheiten angegeben worden sind, so habe ich meistens, z.B. bei Todesfällen, die Schlusskrankheit, oder die hervorstechendste Krankheit im Verzeichnis vermerkt." This translates to: Since many reports concerning vaccine injuries list different and diverse diseases, in the register, I have given the last known or the most prominent disease for the majority of cases, for example, for fatalities. Verzeichnis vermerkt. See also Mayr, *Die Impfgegnerschaft*, 143.
62 Barthes, *Image. Music. Text*, 35.

63 In Barthes original example of the Panzani advert, a tomato represents a tomato, a red pepper represents a red pepper, and so on.

64 Barthes, *Image. Music. Text,* 36.

65 Barthes, *Image. Music. Text,* 36.

66 Bruno Latour, "What Is Iconoclash? Or Is There a World beyond the Image Wars?" in *Iconoclash: Beyond the Image Wars in Science, Religion and Art,* ed. Bruno Latour (Cambridge: MIT Press, 2000), 18.

67 Barthes, *Image. Music. Text,* 46.

68 Lorraine Daston, Peter Galison, "The Image of Objectivity," in *Representations* 40/ 1992: 103.

69 Mary S. Morgan, "Afterword: Reflections on Exemplary Narratives, Cases, and Model Organisms," in *Science without Laws,* ed. Angela Craeger (Durham and London: Duke University Press, 2007), 273.

70 A good example is the Wax Model, Joseph Towne 1850, Collection of the Museum of Health Care, Queen's University, Kingston, item #997002031.

71 See Walloch, *Antivaccine Heresy,* 50.

72 Edward Jenner, *An Inquiry into the Causes and Effects of the Variolae Vaccinae* (London: Sampson Low, 1798), case XVI and Case XVI, "hand of milk maid Sarah Nelmes" (not paginated).

73 Barthes, *Image. Music. Text,* 37.

74 See Martin Dinges, "Medizinkritische Bewegungen zwischen 'Lebenswelt' und 'Wissenschaft,'" in *Medizinkritische Bewegungen im deutschen Reich (ca. 1870–ca. 1933),* ed. ibid. (Stuttgart: Franz Steiner Verlag, 1996), 11.

75 See Barthes, *Image. Music. Text,* 16; Geimer, "Vom Schein," 200.

76 Durbach, *Bodily Matters,* 158. See also Walloch, *Antivaccine Heresy,* 50.

77 See Sharma, *Body,* 124.

78 *Gartenlaube* 32 (1884), 528 (my translation).

79 Notable incidents occurred for example in Montréal under physician William Bessey. See Heather McDougall and Laurence Monnais, "Not without a Risk: The Complex History of Vaccine Resistance in Central Canada, 1885–1960," in *Public Health in the Age of Anxiety. Religious and Cultural Roots of Vaccine Hesitancy in Canada,* eds. Paul Bramadat, Maryse Guay, Julie A. Bettinger, and Réal Roy (Toronto: UofT Press, 2017), 134. See Michael Bliss, *Plague. A Story of Smallpox in Montreal (Toronto: HarperCollins, 1991),* 57–58, and Jennifer Keelan, *The Canadian Antivaccination Leagues 1872–1892* (PhD Thesis University of Toronto, 2004), 191. In Lübeck, 72 out of 251 infants died after having been vaccinated with a contaminated batch of an oral life tuberculosis vaccine developed and first introduced by Calmette and Guerin in 1921. See Eckart Roloff, "Das Lübecker Impfunglück von 1930. Ein Lehrstück der Medizingeschichte," in *Naturwissenschaftliche Rundschau* 69, no. 819 (2016): 461–463; James M. Dunlop, "The History of Immunization," *Public Health* 102 (1988): 204; and also H.L. Rieder, "Die Abklärung der Lübecker Säuglingstuberkulose," in *Pneumologie* 57.7 (2003): 403.

80 See Sharma, *Body,* 118.

81 Sharma, *Body,* 124. See also Walloch, *Antivaccine Heresy,* 28.

82 See Thießen, *Immunisierte Gesellschaft,* 40–41.

83 See Sharma, *Body,* 128–129. Sharma refers here to the aforementioned Kirchner.

84 Sharma, *Body,* 129.

85 Reichstagsprotokolle 09.03.1874, 234. See Thießen, *Immunisierte Gesellschaft,* 45.

86 Thießen, *Immunisierte Gesellschaft,* 47. Thießen in particular refers to a debate in which August Reichensperger points out the social injustices and their conglomeration with religious debates that enter the discussion around punishments for avoiding vaccines.

87 The German word used here is "üblich," in the sense that the material accom-
    panying the petition is "what is to be expected without being especially note-
    worthy," and the term is used repeatedly in connection with anti-vaccine material
    in the protocols of the German Reichstag during the 1870s, for example from the
    Report on Petitions 1877. See *Deutscher Reichstag, Achter Bericht der Kommission für
    Petitionen*, 1877 Aktenstück Nr. 176, 498.

88 The German original reads: "Sollen wir denn die zu Krüppeln geimpften oder die
    Leichen der geopferten Kinder in natura vorführen? Bemühen Sie sich doch
    gefälligst an den Tisch des Hauses und sehen Sie sich die Photographien an."
    Reichstagsprotokolle 6.6. 1883, 2867.

89 Engelmann, *Mapping AIDS*, 3.

90 Engelmann, *Mapping AIDS*, 3.

91 Engelmann, *Mapping AIDS*, 3.

92 It was only in the 1920s that the two opposite sides finally met for repeated talks,
    see Thießen, *Immunisierte Gesellschaft*, 76–77.

93 Paul A.L. Mirus, *Die Impffrage und der Verband deutscher Impfgegner-Vereine*
    (Dortmund: Keßler, 1910), 19 (my translation).

94 Latour, "Bakterien in Krieg und Frieden," in *Bakteriologie und Moderne. Studien zur
    Biopolitik des Unsichtbaren 1870–1920*, eds. Philipp Sarasin, Silvia Berger, Marianne
    Hänseler, and Myriam Spörri (Frankfurt a.M.: Suhrkamp, 2007), 114 (my trans-
    lation). For misconceptions around vaccines see also Durbach, *Bodily Matters*, 164.

95 The metaphor of the "naked eye" (or in German the "unarmed eye"/"un-
    bewaffnetes Auge") appears frequently at this time, e.g. Julius Cohn, "Unsichtbare
    Feinde in der Luft." In *Die Pflanze. Vorträge aus dem Gebiete der Botanik* (Breslau: J.
    U. Kerns Verlag, 1882, 468. See also *Die Gartenlaube* 39 (1866), 611.

96 Martina King and Thomas Rütten, "Introduction," in *Contagionism and Contagious
    Diseases. Medicine and Literature 1880–1933*, ed. Martina King and Thomas Rütten
    (Berlin: De Gruyter, 2013), 7; see also Durbach, *Bodily Matters*, 164 and 168.

97 An example for this argument can be found in Heinrich Molenaar, *Impfschutz und
    Impfgefahren* (München: Melchior Kupferschmid, 1912), 1.

98 Emmanuel Alloa, "Fremdkörper. Fragmente einer Theorie des Eindringlings," in
    Impfen. Pfropfen. Transplantieren, ed. Uwe Wirth (Berlin: Kadmos, 2011), 85 (my
    translation).

99 Roberto Esposito, *Immunitas. Schutz und Negation des Lebens* (Berlin: Diaphanes,
    2004). 73.

100 See Walter Benjamin, *Little History of Photography*, trans. Rodney Linvingstone and
    Others, in *Selected Writings. Volume 2 1927–1934* (Cambridge: The Beknap Press of
    Harvard UP, 1999), 512.

101 This element of fear that is associated with the new medium will continue to
    develop into the well-established notion of film and its uncanny doubles. After
    Freud's epoch-making essay "The Uncanny," Friedrich Kittler and Tom Gunning
    have for example prominently explored the topic.

12 The Politics of Evidence: State Secrecy, Ambiguity, and Counterforensic Practice in "Missing Persons" Cases in Pakistan

Salman Hussain

This chapter explores the politics of evidence in the context of a violence that leaves no proof of its violations. As a spectral form of violence,[1] "disappearances" by state military and intelligence services elude the formal scrutiny of liberal legality under the elusive and often absolute cover of national security. Examining "missing persons"[2] cases in Pakistan, I discuss the use of documentary and visual articles—various types of files, petitions, legal documents, photographs, applications, and other archivable things—as evidence collected and employed by activists and the families of the "disappeared" to uncover violence masked by state secrecy. The ethnographic engagement with this documentary "bricolage"[3] and the practices surrounding it, I suggest, allows for contributing to a counterpolitics of evidence-making by human rights activists. The chapter draws from fieldwork I conducted (2014–2019) among families of the "missing persons": suspected Islamic militants, nationalists, separatists, political activists, and their sympathizers—all allegedly abducted and detained by state military and intelligence services in Pakistan.[4]

My discussion is broadly concerned with documents, photographs, and other types of material things that circulate outside courts and other state repositories and how such materials are retained, transformed, or reconfigured in human rights activism. Many of these documents seem to be part of the state's records alone, while others are obtained from personal sources as well as personal and state sources. The chapter focuses on, what I have called elsewhere, "dossiers of memory."[5] The dossiers assembled by the activists and families of "missing persons" are mobile repositories of various types of fragmentary material evidence collected since the disappearances. Documentary belongings and identification records of the "missing" are frequently confiscated during intelligence and military operations. The dossiers of memory contain legal, personal, and political traces of the disappeared persons—their life-details, photographs, identity documents, and applications (submitted to state officials and circulated publicly

DOI: 10.4324/9781003188612-16

to locate and recover the disappeared)—and they reveal how activists and family members draw on them to reestablish a documentary life of the disappeared, publicly demand justice, and trace and free the "missing."

State Secrecy in Modern Warfare and "Counterterrorism"

Lawfare as well as covert warfare has defined modern wars, and, in particular, the wars against "terrorism" and "rogue" states.[6] In its so-called war on terror and the many violent conflicts that followed in Northern Africa and the Middle East, the United States and its allies have made use of remote surveillance and—often based on evidence collected through that surveillance—have engaged in targeted killings in anticipation of threats against their security interests.[7] Even though launched based on "scientific intelligence" to convince the international community abroad and skeptical publics at home, these new forms of warfare in fact leave behind the least accessible evidence: they are secretive, distant and remotely controlled.

Post 11 September 2001, the counterterrorism operations and legal proceedings against "terrorists" and the states allegedly supporting them were advanced by an outright control over information.[8] The evidence—whether of "weapons of mass destruction" claimed by the United States to be present in Iraq, or that used to categorize captured prisoners of war in Afghanistan and elsewhere as "unlawful combatants" in detention sites like Guantanamo Bay—was kept secret under the cover of national security and state secrecy.[9] Extra-judicial detentions and renditions that followed the attacks, and the acts of torture and even the denial of entry at national borders to foreign nationals, were all justified based on classified information.

Although taking various legal forms in modern statecraft and upheld by courts in Europe and the United States in the years following the attacks of 9/11, the notion of state secrecy can be traced back to *salus rei publicae suprema lex* in Roman law. *Salus rei publicae* expressed the idea that "the safety (or welfare) of the state is the supreme law."[10] According to Arianna Vedaschi, underlying its conception—as a recent Italian Constitutional Court judgement reasoned—is the idea of "the state secret defense" as "an instrument" that aims to "protect the *salus rei publicae* ... linked with national security, which is intended as a 'community's essential and insuppressible interest.'"[11] The court sees the safeguarding of national security as protecting "a collective and general interest" of the community.[12] However, underlying this ambiguous notion of security of "the community" is the fear of an existential threat to the state and its legal order, which then justifies the use of unrestrained violence and violations of basic human rights.[13] Therefore, the Italian Constitutional Court further included in its ruling as permissible the classification of any information (of acts) that endangered the "constitutional order ... [or] actions aimed at overthrowing the government."[14] US courts and other European jurisdictions have also allowed the use of classified intelligence information as evidence in many instances,[15] while judges in Pakistan have examined in their chambers the secret dossiers on "missing persons" presented by the intelligence services.[16] Evidence practices then are asymmetrical; access to information, its nature, admissibility, and use are all led by established legal, official, and scientific practices and institutions of power. Marginalized populations have often been the subject of these knowledge practices, while left out of these institutions. Moreover, in judicial, scientific or other contexts, the demand for "hard" evidence, often aimed at these populations, is itself a "prerogative of power."[17]

I have argued elsewhere[18] that ethnographers are ethically and politically implicated in the politics of evidence, and bear responsibility toward the people engaged in these politics. In conditions of surveillance, censorship, and state violence (and in other "ordinary" contexts), we can't simply engage in the "systematic collection of information,"[19] scientific evidence, and facts to address our research questions. Rather, these practices must respond and take a stance with our research partners' political struggles—as they reconstruct proof of violations and exercise their right to "truth" about them. In this chapter, I further these suggestions by proposing to examine the evidence-collection by the families of the missing as what Eyal Weizman (following Thomas Keenan and Allan Sekula) provocatively calls "counterforensics"—a human rights practice that inverts the "forensic gaze" back upon the state.[20]

Such ethnographic attention to documentary artifacts also allows for identifying modes of speaking to power mediated by "nonhuman" objects. Recent works in ethical and material turns in anthropology have questioned the permanence of the agentive, unitary subject possessing a single voice,[21] and have suggested turning our attention to various otherworldly registers (such as *jinns*[22] and dreamworlds),[23] and material forms[24] by means of which our interlocutors "speak." According to these (re)formulations of the agent, the subjectivities that emerge are refracted through these objects as they perform a crucial social and political role: the agency and the voice—previously assumed to belong solely to humans—are shaped by the encounter between persons, things, and the political and historical conditions of their interactions. Ethnographic practice as a form of counterforensic practice proposes "new ways of conjecturing and operationalizing ethnographic 'facts,'"[25] critically examines the historical context and epistemological genealogy of what is accepted as "scientific" evidence and contributes to the process whereby various types of traces of violations are collected and given a public voice and forum.

Silence, Censorship, and Ambiguity: Enforced Disappearances in Pakistan

Even though many of the "missing persons" are reported by their families to have been detained at check posts (for not carrying their identification documents) and in neighborhood sweeps during combat operations in areas where major counterinsurgency campaigns against militants have taken place,[26] numerous suspects have also been picked up from their homes, shops, offices, buses, hotels, and businesses across the country. Some of these "missing" persons have since been traced to the military and paramilitary-run internment centers (in the northwest of the country), held there with no charge or appearance in the courts. Set up under the Pakistan Protection Act of 2014, the existence of these internment centers and the interminable detention of suspects became known because of the protest and legal activism of the families of those interned there.[27] Information about the internees circulates through rumors and informal connections among the families made possible because of collective protests. The internees are often moved between internment centers and secret houses, and the information about their location is leaked only when a fellow internee is freed and carries messages of others to their families. These messages create new lines of inquiry for the relatives who relentlessly pursue them without having any "hard" proof.

As open violations of the suspected militants' and separatists' constitutional and human rights were (and continue to be) consistently ignored in Pakistan, the practice of disappearing "terrorists" has now extended to political activists, journalists, human rights activists, teachers, and students; the dissenters who dare to protest against, or merely criticize the country's intelligence services, the military's security or foreign policy, or their narratives on militancy and separatism within the country on social media, have become targets of intimidation and surveillance. Impunity from prosecution, and the power to carry out unrestricted surveillance and detentions, have afforded the intelligence services deep reach and influence within Pakistani society and politics.

Secrecy, silence, ambiguity, and denial by the intelligence services are all characteristic of enforced disappearances. In the Pakistani Supreme Court and the Commission of Inquiry on Enforced Disappearances (henceforth, COIOED),[28] formed on the directives of the Supreme Court in 2012 to record and investigate the cases of the missing persons, representatives of the intelligence services consistently deny their having custody of the disappeared. As witnesses of disappearances are intimidated and are fearful of reprisals, they refuse to testify in the courts. Even family members are cautioned by the intelligence services to stay silent and to quietly await the return of their loved ones. Indeed, many families have refused to report a "disappearance," while others have maintained silence as they do not have the resources to carry on legal proceedings in the courts.

"Negative evidence"—lack of "hard" evidentiary material—is invoked by the state's representatives, including its civilian officials (representing the Police and Interior Ministry), who appear in the missing persons cases in the Supreme Court and in the COIOED's proceedings, to respond to judges' inquires.[29] The absence of material proof works in defense of the intelligence services. On the rare occasions when they do present evidence concerning allegations against the disappeared person, it takes the form of secret dossiers and sealed envelopes shown only to the presiding judges and, often, during *in-camera* proceedings[30] in their chambers.[31] It is important to note that the practice of presenting classified evidence in the defendant's absence, and the acceptance of evidence extracted by torture and abuse, was also permissible in the review proceedings of detainees at Guantanamo Bay; the rules of evidence and procedure were set aside in those cases too.[32] State secrecy denies access to, and the scrutinization of, the evidence upon which the missing persons are indefinitely interned on suspicion of militancy and collaboration.

Keeping such information secret allows military and intelligence services to detain and implicate—under the cover of "terrorism," militancy, espionage, rebellion, and separatism—various categories of politically undesirable figures: nationalists, political and human rights activists, journalists, researchers, army deserters, and Islamic militants. Creating ambiguity about the disappeared persons regarding their affiliations and collaborations, maintaining the denial of their detention, and prolonging lack of information further allows for keeping control over who may or may not be categorized as "missing." Against this background, the collection of fragmentary evidence as a "counterforensic" evidence by victims and their families works in the space of the "structural inequality" regarding access to information that only states have access to, and which they refuse to share by invoking national security.[33]

Forensics and Counterforensics

Rooted in the nineteenth century "science" of phrenology, and reinforced by the development of modern photography and biometric fingerprinting, forensic methods continue to primarily serve the dominion of law enforcement and judicial services.[34] Phrenology and other pseudosciences helped to scientifically classify the criminal for the modern state. Colonial practices of record-keeping and photographic archiving (at home and in the colonies) allowed for the recording, classification, and punishment of deviant bodies, and shaped the definition(s) of social and criminal deviance.[35] With scientific developments and devices, such as photography and fingerprinting, the "criminal body" became a site of "interpretation and enforcement" upon which the very authority of law enforcement personnel, like the famous nineteenth century detective, was constituted.[36] These devices aided the discovery of scientific and legal "facts,"[37] and thereby established the authority of the police expert. Popular TV shows, for example, various incarnations of the *CSI* format (Crime Scene Investigation) and those that investigate cold files, continue to provide a popular aura of scientific authenticity and cultural authority to forensics. The field has had an alarming influence on the everyday technologies of surveillance and identification too; finger and iris prints are now linked to national identity cards and passports, and border control.[38] Our digital substitutes, images, and profiles—composed of our digital imprints, physical and virtual movements, and socioeconomic practices—continue to exist and multiply within corporate data and state archives and files. These technologies allow states to survey, surveil, and reinforce their internal and external boundaries.

The securitization of citizenship in Pakistan (materialized by establishing a digital database of citizens in the early 2000s) has taken place in a similar vein. "Missing persons" are taken away for "verification" of their "real" identities, and their families are assured they will be released soon after. The evidence collected by the families is therefore mobilized not only against secretive allegations of collaboration with the militants but to counter misrecognition and misidentification of the missing person, as an illegal, foreign "alien" (read, Afghan) militant. The protesting families thus regularly and publicly display state issued identity documents in their protests.

Forensic sciences have also helped to locate victims of state violence and/ or their (missing) remains, and have played a crucial part in challenging the denial and secrecy of state violence, occasionally helping to bring the perpetrators to justice.[39] Claims made through the use of forensics, however, are assumed to be claims of scientific truth—spoken by various types of material objects that come under the forensic gaze.

Forensics objects do not speak for themselves—they are interpreted by experts and other mediators. Forensic evidence is usually partial, fragmentary, and in need of interpretation; forensic facts have to be created and it's into the space of ambiguity about these "facts" that it becomes crucial to intervene in. Even in the cases of clear human rights violations, forensic evidence has to be "exposed, shown, demonstrated, stated, claimed, proven and made evident to others."[40] Three sites—of collection, inquiry, and presentation—come together to produce forensic facts: the site(s) of investigation where the event of violence took place, the space where the collected material is "processed and composed into evidence," and the forum where that evidence is presented.[41] "Forensic speech," explains Eyal Weizman, is "undertaken as a relation between three elements: an object or a building 'made to speak,' an expert who functions as the translator from the language of objects to that of people, and the forum or assembly in which such claims can be made."[42] However, the relationship between the three components continues to shift: "Objects are animated in the process of presentation; skulls, buildings, and eco systems are referred to as if they were human subjects; the interpreters, meanwhile, are no longer necessarily human experts, but automated or semiautomated technologies of detection, calculation, and imaging, while the forums expand to a multiplicity of modes of articulation."[43] A multiplicity of objects, the entry of different social and political actors—such as human rights activists—who have taken their place beside scientific and legal experts, and the diversity of, often public, sites where forensic facts can be tested, challenged, and debated has changed the relationship between the three traditional aspects of forensic practice and the production of forensic facts. It is in this public and contested context of producing forensic facts that counterforensic practice is situated.

Counterforensics inverts the "forensic gaze" upon the state by drawing from forensics' own public significance (i.e. in terms of presentation and forum) and builds upon its promise of giving voice to material things.[44] It appropriates the original meaning and practice of forensics and yet relocates the field from the legal to the public domain; and from its sole purpose of serving the criminal justice system toward including civil and human rights practice. Drawing from the primary meaning of the term forensics, the term counterforensics can be traced back to its Latin origin, *forensis*, "pertaining to forum."[45] In this regard, Weizman and Keenan explain that Allan Sekula's use of "prosopopoeia—the attribution of a voice [and a face] to inanimate things"—best expresses forensics' counter meaning and practice.[46] "Contemporary modes of prosopopoeia" revive "material objects by converting them into data or images and placing them within a narrative."[47] Interpretation remains crucial to how material objects are imparted a voice: they become "actants" rather than passive objects, as they are brought to life by not only human experts and interpreters but also through various digital and imaging technologies.[48]

Counterforensics thus challenges the state monopoly over the practice and the meaning of, as well as the claims made by, forensics, and importantly, the state control over "information" in modern warfare.[49] Counterforensics use the "means of evidence production to unearth evidence of state violence."[50] It aims to take forensics away from the state, which uses it to legitimize its violence, by producing "scientific facts" about its own actions—often taken against civilians and noncombatants.

The practice of counterforensics—how material objects are turned into data and images, and the images into information that can tell an alternative/counternarrative—makes it an important tool for human rights activists, critics, and even civilians, as they all work together to gather, organize, and analyze information. I will now turn to the files assembled by the families of the missing persons, collected under conditions of secrecy, silence, and denial, and examine how they serve as sources of counter-evidence as the families demand justice and accountability from the state and its judicial system. The files collect various documentary and visual traces of the disappeared, and a critical reading of these bits of information provide insight into the events of the disappearances and the claims of inculpability made through these paper facts. Attention to "traces," in other words, require a rereading of existing objects, new ways of assembling and interpreting "ethnographic 'details.'"[51]

Files: Documentation and Documentary Remains

Files are ubiquitous mobile objects accompanying the family members of the missing persons, regularly found at protests, courts, and the offices of human rights organizations. They are, I suggest, "evidence assemblages"—made of petitions, legal documents, photographs and other images, and of witness testimonies/testimonial evidence—in the missing persons cases.[52] Files are displayed by the families as visual evidence toward the identification of a disappeared person and, most significantly, as evidence itself of the disappearance. The documents collected in them and displayed publicly are counted upon by the families to advocate on behalf of the missing persons. Crumpled, aged documents, copies of the copied originals, folded, creased, and faded photographs, state issued identity cards and other identifying documents, applications and petitions addressed to the civilian and law enforcement agencies, and copies of court petitions and orders are all put together in these files and are assumed to speak on behalf of their bearers—the majority of whom are illiterate and live at a spatial and cultural distance from the labyrinths of state bureaucracy and its arcane legal system.

As "ethnographic artifacts," Annalise Riles argues, documents raise questions for ethnographers about agency, authorship, and responsibility; they demand a response from ethnographers.[53] Rather than objects to be assessed simply for their own veracity, or the veracity of persons or events that they refer to, I suggest, documents tell narratives about their production and circulation, and the effects they produce on their carriers. Files and the objects making them up are "saturated with meaning and power,"[54] condensed traumas, events of suffering and narratives of (alternative) histories within them. They have living relationships with their bearers and are composed of a "bricolage"—an assortment created out of various available objects and ideas—of visual and documentary fragments.[55]

In South Asia, documentary or paper *raj* (rule) was the binding technology of governance that cemented the colonial state and, following independence, of the postcolonial state. Stamps and seals, stamped paper, and many other state practices (e.g. introduced through acts such as the Stamp Act of 1899) were brought about and divorced the official from other forms of documents.[56] The copies of documents originally issued by state departments and courts continue to be duly signed and stamped, and without these marks of authentication they apparently possess no official value, since unsigned and unstamped copies fail to mediate or intervene on behalf of the bearer. Signature, seal, and stamp attest to the authenticity of a paper, turning it into an official document signified with state power.[57] Documents represent and declare the administrative, the bureaucratic, and the legal authority of the state, and forging them carries severe punishments.

It is important to note here that forging evidence—in documentary and other forms—was a major concern of the colonial judicial system. The colonial state had a deep distrust of the locals and their presumed tendency to manipulate the courts in personal disputes; therefore, the Indian Penal Code and the Code of Criminal Procedure carried "extensive provisions on perjury, forgery, and false charges."[58] Various methods of authenticating documents submitted to and issued from the court and state bureaucracies were introduced, in order to protect them from potential manipulations outside of their official networks of circulation and use.

At the courts and in state offices it is common to find bulky files in the arms of lawyers, clients, petitioners, and on the desks of the high and low bureaucrats. These files accumulate documents—originals as well as copies—while cases slug through the judicial system and/or state bureaucracy.[59] However, there is an intended dearth of documents in the missing persons cases in the higher courts, as well as in the proceedings of the COIOED. Documentary proof about the missing persons and their disappearances is simply absent, and this absence—generative of ambiguity about the disappearances—is in turn made use of by the families and activists by assembling their own evidence regarding the disappeared persons. Meanwhile, access to the COIOED and its proceedings is restricted and only the family members invited for a hearing may attend the proceedings. Moreover, in the courts and in the COIOED, the intelligence services maintain a complete denial, providing no written affirmation or refutation of any detention. They keep silent because a documentary trail of the disappearance would provide evidence of the detention and could therefore serve as grounds for litigation against the security services. The higher courts have occasionally used such documentary evidence to hold the military and intelligence services responsible for the "disappearances."[60]

The files assembled by claimants in the missing persons cases manifest their age and the distance they have traveled—spatially and temporally—by the documents they have accumulated since the disappearance. The thicker the folder, the longer and more unresolved the disappearance. Court orders and petitions, applications made to the internment centers (where many of the "missing" are allegedly held), national identification cards and other forms of identification, and First Information Reports collected by the families over the years of the disappearance, all serve as documentary trails establishing evidence of the disappearance. Rather than merely copies of the real documents—as most of these indeed are—they are documentary forms that, in replication, escape their legal official networks and, by their circulation and multiplication, blur the legal and nonlegal spheres, significations, and usages.

Like many of the missing persons, Baz Muhammad's file and his photograph were carried by his brother, who had traveled from Mohmand Agency (located eight to ten hours away) to Islamabad to attend a protest organized by the Defense of Human Rights Pakistan (henceforth, DHRP) in August 2018. After the protest, we met at the DHRP's office, where he had brought all the documents he had collected over the years to file a petition in the Supreme Court.[61] Muhammad, his brother stated, had disappeared six years ago. His file disclosed that Muhammad was picked up from Islamabad but had been transported to an internment center in Mohmand Agency, in the former Federally Administered Tribal Areas (FATA).[62] Muhammad had been "missing" since May 2012. He was a day laborer and a vegetable hawker, working at the main vegetable market in the capital, and disappeared from there. In a form submitted to the National Crisis Management Cell, his brother admitted Muhammad had been to Afghanistan four years prior to his disappearance. He claimed Muhammad hadn't been involved with any militant organization since his return from Afghanistan and subsequent move to the capital. Among other documents, Muhammad's file contained his crumpled photograph, copies of various legal orders (passed by the Peshawar High Court), and a paper slip stating his tribal identity and address.[63] The most important document in his file was the copy of an official letter signed and issued by an officer in the Mohmand Rifles (a unit of the paramilitary force, Frontier Corps) and addressed to the Commissioner of Peshawar. The letter asked the Commissioner to arrange a meeting between the "internee" (i.e. Baz Muhammad) and his family. As a crucial piece of evidence in locating Muhammad, this letter affirmed that he was indeed in military custody and challenged the intelligence and military services' claims that they were not involved in Muhammad's disappearance, alleging he had in fact fled, once more, to Afghanistan. Since his last meeting with his family in 2016 at an internment center, Muhammad was, once again, considered "missing." Intertextuality and the chain of documents within the file establish the facts of the event of disappearance and what has followed. The file form allows for the collecting and condensing of events, actions, and images into narratives.

Noor Khan was also picked up by the military forces during their counterinsurgency operations in South Waziristan. His brother, whom I too met at the DHRP, told me that Noor had been "missing" since 2009. Through different informal channels and subsequent rumors, his brothers and cousins had been searching for him at various internment centers (in the former FATA region) and elsewhere in the KPK Province. Petitions to commissioners of the area where these internment centers were located, an aged photograph attached to one of the applications, and a petition in the Peshawar High Court made by Noor's mother seeking permission to meet him were all part of Noor's file; its latest addition was the application addressed to the DHRP to represent Noor's case at the Supreme Court of Pakistan. The day we met, his brother had brought his photographs and the documents he had collected to the DHRP since Noor's disappearance. Petitions to the courts and applications to the state officials had all failed to locate him. Noor had been untraceable for the last ten years.

The disappearances are all marked by the absence of procedural documentation and material and visual evidence that could serve as a proof of unlawful custody. They exist at the "threshold of detectability"—things that hover between being identifiable and not.[64] They take place at the "juridical threshold,"[65] and mark their spectral presence by documentary absence, denial, and ambiguity. The dossiers created by the families of the disappeared challenge the legal threshold of disappearances, reestablish their documentary life, and serve as a counter biographical archive. These files do not only have legal, but also emotional, material, and narrative values for their bearers, and these attachments feed their protest politics, activism, and the demand for justice. The photographs of the disappeared persons— always present in these files—are a key part of their documentary remains, and act as a visual as well as affective articles of evidence.

Photographs: Frames and Counter Frames of War

The dossiers of memory contain the photographs of the "missing persons," silently displayed at the protests by their families. The photographs take on archival, evidentiary, and performative roles, and in the body's absence, they condense, advocate, and are presumed to speak on behalf of the missing person.

Photography also played a crucial role in the development of nineteenth century forensics. Allan Sekula argues that as a visual technology, photography helped establish a generalized typology of individuals in an archivable form. In relation to this typology's inverse, a particular deviancy of individual types was crafted through its "photographic portraiture."[66] It did so as the "criminal identification photograph" became essential "to the process of defining and regulating the criminal."[67] It was against the background of a larger "universal" archive of photographic classification of things and persons, that the "zones of deviance and respectability could be clearly demarcated."[68] Based on a comparative difference made possible by compiling images, criminals' visual classification emerged in relation to the respectable "law-abiding body" of the bourgeoisie.[69] "Bourgeois respectability" desired "some identifying sign"—the practice of mutilating or marking was not acceptable to bourgeois sensibilities anymore—for the working-class repeating offenders.[70] Photography helped read the body, its external signs, contours, shapes, colors, and particularly the face and the head, which, according to the pseudoscience of phrenology and physiognomy, reflected one's inner self.

Photography made possible the creation and development of the state archive, as a modern "encyclopedic repository of exchangeable images."[71] It became part of the larger "ensemble" of a "bureaucratic-clerical-statistical system of 'intelligence'"—"a sophisticated form of archive," whose "central artifact" was not the camera but the file and "the filing cabinet."[72] Information collected, condensed, and secured could not but become secret through these technologies of recording and cataloguing. Criminology and statistics joined together as crime became a statistical matter of knowledge-politics-science.

Noted French criminologist Alphonse Bertillon's system of "criminal identification" drew from "'microscopic' individual record" and "macroscopic aggregate" in order to combine "photographic portraiture, anthropometric description [based on the study of the measurements and proportions of the human body], and highly standardized and abbreviated written notes on a single fiche, or card."[73] Bertillon then "organized these cards within a comprehensive, statistically based filing system."[74] The professionalization of the police, investigative work, and the development of criminology as a science took place in tandem with the emergence of social statistics. The various typologies of criminal bodies collected and filed not only resulted in the establishment of an archive of criminal deviancy but also gave birth to the social statistical notion of the "average man" who stood in opposition—within the generalized system of classification—to the deviant criminal.[75] Biological and racial determinations of crime started to rely upon photography, which could now capture head and facial features and allow for comparative measurements. Together, these aspects served the growing science of criminology by helping to archive, classify, and regulate the "criminal types."[76]

Photography fashioned a legal-political panopticon based on a technology of documenting and an "ethics of seeing."[77] An evidentiary value was attached to the photograph; "only the photograph could ... claim the legal status of a visual document of ownership."[78] The photograph established a new kind of "legalistic truth."[79] Relying upon the primacy of the visual senses, photography claimed to represent "the real" and "the authentic," as the declaration of *being there* could be easily and confidently captured by the photograph.

If war is a "regime of truth," photographic and other forms of visual evidence shape war-making and the responses to its many forms of violations.[80] State assertions regarding the torture and abuse of prisoners of war and alleged "terrorists," drone attacks and targeted killings, and the attacks on civilian areas—leading to the deaths of civilians and noncombatants—all rely upon the presentation of visual evidence. Counterforensics, as an activist and human rights practice, acts within the domain of the fundamental claims and counterclaims that often form the basis of justifying war and its violations.[81]

Photographic evidence is utilized both to shape and disrupt the state framing of war. The visual evidence of war, such as records of torture, abuse, and the killings of soldiers, "terrorists," and civilians, molds the "frames of war—the ways to control and heighten affect in relation to the differential grieveability of lives."[82] Public outcry over the torture and abuse of detainees at US-run prisons in Abu-Gharib, Iraq, and the inhuman conditions of imprisonment at Guantanamo Bay, emerged only once the photographs of the prisoners' sexual and psychological abuse were made public.[83] The photographs of the sexual and psychological torture and abuse of prisoners at Abu-Gharib, and of hooded, shackled prisoners at Guantanamo Bay prison provided, as Judith Butler suggests, another "frame" within which to view the wars on "terror."[84]

Over the past two decades, the severed heads and headless bodies of suicide bombers and militants, and the flag-wrapped coffins of martyred soldiers and officers, became the visuals of counterinsurgency for Pakistanis. If the visuals of the latter were broadcasted to carefully choreograph the military's sacrifices for the nation, the former were often presented as faceless, disfigured figures that represented the specter of violent militancy brewing in the "lawless," troubled border regions of the country. Security forces displayed the headless "terrorists" and/or their bullet-ridden cadavers following suicide attacks and encounters with them. These images were often captured and broadcasted by the media, which then played a part in affectively managing the public response(s) to war and its violence.

Images of the coffins of soldiers and officers killed in military operations against militants, wrapped in the national flag, frequently appear on Pakistani news channels. The fallen men and those who have survived, albeit severely wounded and/or maimed, are commemorated on national holidays celebrating independence, the defense forces, and the wars fought with India in 1965, 1971, and 1999. The dead are also remembered during the sacred days of Ramadan. These programs are broadcasted on almost all TV channels. The family members of the martyred speak of and on behalf of those they have lost. These visual displays of emotion, sacrifice, grief, and national pride invite the viewers to bear witness to the loss and pain that the families have suffered to protect the nation. These visual spectacles further help the military to manage its narrative of *shahadat* (martyrdom), challenged by the militants who professed to fight foreign troops (in neighboring Afghanistan) and their proxies (in Pakistan), in the name of Islam. Differential grievability of life determines who is remembered, who is forgotten, and who is made to disappear from the official memory of war.

In a counterforensics narrative however, the visual documentation of the missing persons humanizes those suspected of militancy and those disappeared and interned in remote areas of the country. The dossiers of the missing contain their photographs taken in professional studios—the face-captures—as well as a disparate variety from personal collections. Most of these studio photographs are headshots captured against a blank background, devoid of emotions and expressions, as modern bureaucracies demand of their citizens for identification. Their earlier forms, as explained, were used to establish a visual archive of criminals separating them from the respectable bourgeoisie and abstracting the images, and the bodies they represented, from their historical and political contexts.

The photographs of the disappeared demand a political and ethical response. Disappearances are often followed by the threats to the families of missing persons—they are intimidated and silenced. Rather than speaking themselves, the families therefore quietly display the photographs at their protests and allow them to represent the missing persons. The photographs, even though providing a "mute testimony,"[85] are not simple conveyors of meanings or of communications. They take on a performative function, reminding the onlookers that "a photograph is a pseudo-presence and an absence."[86]

Diane Taylor has also observed the use of similar kinds of photographs in the protests of the Mothers of the Plaza de Mayo in Argentina. Such a framing, she states, "allows for no background information, no context, no network of relationships"; these images "appear to be artless and precise. Yet they are highly constructed and ideological, isolating and freezing an individual outside the realms of meaningful social experience."[87] The facial captures—without expressions, jewelry, or other objects that might distinctly mark a person—catalogue and archive individuals as decontextualized images: parts of a universal archive in which citizens as images are substitutable.

Photographs—the bureaucratic face-captures, decontextualized of personal signs and affects—are reframed and resignified by the families of the disappeared in their dossiers and protests. In the protests, I often found that the family members, friends, and activists lined up and held victims' photographs, for spectators, journalists, and for each other, as they demanded the release of the missing and accountability for those killed, tortured, and interned. For example, Noor Khan's brother and cousins displayed his old photographs, riding a horse or just posing in his house, taken years before his "disappearance." Even though, like Noor, many of the men were barely visible in the blown-up, laminated photographs, the act of holding them up distinguished one from the other missing—contextualizing each one as their lives and identities were flattened by the ambiguity and doubt created by their abductors or under the bureaucratic burden of the inquiry commission, the COIOED. The protestors, moreover, presented the photographs and documents to the activists and journalists covering the protests to speak/advocate on behalf of not only the missing but also the protestors as well. The fetishization of these objects shaped the identities of the missing but were also assumed to break the silence over their erasure.[88] The protesting bodies that held these photographs silently served as archives of another kind: by embracing and wearing the photographs around their necks, the protestors turned "their bodies into archives, preserving and displaying the images that had been targeted for erasure."[89] The photographs declare that the bodies—the absent and the present—are joined "genetically, affiliatively, and … politically."[90]

While the higher courts appear powerless in locating the disappeared persons, the COIOED keeps on delaying its proceedings. As the censorship of unlawful detentions and disappearances by state intelligence and military services tacitly remains in effect, the abductions and detentions continue to take place. In this context of judicial paralysis, the dossiers and photographs provide counterforensic information that bolsters the demand for state accountability and works to reframe the memory of war and militancy in the country. War-making against terrorism is recontextualized by public displays of grief and suffering, which uncover the violence of the state's extra-legal sovereignty and the various thresholds, judicial and territorial, upon which it rests and justifies its violence.

Evidence, Power, and Ethnography

I have suggested that together, the visual and material artifacts collected in the dossiers of memory serve as repositories of (counter) evidence regarding the spectrality surrounding disappearances, as well as the allegations and absence of information about them. The dossiers are repositories to be explored—to unpack events of state violence, unlawful detentions and torture, and thereby provide evidence that questions the official truths on "terrorism," and of those who are declared "terrorists" and "enemies of the state" whose fate is sealed under state secrecy.

My discussion broadly suggests that evidence practices emerge in and are shaped by their political and historical contexts. The hegemonic forms they take over the long durée give them an aura of legitimacy as well as authority as objects of state fetishism.[91] As I have examined above, forensic science, visual archive (of the criminals) and the objects that constituted them—files, documents, and photographs—emerged historically and became "rational" tools of investigation and of maintaining law and order by "scientific" and "humane" means. As Michel Foucault has shown in his work, modern power and sovereignty relied for their effect upon forms of knowledge embedded in social and scientific practices.[92] Evidence practices therefore are inherently intertwined with how power is exercised, but, also, contested through them. That is, they shape subjects but also provide means for them to resist the conditions imposed on them through these forms.

A counterforensic practice backed by ethnographic fieldwork helps assemble a counternarrative on violent events and official "facts" that allude inquiries by formal, legal means. Counterforensic practices work to expose, prove, and demonstrate human rights violations and thus help uncover the voices of those marginalized by state violence and silenced by censorship. Even the most apparent violations of human rights have to be documented, witnessed, and corroborated for observers, and "the discourse of human rights," as Thomas Keenan argues in his commentary on counterforensics, "seems to turn fundamentally on this question of evidence—its discovery or production, its presentation, and its reception."[93] The evidence of state violence and human rights violations *has* to be "exposed, shown, demonstrated, stated, claimed ... and made evident for others."[94]

Ethnographers can play a crucial role in how "evidence" is identified and collected and, more importantly, in determining what qualifies as material evidence, produced through various archival, affective, and performative practices. If counterforensic practice lies at the tense threshold between state secrecy and the making and declaring of public records, ethnographers as witnesses advance the ways of listening and seeing that uncover voices and subjectivities—albeit partial and dispersed—that would otherwise remain invisible and lost under the hegemonic burden of official discourses and facts. However, the marginalized voice that speaks through material and visual objects of evidence is neither fully articulated—it contains contradictions and gaps in the narratives of suffering and trauma—nor does it possess a unitary subjectivity. Fragmented subjectivities are then displaced onto—and articulated through—documents, photographs and other objects. Various forms of ethnographic objects help us trace and interpret social, political, and judicial processes, and understand how these objects, files, documents, and photographs set them in motion, and resist official moves of erasure and denial.

Acknowledgments

This chapter draws from a research project supported by The Social Sciences Research Council (International Dissertation Research Fellowship) and Social Sciences and Humanities Research Council of Canada (Doctoral Fellowship). Earlier versions were presented at Practicing Evidence—Evidencing Practice: How is (Scientific) Knowledge Validated, Valued and Contested? Munich 2020; The Use of Law by Social Movements and Civil Society, International and Interdisciplinary Symposium, Brussels 2018; South Asia Graduate Student Conference, University of Chicago, Chicago 2017; and the Annual South Asia Conference, Madison 2017. I thank all those who raised questions and critiques at these venues. I also thank editors of this volume, Stefan Esselborn and Sarah Ehlers. I am especially grateful to Malcolm Blincow for his comments on earlier versions of this chapter. I am in debt to those men and women who shared their stories with me for this project.

Notes

1 Begona Aretxaga, *States of Terror: Begoña Aretxaga's Essays* (Reno: Center for Basque Studies, University of Nevada 2003).
2 "Enforced disappearances" and "Missing Persons" are terms used interchangeably by human rights activists and by journalists to refer to extra-judicial abductions and detentions in Pakistan.
3 Levi-Strauss, *The Savage Mind* (London: Weidenfeld and Nicolson, 1962), 18.
4 Social media has played an increasingly important role in highlighting disappearances, but campaigns by activists and their allies have been mostly, albeit not always, mobilized for the release of journalists and activists connected with urban civil society.
5 Salman Hussain, "Violence, Law, and the Archive: How Dossiers of Memory Challenge Enforced Disappearances in the War on Terror in Pakistan," *POLAR: Political and Legal Anthropology Review* 42, no. 1 (2019), 53.
6 Salman Hussain, "The 'Ethical' Framework for Counterinsurgency: International Law of War and Cultural Knowledge in the U.S. Army and Marine Corps Counterinsurgency Field Manual," *Anthropologica* 57, no. 1 (2015): 105–113; Claire Finkelstein, Jens David Ohlin, and Andrew Altman, eds., *Targeted Killings: Law and Morality in an Asymmetrical World* (Oxford: Oxford University Press, 2012).
7 Hussain, "Ethical Framework"; Finkelstein, Ohlin, and Altman, *Targeted Killings*.
8 Michael Hafetz, *Habeas Corpus after 9/11: Confronting America's New Global Detention System* (New York: New York University Press, 2011).

9 See Munir Ahmad, "Resisting Guantanamo Bay: Rights at the Brink of Dehumanization," *Northwestern University Law Review* 103 (2009): 1683–1764.

10 *Oxford Dictionary of Law,* 9th ed., ed. Jonathan Law (Oxford: Oxford University Press, 2018), s.v. "*salus rei publicae suprema lex.*"

11 Arianna Vedaschi, "The Dark Side of Counter-Terrorism: Arcana Imperii and Salus Rei Publicae," *The American Journal of Comparative Law* 66, (2018): 896.

12 Vedaschi, "Dark Side of Counter-Terrorism," 896.

13 On existential threat to the state and exceptional violence, see Carl Schmitt, *The Concept of the Political* (Chicago: Chicago University Press, 1932).

14 Vedaschi, "Dark Side of Counter-Terrorism," 877.

15 Vedaschi, 880–881.

16 This information was shared with me by lawyers representing the families in the Supreme Court.

17 Webb Keane, "The Evidence of the Senses and the Materiality of Religion," *Journal of the Royal Anthropological Institute* 14, no. 1 (2008): 116.

18 Salman Hussain, "Witnessing 'Imperfect Victims': Inculpability, Collaboration, and the 'Suffering Subject', *American Ethnologist*, no. 49, 1 (2022): 92–103.

19 Joanne Rappaport, "Rethinking the Meaning of Research in Collaborative Relationships," *Collaborative Anthropologies* 9, no. 1–2 (2016): 1.

20 Eyal Weizman, *Forensic Architecture: Violence at the Threshold of Detectability* (New York: Zone Books, 2017), 9.

21 Saba Mahmood, *Politics of Piety: The Islamic Revival and the Feminist Subject* (Princeton: Princeton University Press, 2005); Keane, "Evidence of the Senses"; Didier Fassin, "The Ethical Turn in Anthropology: Promises and Uncertainties," *HAU: Journal of Ethnographic Theory* 4, no. 1 (2008): 429–435.

22 In Islamic cosmology *jinns* are supernatural beings who occupy an important place and a role on the spectrum of creation spanning angels to humans.

23 See Stefania Pandolfo "'Nibtidi mnin il-hikaya [Where are we to start the tale]' Violence, Intimacy, and Recollection," *Social Science Information* 45, no. 3 (2017): 349–371; and Amira Mittermaier, *Dreams That Matter: Egyptian Landscapes of the Imagination* (Berkeley: University of California Press), 2010.

24 On documents, see Mathur, *Paper Tiger*, 2015; and Bruno Latour, *The Making of Law: An Ethnography of the Conseil d'Etat* (Cambridge: Polity, 2002).

25 Valentina Napolitano, "Anthropology and Traces" *Anthropological Theory* 15, no. 1 (2015): 47–67.

26 Especially in the area previously known as FATA (Federally Administered Tribal Areas), the Swat valley and its adjacent areas.

27 The torture of detainees, many of whom are among "the disappeared," by the United States and its allies is a common practice. In 2010, in the United Nations Rapporteur on enforced disappearances, Martin Scheinin, reported that since 11 September 2001, systematic human rights abuses have taken place in covert operations against terrorism across the globe. See Martin Scheinin and Manfred Nowak. *Joint Study on Global Practices in Relation to Secret Detention in the Context of Countering Terrorism* (Geneva: UNHRC, 2010).

28 As of 31 January 2021, COIOED reported it had received 6,921 complaints and had "disposed" of 4,798 cases. The Defense of Human Rights Pakistan (DHRP) disputes these figures and contends that under the term disposed, cases are ignored and discarded; that is, the missing are neither freed nor even located, and the cases are considered resolved simply because the intelligence services submit that they are not holding the person.

29 Weizman, *Forensic Architecture*, 18.

30 In-camera proceedings refer to case proceedings that take place in judges' chambers or in absence of an audience in a courtroom.

31 Shahzad Malik, "Col. Rtd. Inam Rahim: The Missing Persons Lawyer Returns Home," *BBC Urdu*, 24 January 2020, https://www.bbc.com/urdu/pakistan-51232304.

32 Ahmad, "Resisting Guantanamo Bay," 1712–1719.

33 Weizman, *Forensic Architecture*, 30.

34 Ian Burney and Christopher Hamlin, eds., *Global Forensic Cultures: Making Fact and Justice in the Modern Era* (Baltimore: John Hopkins University Press, 2019).

35 Allan Sekula, "Body and the Archive," *October* 39 (1986): 3–64.

36 Ronald R. Thomas, *Detective Fiction and the Rise of Forensic Science* (Cambridge: Cambridge University Press, 1999), 1.

37 Carlo Ginzburg, "Morelli, Freud and Sherlock Holmes: Clues and the Scientific Method" *The Society for the Study of Labour History* 39 (Autumn 1979): 5–36.

38 Mark Maguire, "The Birth of Biometric Security," *Anthropology Today* 25, no. 2, 9–14.

39 See Rojas-Perez, *Mourning Remains*; Francisco Ferrándiz and Antonius C.G.M. Robben, eds. *Necropolitics: Mass Graves and Exhumations in the Age of Human Rights* (Philadelphia: University of Pennsylvania Press, 2017); and Adam Rosenblatt, *Digging for the Disappeared: Forensic Science after Atrocity* (Stanford: Stanford University Press, 2015).

40 Thomas Keenan, "Getting the Dead to Tell Me What Happened: Justice, Prosopopoeia, and Forensic Afterlives," *Kronos* 44, no. 1 (2018), 109.

41 Weizman, *Forensic Architecture*, 67.

42 Weizman, 67.

43 Weizman, 67–68.

44 Weizman, 9.

45 Weizman, 65.

46 Weizman, 65; Keenan, "Getting the Dead," 107.

47 Weizman, 65.

48 Weizman, 68.

49 Weizman, 64.

50 Weizman, 64.

51 Valentina Napolitano, "Anthropology and Traces," 47.

52 Weizman, 134.

53 Annalise Riles, *Documents: Artifacts of Modern Knowledge* (Ann Arbor: University of Michigan, 2006), 23.

54 Riles, 23.

55 Levi-Strauss, *Savage Mind*, 18.

56 Shyrimoyee Ghosh, "'Not Worth the Paper It's Written on': Stamp Paper Documents and the Life of Law in India," *Contributions to Indian Sociology* 53, no. 1 (2019): 19–45.

57 Cornelia Vismann, *Files: Law and Media Technology* (Stanford: Stanford University Press, 2008).

58 Mitra Sharafi, "The Imperial Serologist and Self Harm: Blood Stains and Legal Pluralism in British India," in *Global Forensic Cultures: Making Fact and Justice in the Modern Era,* ed. Ian Burney and Christopher Hamlin (Baltimore: John Hopkins University Press, 2019), 61.

59 Navanika Mathur, *Paper Tiger: Law, Bureaucracy and the Developmental State in Himalayan India* (Cambridge: University of Cambridge Press, 2015); Mathew Hull, *Government of Paper: The Materiality of Bureaucracy in Urban Pakistan* (Berkeley: University of California Press, 2012).

60 For example, in the famous Mohabat Shah Case, in its judgment on 10th December 2013, the Supreme Court declared that the Army had unlawfully removed 35 missing persons, traced to an interment center in Malakand Garrison, against court orders, and out of whom only seven had been produced before the Court (H.R.C 29388-K, Supreme Court of Pakistan).

61 Located in Rawalpindi, the Defense of Human Rights Pakistan's (DHRP) office was established (in 2006) by Amina Janjua and other claimants in the missing persons cases. DHRP records all reported cases of disappearances and it also appears on behalf of the claimants in the Supreme Court.

62 These centers were established by military forces in their counterinsurgency operations between 2007 and 2009.

63 Tribes are assumed to be located, fixed in their areas, following the colonial system of governance in the tribal-North-Western and North-Eastern border areas of Pakistan. On colonial resettlement of tribes, see *The Great Agrarian Conquest: The Colonial Reshaping of a Rural World*, Neeladri Bhattacharya (Albany: SUNY Press, 2018).

64 Weizman, *Forensic Architecture*, 20.

65 Weizman, 20.

66 Sekula, "Body and the Archive," 7.

67 Sekula, 7.

68 Sekula, 12.

69 Sekula, 13.

70 Ginzburg, 25.

71 Sekula, "Body and the Archive," 15.

72 Sekula, 15.

73 Sekula, 18.

74 Sekula, 18.

75 Sekula, 19.

76 Sekula, 18.

77 Susan Sontag, *On Photography* (New York: Dell Publishing, 1977), 3.

78 Sekula, "Body and the Archive," 6.

79 Sekula, 6.

80 Allen Feldman, *Archives of the Insensible: Of War, Photopolitics, and Dead Memory* (Chicago: Chicago University Press, 2015), 2.

81 The US campaign of drone attacks in the tribal areas took advantage of the visual and geographical threshold. Lack of evidence of the destruction and killing of civilians in these attacks kept the attacks disputed. However, once the witnesses of these attacks came forward and produced visual evidence of civilian deaths, the denial was changed to silence.

82 Judith Butler, *Frames of War: When Is Life Grievable* (London: Verso, 2009), 26.

83 Seymour Hersh, "Torture at Abu Ghraib: American Soldiers Brutalized Iraqis. How Far up Does the Responsibility Go?" *New Yorker*, 30 April 2004, https://www.newyorker.com/magazine/2004/05/10/torture-at-abu-ghraib.

84 Butler, *Frames of War*, 26.

85 Sekula, "Body and the Archive," 6.

86 Sontag, *On Photography*, 16.

87 Diana Taylor, *The Archive and the Repertoire: Performing Cultural Memory in the Americas* (Durham: Duke University Press, 2003), 176.

88 Gastón Gordillo, "The Crucible of Citizenship: ID-paper Fetishism in the Argentinean Chaco," *American Ethnologist*, 33, no. 2 (2019): 162–176.

89 Taylor, *Archive and the Repertoire*, 177–178.

90 Taylor, 177–178.

91 Gordillo, "The Crucible of Citizenship," 1.

92 See Foucault's early works, such as *Discipline and Punish: The Birth of the Prison* (New York: Vintage Books, 1975).

93 Keenan, "Getting the Dead," 110.

94 Keenan, 109.

13 Digital Ethnographic Art(i)Facts as Evidence: Anthropological Entanglements between Techne and Episteme

Anna Apostolidou

Evidence and Fabrication in Anthropology

The designation of convincing and trustworthy evidence has caused considerable turmoil throughout the history of anthropology as a discipline, especially following the intense discussion provoked by the 1980s wave of cultural critique. Practices of anonymization, interpretation, and contextualization in conjunction with questions about authorship and the limits of the textual representation of ethnography have been at the center of academic attention ever since, often calling into question the very legitimacy of anthropology as a social science. In addition, anthropology's exceptionally varied nature, ranging from social statistics to the study of affective practices or material objects, constitute a chameleonic disciplinary terrain, wherein the scrutiny of what counts as evidence each time is of great importance. The shift from what we know to how we came to know it has been at the heart of such disciplinary introspection for quite some time,[1] illuminatingand problematizing, among other things, the literary aspects of anthropological work that are apparently always at play.[2]

This introspection has reappeared even more forcefully in light of the digital turn in anthropology[3] as the modalities available for recounting ethnography have undergone profound technical, theoretical, and political reconfiguration. In the twenty-first century, with the escalating dissemination of digital technologies, a growing number of ethnographers have been actively pursuing alternative ways to register these diverse anthropological ways of knowing in nontextual and often nonevidence-based representational modes, in an effort to do justice to the experience of fieldwork and the complex dynamics of sociocultural activity that usually get obscured or omitted from the standard academic text.[4] Departing from the exciting affordance and the inherent creative and generative qualities of multimodal representation that have become temptingly available—especially to younger scholars—a timely question has returned: What is (ethnographic) evidence and to what extent are we allowed to fabricate it in order to make the

DOI: 10.4324/9781003188612-17

ethnographic narration more accurate and at the same time more accessible to the wider public?

The idea of fabrication—both as a knowledge generative process and as a political positionality—inevitably raises ethical questions about representation and the politics of truth. In the context of a discipline as fragile to the critique of objectivism as anthropology, these questions immediately become a locus of interrogation with regard to the *scientific* properties of social science and the artistic liberties of ethnographic writing.[5] The discussion that follows maintains that the persistent issues of writing-up the research field and articulating meaning in anthropology may be reframed in an expanded understanding of research ethics that leaves room for imagination, omission, and even "tampering with evidence" as part of a viable ethnographic practice; one which concentrates on the detailed portrayal of multifaceted cultural information that comes from the accounts of various field agents and cultural contexts. The synthetic quality of this information is often hard to convey when one is obliged to stick to the stories of particular persons, to compare their experiences, and to discuss them vis-à-vis other ethnographic and theoretical material. It is often the case that, after post-fieldwork analytical work is complete, the merging of standpoints, experiences, and information may lead to the fabrication of incidents, personas, or interactions, which may deliver an equally, if not more truthful account of the issues raised by the ethnographic material.[6] This is especially so when studying phenomena that engage with liminal acts—legally or morally, according to local code—or when stigma and shame are incurred by certain social practices.[7]

The following discussion draws on the research project "Ethnography and/as Hypertext Fiction: Representing Surrogate Motherhood,"[8] and focuses on the emerging social practice of surrogate motherhood in order to critically address digital ethnographic textuality. The research was implemented through collaborative research methodology (duoethnography)[9] and involved a small-scale ethnographic study of surrogate motherhood in contemporary Greece, mainly via interviewing and collecting objects and materials from social parents we encountered about their experience with infertility, and their having a child through the processes of surrogacy (participants who agreed to participate included single gay fathers, married mothers, and heterosexual couples). The initial research plan aimed to interview an equal number of surrogate women who have carried these parents' babies, but despite persistent and systematic pursuit, the research team did not manage to locate and engage in ongoing discussions with more than one surrogate, which—as contextual evidence suggests—is very revealing of the positionality and disempowerment of women who act as surrogates, as well as of the social stigma that their choices often incur.[10] A principal concern of the research revolved around practices of ethnographic writing of surrogacy, and sought to experiment with the proposition that research findings may be presented in the form of fiction, including poetry, short stories, animation, memes, graphics, and audio testimonies that exceed the standard text-based academic discourse. In our understanding of the research material, we were faced with two distinct forms of evidence at work, which were intrinsically related, but not identical: as in every anthropological study, on the one hand, there is the evidence that anthropologists collect in the field to match with theory and arrive at their conclusions, and on the other the evidence they present to others (i.e. intersubjective evidence). The argument made for fabrication in this particular case is primarily directed at the latter set of evidence, and the liberties we took to fictionally reframe them, nonetheless bearing in mind that the very acts of collecting and sorting evidence are equally constitutive (indeed fabricative!) of the field itself.

By employing an audiovisual fictional approach, we focused on the narrative processes of our informants as they were sharing their experiences and stories with us, and concentrated on the affective and imaginative features often entailed in the processes of pursuing parenthood in nonnormative ways. In addition, we decided that the strong affinities as well as the telling disparities of informants' accounts supported our choice to produce[11] a set of fabricated stories and viewpoints that would be placed in a digital ethnographic artifact,[12] along with (and indiscriminately from) collected material and recorded testimonies (ranging from audio recordings, videos, and photos to notes, academic papers, collages, and interview excerpts) (Figure 13.1).

The overarching stake of this methodologically unorthodox processing of data which derived from a conventional ethnographic study was the fashioning of ethnographic "texts," which may facilitate nonlinear readings of fictional ethnography and, most importantly, yield uninhibited access to nonexpert readers. To that end, we mapped out the areas of theoretical interest that support new modalities of ethnographic "writing," which are phenomenologically relevant[13] to the object of representation and at the same time, present affinities with new media, feminist thought, and experimental methodologies.[14] This approach stems from emergent writing methodologies in feminist studies, which have recently informed the blending of writing styles and have helped legitimize in academic discourse[15] those alternative writing modes that embrace imaginary and poetic spaces of registering social reality, a powerful example of this tendency being the ethno(bio)graphic memoir.[16] In our case, this was performed by analytically correlating the practices of new reproductive technologies to practices of new media technologies,[17] and by locating conceptual and methodological overlaps between these two otherwise distinct and seemingly irrelevant research territories.[18] Our purpose was to render ethnography as emotionally engaging and audiovisually stimulating as possible while keeping contextual and personal information that derived from fieldwork as our principal locus of reference. We find that this is especially relevant to researching queer parenthood and assisted reproductive strategies, in the sense that the latter directly challenge the normative technologies of procreation just as digital multimedia writing upsets academic prose. Nonetheless, this experimental methodological stance urgently raised the question as to whether practices of fabricating evidence can be registered as legitimate practices of producing scholarly evidence in the academic culture of contemporary anthropology.

Letting Go of *the Truth*?

In one of his famous aphorisms Nietzsche asks:

> How could anything originate out of its opposite? for example, truth out of error? or the will to truth out of the will to deception? or selfless deeds out of selfishness? or the pure and sunlike gaze of the sage out of lust? Such origins are impossible; whoever dreams of them is a fool, indeed worse; the things of the highest value must have another, peculiar origin—they cannot be derived from this transitory, seductive, deceptive, paltry world, from this turmoil of delusion and lust... This way of judging constitutes the typical prejudgment and prejudice which give away the metaphysicians of all ages; this kind of valuation looms in the background of all their logical procedures; it is on account of this "faith" that they trouble themselves about "knowledge," about something that is finally baptized solemnly as "the truth."[19]

In the course of his notably anti-metaphysical project, Nietzsche foresees the rise of "the philosophers of the dangerous perhaps," those who would not duel upon "the belief between the opposition of values." His pondering sets the scene for a metaphysics of truth that has affected the rise and consolidation of the social sciences in the West. He aptly portrays the impossibility of generating "truth" out of deception; thus cementing the antithetical balance between truth and fiction—also perceived as science (*episteme*) versus art (*techne*)—and echoed in the binary logic of structuralist anthropology.

During the 130 years that have passed since Nietzsche's formulation, the problem of anthropological knowledge (as grounded in empirical reality) has been seen to heavily rely upon evidence and context in order to account for the cultural understanding of the Other. And even though evidence connotes a certain degree of positivity—as anthropological theory contests—it cannot in any form be freed from human intention. So, how do we deal with the question of truth[20] at a time when academic anthropological knowledge—often engaging with theoretical abstraction and contained in texts written and accessed by a limited audience—has become completely inaccessible to the very persons qualified to verify its gravitas, that is, the field interlocutors? Even if this inaccessibility was a "forgivable" trend in the early, somewhat naïve, years of colonial anthropology, in our days it cannot stand the decades of critique that have highlighted this matter as an ethical priority of the discipline. How then may we attempt to reconcile academic jargon and text-based narration that is

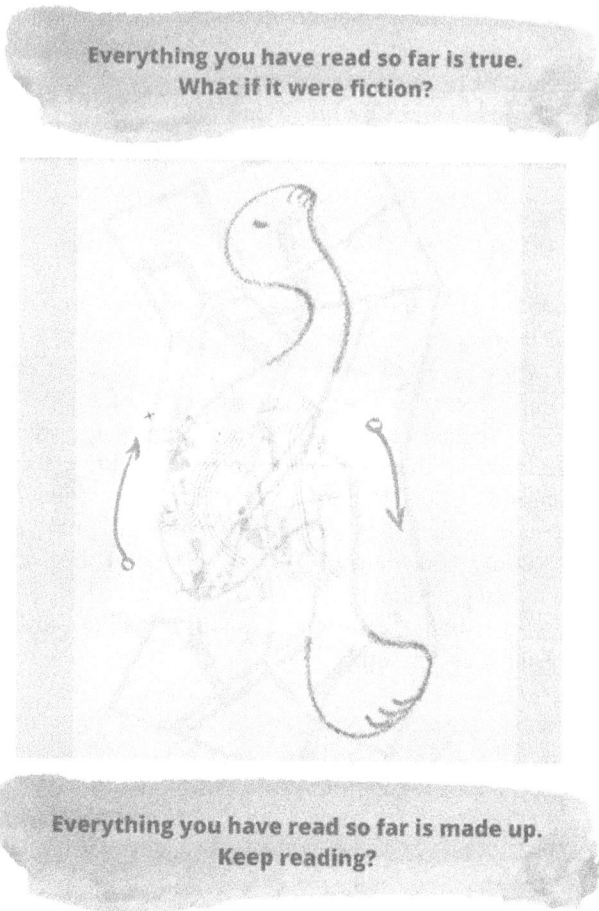

Figure 13.1 Hints of tampering with truth (from digital artifact).

reliant on specific learning practices and institutions, with the stories and experiences that arise from the field?

Perhaps a viable tentative answer could be to draw our attention to the loci
and modalities of knowing that render ethnographic truths legible to wider
audiences and ultimately, manage a degree of accuracy that cannot be
measured by current canonical academic means. Yet, if we are to follow this
line of thinking, we are presented with a twofold predicament: a heightened
responsibility to accurately portray the cultural context of the field and its
agents and, on the other hand, the necessity to forge new ways of conveying
cultural meaning, compatible with the perceptive and narrative practices of
contemporary audiences, which more often than not include the persons and
groups that trusted us with this information in the first place.

With respect to the former predicament, anthropologist Kirsten Hastrup draws on her own fieldwork in Iceland in order to discuss the concept of "getting it right," and makes the distinction between knowledge and evidence in anthropological theory and practice.[21] In her view, in a postpositivist era "the question of evidence is acute if anthropology shall aspire to anything but reporting quaint stories from strange places."[22] The author points to a new awareness of the anthropological mode of knowing about the world and of "getting it right"; this in turn implicates practicing anthropologists deeply in the narrative ethics upon which anthropological authority rests. She views knowledge as organized information, both reductive and selective, a modality that heavily rests on the modes of knowing and of interpreting, both of which are disciplined and already culturally mediated:

> The question of evidence is far more complex than the claim to positivity suggests, because it is enfolded within the relational nature of anthropological knowledge that—epistemologically—precludes the use of evidence as an independent measure of validity. Another complication is found in the fact that so much evidence in anthropology is circumstantial or inferential, and relates to sensations, silences, deceptions, and moods. These sensations are not external to categories but inform them deeply.[23]

Inevitably, the evidence we as anthropologists provide in our ethnographies is always evidence of our own mediation with the object of study, therefore rendering arbitrary any claim to reliable representation from the outset. Therefore, the main problem that presents itself is how to translate "evidence for me" into "evidence for others," if the evidence in question is based on my own (necessarily subjective) impressions. As Hastrup contends, "the objective of the anthropological narrative (of whatever scale) is to provide a mode of imagining how individual actions and collective illusions are interlinked."[24] Accordingly, the predicaments we found ourselves in while researching surrogate motherhood related to questions such as "which standpoint are we going to put first?" and "which data [evidence] should we use and in which hierarchical order or viewpoint?" and often "how do we produce data which is not even there?" How do we measure and produce data from the loudly silent surrogates, particularly when the evidence we gathered comes on their behalf and not directly from them? This led the research team to make strenuous efforts to position themselves ethically, again and again, in relation to the object of study as a whole and not solely to the persons we physically encountered in the field or the circumstantial evidence we collected.

In addressing these ethical queries, Hastrup's proposition does not take away the responsibility of the anthropologist vis-à-vis the social conditions studied. It does however shift the ethical focus toward an accurate mobilization of the anthropologist's imagination in rendering legible the performative and relational modalities of knowing. Such an endeavor entails the promulgation of our ethnographic imagination and introduces a "dicey" narrative ethics that largely rests on intuition, presumption, and ultimately fabrication of the elements that emerge from fieldwork. This process of fabricating—in the sense of manufacturing or inventing something (ποιείν)—is not a surprising idea for anthropology. Rather, this positionality evokes James Clifford's famous 1986 work that views ethnography as a semifictional genre in which allegory and the "poetic dimensions," or "plural poesis," of the text are of paramount importance.[25] The ethnographic text has since been introduced as an incomplete "true fiction"—a turn to subjectivity which "not only undermined the 'scientific' claims of anthropology, but also further opened its scope towards the representation of everyday life in a reflective manner as a poetic mimesis of reality."[26]

Departing from a reworking of this seminal position, and making use of the technical means offered by online technologies, the research team elected to produce fictive elements of the interlocutors' accounts, by engaging with the economies of emotive truths of the field and digging out the poetic form of their testified experience. Some informants contributed to this with their own written accounts (poems, short stories, diaries), but most of the fictionalization took place after the analytical phase, set forth by the ethnographers' jointly listening and reexperiencing the field interlocutors' accounts. "Getting it right" in this scope did not necessarily entail telling the truth or presenting the field evidence in all clarity, but rather merging the information with the imaginative propensities of fiction in order to accurately render it comprehensible to the cultural context from which it sprang.

Turning to the second predicament mentioned above, recent anthropological approaches challenge us to reconsider the relationship between anthropological texts and the worlds they purport to describe, and to conceptualize this relationship in terms other than those of representation. The fictional rendering of ethnographic data interrogates the very limits between objectivity and subjectivity—as imprinted in the contrast between "scientific" and "literary" language, anthropology and literature—and explores the possibility that the world of the imaginary (fiction writing) may be more likely to help us experience and understand the real (ethnographic findings). It is therefore important to place this endeavor in its historic context.

The interrelation between anthropology and fiction has a long genealogy of shared concerns, insights, and problematics, with the tendency to note the fictional character of even the most well-documented ethnographic accounts and the evident or subtle authorship that telling the story of the Other inevitably entails. Before Clifford's 1986 *Writing Culture,* the terms ethnography and fiction had been jointly discussed and/or juxtaposed in several overlapping bodies of theoretical and historical scholarship; ranging from anthropologists writing both ethnography and fiction, to experimental modes of feminist writing, and to anthropologists treating the work of particular novelists as ethnographic sources.[27] The distinction between ethnographic truth and fabrication has been at the core of this discussion and has troubled many generations of anthropologists striving to achieve an honest balance between fieldwork and textual ethnographic realism.[28] However, recent work—into whose proximity our research project also falls—pushes for a further reconceptualization of ethnographic evidence that resists the clear-cut distinction between collected and creatively manufactured data, so long as the end narration of the fictional ethnographic artifact holds the "real" as a locus of inspiration, and merges the collected data in a culturally informed performative narrative. This trend, as recently acknowledged by the history and the philosophy of science, points out that "narratives often rely on an appeal to ambiguity about causality and temporality in giving accounts of the scientific phenomenon at hand,"[29] which is a position that may also assist social researchers when having to recount complex cultural phenomena and stay tuned to conflicting narrations by field interlocutors. In attempting to give back to anthropology "the emotional quality of social life: the imponderabilia" that get taken away by standard ethnographic writing, pleas have been made for the inclusion of emotion in ethnography as a way of purging the "real" and of transcending the very limits of ethnography.[30]

In this vein of inquiry that stems from deep disciplinary transformations, in many cases blurred genres of writing have been proven to be more powerful in terms of both popularity and reading experience than conventional ethnography.[31] Renato Rosaldo's student, Sasha Su-Ling Welland, produced a blended memoir on the topic of the Chinese women's movement, which stands midway between genealogy, literary criticism, and theoretical reflection, and is articulated through nontextual ethnographic writing, showcasing fieldwork by use of diverse media forms (e.g. photo essays, crowd-sourced videos, digital archives, video sculpture installations, and so on).[32] A predominant urge evident in this and other similar examples of recent anthropological work is the premise that scholarship should reflect the aesthetic richness of the worlds it represents.

Faced with the current normative figuration of social sciences as mechanisms of "achieving evidence" (from research design to data analysis) we argued that the reading practices of our times—together with the deep crisis that the humanities and social sciences face (one of credibility, departmental shrinkage, and lack of funding, among other things)—call for alternative and at times unsafe and precarious methodological standpoints and techniques in order to maintain our work's social relevance and epistemological pertinence. Indeed, as we strive to perform an orthodox management of research data and to address our texts, papers, and dissertations to strictly demarcated disciplinary and institutional audiences, we may in fact generate less accurate depictions of the phenomena under study.

In this respect, allowing space for the imagination comes close to the pursuits of biofiction, a recent trend in literature which takes a real biography as its point of departure, and is powered by what Colm Tóibín has called "the anchored imagination."[33] Engagement in digitally mediated ethnography grants the fictional narrative a certain ambiguous and fragile credibility. Furthermore, work on alternative representational politics, such as the ethno-graphic novel and informants' response to fictionalized parts of their experience has indicated that "as a distillate or reconstructed realism, truth was less about getting to authenticity but appreciating the possibilities to imagine and re-create a sense of authenticity."[34] This insight aligns with feedback that came out of the study of surrogacy in our project, verifying the methodological strategy of fabrication as a holistic rendering of the experience of the individuals, instead of a strict commitment to factual accuracy or representational realism. After presenting the digital artifacts to field interlocutors, as well as expert readers, digital designers, art practitioners and women and men who have dealt with infertility, we noted that the emotive properties and the condensed narrative manner of stories as employed in the artifact were well received and triggered memory, comparison, and self-reflexivity over cultural standards and other agents' experiences that had not been there before. Given the often intimidating and variously inaccessible character of academic writing and its monolithic modality (written text), we presume that these responses would not have been possible if the ethnographic product had been generated in the traditional style of presenting "evidence," for example, as a book or journal article.

After seeing the project's digital, multimedia rendering of the ethnographic text, Alex, a single woman who had made two attempts to have a child through surrogacy, reflected, "I had never thought [of?] the position of the surrogate, it was only my own problem that I could grasp," while Tania, her partner, commented, "It [the artifact] made me cry, it is so rich and precise.... Although I could not see the stories of specific individuals, I got the whole array of things that get involved in this process.... Especially the medical, the legal, and the internal spaces [in the digital artifact] are extraordinary. I realize the work you had to put in it to get it all right." Jim, a middle-aged man who works in the development of digital museum exhibitions, also attested, "It would be impossible to convey this powerful silence of the women who are surrogates.... Not a movie, not an installation, not a text, and surely not a novel. A *tour de force* of imagination, but I am not sure how anthropologists are going to see it." We acknowledge that probably the most crucial testimonies in regard to "grasping the fieldwork truth" would come from the surrogates themselves but, unfortunately, with the exception of a woman who carried her sister's twins, none of them consented to officially participate in interviews for this study.

Digital Ethnographic Art(i)Facts as Evidence

The full breadth of this discussion, in the complexity of multimodal forms afforded by digital technologies, may only be plausible in the era of ever-growing online performances of sociality. Commenting on these recent trends, George Marcus identifies two tendencies in modern methodological experimentation in anthropology. Firstly, the emergence of new ecologies in the forms of scholarly communication, wherein the book and other relevant texts (such as the scholarly article) are losing ground to other, more embedded and accessible expressions of cultural meaning. Secondly, he perceives an "involution of form" in the writing of ethnographic accounts, as theoretical influences are settling in and become "dictating" writing practices—although he points out that some of these experimental forms are often marked by tendencies of excess in descriptive and theoretical ardor. For Marcus, the emergence of alternative forms of articulating thinking, ideas, and concepts will be performed in "third spaces," such as archives, studios, labs, "para-sites," and so on.[35]

One of these alternative forms is the digital ethnographic artifact—a notional and tangible object that stands both etymologically and epistemologically midway between art and fact. This was viewed in our work as a tentative resolution of the tensions between a politics of truth and an aesthetics of fabrication and storytelling. It also presents interesting disciplinary affinities with certain aspects of the history of representation in anthropology. Ethnographic artifacts—both a remnant of colonial commodification and a material culture testimony—inhabit a prominent position in the early times of ethnographic preoccupation with the Other.[36] In addressing Alfred Gell's treatment of artworks and artifacts alike as "embodiments or residues of complex intentionalities," Foster posits that ethnographic artifacts were being displayed as artworks in natural history or art museums, via installation practices which drew on techniques characteristic of not only natural history museums but also of commercial urban window displays.[37] Reflecting on the history of the artifact and its political gravity in the colonial project, we might consider a forceful appropriation of its properties in the digital realm. In a sense, there are now means available to subvert the century-long paradigmatic function of artifacts as crutches for the colonial project, and attempt to create digital artifacts that offer accessible ethnographic data back to the informants in a more democratic and horizontal way. In positioning ourselves as agents of well-documented politics of representation in the production of anthropological insight, the digital artifact is proposed as a solution for rendering layered meaning, and for evading the dichotomy between exclusively textual academic narration of ethnographic fact and untenable fiction.

Taking away the artifact from the territory of commodity or museum display and placing it in the repertoire of ethnographic digitality, it regains its properties as an incubation of complex and contrasting meaning; one which escapes imperialist motivation[38] and also attempts a convergence between art (techne) and ethnographic fact. Techne (τέχνη), both as art and as technology, is in this context not necessarily expelled from anthropological renderings of the truth but could be viewed as intrinsically related to the act of meaning-making and representation in anthropology, thereby collapsing the central epistemological division between science and art that has dominated western thought. Techne here refers to art or craft, to human action that engages with the world and thereby results in a different world. Techne is not just knowledge about the world, what Greek thought termed episteme; it is intentional action that constitutes a gap between the world as it was before the action, and the new world it calls into being.[39]

This generative propensity of art-as-fabrication—which has recently become available to anthropologists who deal with digital technologies—strongly challenges the hierarchical juxtaposition between "hard"/tangible and "soft"/intangible evidence and data, and pushes for the merging of the "truth" and "reality" of fieldwork experience with the analytical, fictional, and reflexive imaginative engagement with this experience. As Lorenzo Brutti posits, when addressing visual anthropology as a *terra nullius* between science and art, "aesthetics may help science in furnishing a more truthful description of an ethnographic context. Art can help anthropology to be more sensitive."[40] Since academic writing has often failed to do justice to its research subjects or adequately engage with nonacademic audiences, research-based fiction may provide meaningful answers to the pursuit of a wider politics of representation and knowledge dissemination.

It has been over twenty years since Lila Abu-Lughod identified three modes of "writing against culture" and experimenting with narrative ethnographies of the particular: first, contesting the terms of theoretical discussion as the way in which anthropologists engage with each other; second, reorienting the problems or subject matters of anthropology with a focus on the thematic interconnections and the immersive methodology we encounter in the field; and third, producing ethnographies of the particular as a means to subvert the process of othering, homogenizing, and timelessness that has been overwhelming in twentieth century writing. As she notes, "by bringing closer the language of everyday life and the language of the text" we may really address the particularity of social experience and refute academic generalization.[41] Contemporary anthropology has a chance to answer to the multimodal language of everyday life through ethnographic artifacts, which have the potential to become a new "language of the text" and thus register feminist anthropologists—initially excluded from *Writing Culture*—in what she terms "textual innovation." Attention to detail, to the particularity of perspectives and the lived experience of specific actors is not a theoretical proposition and cannot be conveyed through standard academic language. It requires the fostering of alternative representational and analytical ethics, which, unorthodox and subversive as they may appear, greatly coincide with current shifts in everyday language, communication, and social connectivity. Returning to Hastrup:

> To speak of a narrative ethics, however, is not simply to acknowledge that connections between social facts and larger frames cannot be "constructed" at random.... Practicing anthropology implies a "using" of other people's understanding to further an anthropological understanding that is

narratively mediated. In writing, anthropologists make connections and sort out hierarchies of significance that cannot bypass local social knowledge even while transcending it. The ethical demand is to "get it right", not in any ontological sense, but in being true to the world under study and to the epistemological premises of anthropology.[42]

In this sense, fictional digital artifacts direct their evidence equally towards a pursuit of knowledge and an aesthetic project, which further accentuates the mixture of genres in representational modalities.

Ethnography is itself often a critique of the way in which knowledge is commonly produced and communicated within social science research. Whatever the representational means selected, ethnography remains a strenuous "exercise in being truthful about the distance we travel from research questions to finished manuscript, with all its doubts, epiphanies and improvisations."[43] To that end, the inclusion of practices of imagination, fantasy, fiction-making, and fabrication in the ethnographic account may ultimately provide a deeper sense of doing justice to the recordable and inconspicuous aspects of the field.

In Defense of Poetry

In addressing the representational predicament set forth by the combination of digital affordances and truthful ethnography, the production of poetic texts—in the widest possible definition—could be introduced as an ethical mode of providing evidence.[44] If stories have always been at the core of the anthropological project, our gesture of presenting ethnography in digital media and fictional works aims at accentuating this centrality, and furthermore restoring the epistemological and philosophical character of *knowing* that anthropology supports. As Symons and Maggio speculate, with reference to Michael Taussig's lyrical writing style and visually stimulating ethnographic rhetoric:

> When Taussig asks "is it not the ultimate betrayal to render stories as 'information' and not as stories?" he warns against over-focusing on stories as data producing mechanisms. In a story's transformation into ethnographic data and the storyteller into informant, "the philosophical character of the knowing is changed. The reach and imagination in the story is lost."[45]

In addition to preempting this fundamental loss, poetic accounts are—in certain areas of anthropological study—far more accurate than ethnographic realism. Rosaldo's recent defense of poetry in accounting for personal turning points is revealing. In "Notes on Poetry and Ethnography," he makes the case for poetry as a rigorous form of cultural inquiry: "Poetic exploration resembles ethnographic inquiry in that insight emerges from specifics more than from generalizations," he writes. "I strive for accuracy and engage in forms of inquiry where I am surprised by the unexpected."[46]

Particularly since the revolution in writing modalities enabled by digital tools and media, ethnography has been at the core of questioning about the limits of "objective representation." Peter Pels' approach to the philosophy of science places our era in a postsubjectivity ethnographic rationale, and borrows a critique of objectivity that recalls Daston and Galison's "trained judgment," defined as directed at the "users" of knowledge.[47] Along the same lines, Elizabeth Chin argues that using fiction is a sharp way to explore social facts. If we are to borrow her idea of a "laboratory of speculative ethnology"[48] and turn to the Greek word for poetry, ποίησις/*poihsis*—which actually translates to "making/creating/manufacturing"—this allows for a focus on the critical functions of ethnographic and poetic texts,[49] which may in turn shed light on underlying assumptions and implications of the ethnographic imagination and writing. Taking this enlarged articulation of poetry, we follow Rosaldo in proclaiming that "the ambition of poetry is … to *be* the event itself."[50] Consequently, in many instances in the artifact that is our digital rendering of the ethnography, we articulated the event itself with the use and depiction of poetic language (Figure 13.2).

Figure 13.2 Animation and visual poetry crafted out of interviews.

In working with parenthood via surrogacy, the component of the bodily labor that is central yet silenced in our informants' accounts has been shown to be affectively understood by modes of poetic representation rather than discussion, interviewing, or reading academic books and papers. Poetry writing as ethnography has, in the context of digital narrative tools (such as hypertext or visual poetry), proved to be a refined, nuanced, and useful critical-creative communicative form of scholarship and research method in the geohumanities.[51] Leonora Angeles also problematizes the concept of empathy and the use of (auto)ethnographic poetry in articulating the doubts, dissonance, and disjuncture produced from inter- and intracultural differences, as they are generated from the study of (dis)embodied and (dis) emplaced liminal experiences, such as surrogate motherhood.[52]

Other ethnographers go so far as to argue that poetic representation in research supports "new ways of seeing and understanding phenomena."[53] Women writers have a strong presence in this trend, as women have long been the object of literary representation on the basis of gender difference, and gender is in itself a lens for reading literature.[54] When it comes to accounts of motherhood and mothering, poetry has been indicated to act as a prominent expressive means, both in representing intersecting concerns such as gender-race-ethnicity,[55] and as a gendered political stance, which at times is more effective than academic prose.[56] It has furthermore been labeled as a "breathing room," a locale to resist conventions of analytic voice and textual, spatial, and existential "captivities," when it comes to ethnographically depicting human suffering.[57] This modality of storytelling has often been connected to freedom, as it allows for playful abstraction and emotive mobilization, freed from the constraints of conventional discursive schemata. Marc Augé acknowledges literary "play" as an obligatory preliminary to any development of philosophical thought, and comments on the intellectual freedom which it institutes in relation to established cosmologies—a freedom which moreover assumes the existence of writing as a guarantee of memory and a support for argument.[58] These recognitions also entail a methodological shift. As anthropologists in the digital era enter never before imagined quests of knowledge acquisition and representation, so, too, must their instruments be modified in accordance to these transformations.

Why Experiment at All?

Putting aside personal inclinations and aesthetic preferences in defending fictional digital ethnography, a question still persists: Why experiment with ethnographic methodologies? Is it merely a challenge of canonical discourse, or a sense of fatigue with reading, crafting, and correcting linear, serious-appearing, and dense ethnographies that satisfy institutional requirements? Is it a quest for wider readership, more connections, and the possibility of more democratic access to contemporary research? Principally, it may be about

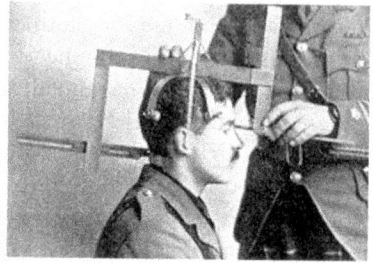

Figure 13.3 Fabricated analogies: depicting a genealogy that ranges from nineteenth and twentieth century anthropometric practices of "primitives" and "criminals" to twenty-first century biopolitical measurements of surrogates (short video and photograph embedded in the artifact).

being able to give research back to the people that allowed its emergence. Since academic writing often fails to engage with the ways of apprehension (in its dual meaning: both comprehending, and fear) of nonacademic audiences, digital fictional artifacts may be a forceful rejoinder to a wider politics of representation and sharing of research findings.

Even though fiction writing may be viewed as promoting the singularity of authorship on the part of the ethnographer, the replacement of academic jargon with literary fiction might actually diminish the effect of this authoritative voice and reshift the focus on to the research subjects' experiences and reception modalities. The multivocal ethnography achieved through the use of digital media and artefacts has been shown to enrich and nuance the concept of the ethnographic text itself[59] and to provide fresh new discursive ethnographic territories that "accurately" reflect new sensibilities and manners of registering the "nontellable" of social reality (Figure 13.3). This idea is admittedly situated on the verge of a cultural shift, when normative ideologies and "innovative" practices forge a discursive terrain of tension, convergence, and the potential for articulating new gendered and cultural meanings. If writing inevitably relates to privileged and subjugated knowledge—established "truths" and emergent "antilogues"—actively addressing the question of how our choice of writing modalities shape anthropological knowledge becomes urgent, as it strives to embrace temporal and spatial locations, activist proclamations, and an aesthetic of resistance. When Giovanni da Col addresses the two fundamental turns of recent years—one ontological and the other ethical—he underlines the conditions of uncertainty and unknowability that both researchers and their subjects face in critical times.[60] He argues that intersubjective relations in the field are the principal point of our political engagement with knowledge. For Bob White and Kiven Strohm,

> the ethics and politics of ethnographic knowledge take place primarily in the field and not in particular products or political postures. It is from this vantage point that we seek to critically examine how anthropology knows and, by extension, what is political about anthropological knowledge.[61]

In my view, the imprint of this political engagement in the final artifact or text is equally crucial so that this knowledge may be available, accessible, and open to critique by our field interlocutors as well as wider audiences.

As anthropologists, we are used to splitting our attention across two usually antithetical directions of audience reception: on the one hand, to the local actors who have helped us shape our perspective on any given topic and, on the other, to our academic and institutional interlocutors. This antithesis immediately generates two sets of receptive modalities; the latter conventionally more prone to objectivity-oriented discourse, and the former set to engage larger audiences through fiction-based storytelling means. As demonstrated, the research project "Ethnography and/as Hypertext Fiction: Representing Surrogate Motherhood" embraces the argument that reality can only be phenomenologically perceived, and thus never truly objectively observable. This is not the result of a wish to bypass reality as a mere fabrication or to release the creative instincts of the ethnographer untamed. Rather, the project is a meticulous effort to engage fieldwork interlocutors and fellow researchers in an intellectual, sensorial, and emotional quest, which allows space for the imaginary, the mythical, the irrational, the playful, and the fictional, in order to account for reality at least to the same extent that orthodox, recordable, and analyzable data are considered to be "'reflective" of the "real."

In acknowledging anthropology's ultimate aspiration to interpret the world through tampered truths and made-up lies, it is useful to remember Clifford Geertz's view that "reality privileges no particular idiom in which it demands to be described—literally, positivistically, or without fuss."[62] As one of the first anthropologists to bring increased attention to the fictive character of ethnography, Geertz portrays ethnographies as made and fashioned things, rather than simple reports of things as they are. However, he does not consider them untrue accounts. Thus, if a principal goal of ethnography is for the informants, the interlocutors, the subjects to recognize themselves in the final "texts," then our efforts should lie in manufacturing a complex text, or a digital artifact, that stays true to the experience and narration of the individuals who entrusted us with their stories—even if this may only be achieved by consciously producing "tampered" ethnographic data. This might be the ultimate form of knowledge validation even if the point-by-point characteristics of the ethnography (i.e. the evidence provided) fail to portray the words and actions of specific research informants in all exactitude. A greater degree of interference with evidence may paradoxically diminish scholarly authorship (in regard to form, vocabulary, and content), and bring forth more congenial ways of interpreting and portraying social reality.

A final note in defense of "evidence" is important at this concluding point: even though I strongly argue for a poetic, multimedia, nonlinear, "fabricated" ethnography, it should be clear that this only refers to the final outcome of a research project, that is, the write-up of ethnography. In my view, in order to let go of evidence and rely on imagination and creativity, one needs to have immersed oneself in the social reality being studied, and to have kept trailing the facts as well as their own "re-actions" to them through the long-term act of immersive systematic fieldwork. After all, "an ethnography is ultimately a story that is backed up by reliable ethnographic data and the authority that comes from active ethnographic engagement."[63] Much like music, where one needs to have been attentive to all detail and mastered all rules included in the score in order to improvise, so in fabricating multimedia stories as evidence we need to remain committed to all the information—intellectual, sensory, imaginative—that we have collected and coproduced in the field and also to the rigorous analytical organization of incoming data with reference to their wider discursive and cultural context. Only then might we enter a notional and disciplinary territory that actively subverts academic orthodoxies while maintaining a strong anchor on the recognition of social reality as a convergence of fiction, incomprehensibility, truth and societal trust.

Acknowledgments

I would like to thank the editors for the attention and time they devoted to my manuscript and for their detailed comments for which I am truly grateful. My gratitude also goes to field interlocutors and colleagues, who contributed to the adventure of this research project.

Notes

1 Kent Maynard and Melisa Cahnmann-Taylor, "Anthropology at the Edge of Words: Where Poetry and Ethnography Meet," *Anthropology and Humanism* 35, no. 1 (2010): 2–19.
2 Nigel Rapport, "Representing Anthropology, Blurring Genres Zigzagging Towards a Literary Anthropology," *Mediterranean Ethnological Summer School* 3 (1999): 29–48.
3 Sarah Pink et al., *Digital Ethnography: Principles and Practice* (Thousand Oaks, CA: SAGE, 2015).

4 For example, see the works by Grimshaw "The Multipart Film: Extended Anthropological Analysis and Non-textual Forms" (2021), Dattatreyan and Marrero-Guillamón's "Introduction: Multimodal Anthropology and the Politics of Invention" (2019) and Cox, Irving, and Wright's seminal book *Beyond Text? Critical Practices and Sensory Anthropology* (2016).

5 See the edited volume *Anthropology Off the Shelf: Anthropologists on Writing* (Waterston & Vesperi, 2011).

6 For the construction of fictional realities as a means of gaining knowledge to parts of "real" reality that are not easily accessible otherwise (i.e. the multiplicity of human intentions and perspectives, or the inherently contingent future) in modern literary fiction and prob-abilistic calculations, see the work of Elena Esposito (2007) on the fiction of probable reality, where she employs the Luhmannian concept "Realitätsverdopplung" (a "re-doubling of reality"), which greatly coincides with my own understanding of "fabricating evidence." Special thanks to Stefan Esselborn for pointing out this analogy.

7 For practices of digital fictionalization vis-à-vis methodological and fieldwork impediments see: Anna Apostolidou, *Re/producing Fictional Ethnographies: Surrogacy and Digitally Performed Anthropological Knowledge* (New York: Palgrave Macmillan, 2022).

8 The three main research aims of the project "Ethnography and/as Hypertext Fiction: Representing Surrogate Motherhood" (HYFRESMO), undertaken at the Social Anthropology Department of Panteion University of Social and Political Sciences (Athens, Greece), and funded by the Hellenic Foundation for Research and Innovation & the General Secretariat for Research and Technology, were (1) to identify the cultural understanding of surrogacy practices in Greece, (2) to fictio-nalize fieldwork data, generating multimodal digital media that would accurately portray interlocutors' experiences and testimonies and (3) to create a digital eth-nographic artifact that would render ethnography accessible to wider audiences. The research team comprised Anna Apostolidou (Principal Investigator), Ivi Daskalaki (Postdoctoral Researcher), Maria Niari and Ilona Lasica (Research Assistants), Athena Athanassiou and Eirini Tountasaki (Advisors).

9 Anna Apostolidou and Ivi Daskalaki, "Digitally Performing Duoethnography: Collaborative Practices and the Production of Anthropological Knowledge," *Entanglements* 4, no. 1 (2021): 51–69.

10 Ivi Daskalaki and Anna Apostolidou, "Digital Encounters of Surrogacy: Nodes of a Fictional Ethnography," *Journal of Contemporary Ethnography* 50, no. 4 (2021).

11 In the context of creationist capitalism, as deployed in virtual spaces, anthropological labor and knowledge production are viewed in terms of creating rather than working (cf. Boellstorff, 2015).

12 The artifact can be accessed at https://prezi.com/view/nEhTIKCs7yE1o0sd4TyK/.

13 Sarah Ahmed, "Orientations Matter," in *New Materialisms: Ontology, Agency, and Politics*, ed. Diana H. Coole and Samantha Frost (Durham: Duke University Press, 2010), 234–258.

14 Maria do Mar Pereira, *Power, Knowledge and Feminist Scholarship: An Ethnography of Academia* (London: Routledge, 2017).

15 Livholts, Mona, "Introduction. Contemporary Untimely Post/Academic Writings: Transforming the Shape of Knowledge in Feminist Studies," in *Emergent Writing Methodologies in Feminist Studies*, ed. Mona Livholts (London: Routledge, 2012), 1–24.

16 For an example of this experimental modality, see Ulrika Dahl, "The Road to Writing: An Ethno(bio)graphic Memoir," in *Emergent Writing Methodologies in Feminist Studies*, ed. Mona Livholts (London: Routledge, 2012), 164–181.

17 Anna Apostolidou, Ivi Daskalaki, and Maria Niari, "Anthropological Intersections be-tween New Reproductive Technologies and New Digital Technologies," *TECHNO REVIEW: International Technology, Science and Society Review* 9, no. 1 (2020): 49–59.

18 For an elaborate discussion see Anna Apostolidou, Re/producing Fictional Ethnographies: Surrogacy and Digitally Performed Anthropological Knowledge (New York: Palgrave Macmillan, 2022).

19 Friedrich W. Nietzsche, *Beyond Good and Evil,* Trans. Walter Kaufmann (New York: Vintage, 1966).

20 For a masterful analysis of the ethics of truth in contemporary anthropology see: Henrietta Moore, "The Truths of Anthropology," *Cambridge Anthropology* 25, no. 1 (2005/06): 52–58.

21 Kirsten Hastrup, "Getting It Right: Knowledge and Evidence in Anthropology," *Anthropological Theory* 4, no. 4 (2004): 455–472.

22 Hastrup, "Getting It Right," 455.

23 Hastrup, "Getting It Right," 460.

24 Hastrup, "Getting It Right," 467.

25 James Clifford, "On Ethnographic Allegory," in *Writing Culture: The Poetics and Politics of Ethnography,* ed. James Clifford and George E. Marcus (Berkeley, Los Angeles, London: University of California Press, 1986), 98–121.

26 Michelangelo Paganopoulos, *In-between Fiction and Non-fiction: Reflections on the Poetics of Ethnography of Literature and Film* (Newcastle upon Tyne: Cambridge Scholars Publishing, 2019), 4.

27 Kirin Narayan, "Ethnography and Fiction: Where Is the Border?" *Anthropology and Humanism* 24, no. 2 (1999): 135.

28 This paper addresses research in what is largely considered social and cultural anthropology. The paradigm of evidence-based work is still dominant in other areas, such as forensic anthropology (cf. Ecks, 2008; Steadman, 2015).

29 For an overview of a wide array of examples that upset standard assumptions of a fundamental opposition between narrative and science, see Mary Morgan and Norton Wise, "Narrative Science and Narrative Knowing. Introduction to Special Issue on Narrative Science," *Studies in History and Philosophy of Science Part A* 62 (2017): 1–5.

30 Andrew Beatty, "How Did It Feel for You? Emotion, Narrative, and the Limits of Ethnography," *American Anthropologist* 112, no. 3 (2010): 430–443.

31 Ruth Behar, "Ethnography in a Time of Blurred Genres," *Anthropology and Humanism* 32, no. 2 (2007): 145–155, offers suggestions for reading ethnographies with an eye toward learning how they were written and what literary feats they accomplished. She argues that the line between fiction and nonfiction has become increasingly indistinct; so, it is important to reassess the possibilities and limits of ethnography as a literary genre if we are to understand the idiosyncrasies of its "art." "Ethnography in a Time of Blurred Genres." See also Visweswaran, *Fictions of Feminist Ethnography.*

32 For details and examples of this project, see Jonathan Hiskes, "Ethnography Unbound: Experiments in New Scholarship." See also Laterza, "The Ethnographic Novel: Another Literary Skeleton in the Anthropological Closet?" (2007).

33 Bethany Layne and Colm Tóibín, "Colm Tóibín: The Anchored Imagination of the Biographical Novel," *Éire-Ireland* 53, no. 1 (2018): 150–166. On biofiction, its formal innovations, its play with the boundaries of fact and fiction, and its intrinsic relation with fact and fiction in anthropology, see Stefan Helmreich and Caroline Jones, "Science/Art/Culture through an Oceanic Lens." *Annual Review of Anthropology* 47 (2018): 97–115.

34 Benjamin Dix, Raminder Kaur, and Lindsay Pollock, "Drawing-Writing Culture: The Truth-Fiction Spectrum of an Ethno-Graphic Novel on the Sri Lankan Civil War and Migration," *Visual Anthropology Review* 35, no. 1 (2019): 107.

35 George E. Marcus, "The Legacies of Writing Culture and the Near Future of the Ethnographic Form: A Sketch," in *Writing Culture and the Life of Anthropology,* ed. Orin Starn (Durham: Duke University Press, 2015), 36.

36 For a detailed account of anthropological artifacts as evidence in the context of Early Americanist Anthropology, see Rodriguez's contribution in this volume.

37 Robert J. Foster, "Art/Artefact/Commodity: Installation Design and the Exhibition of Oceanic Things at Two New York Museums in the 1940s," *The Australian Journal of Anthropology* 23, no. 2 (2012): 129–157.

38 Janet Owen, "Collecting Artefacts, Acquiring Empire: Exploring the Relationship between Enlightenment and Darwinist Collecting and Late-Nineteenth-Century British Imperialism," *Journal of the History of Collections* 18, no. 1 (2006): 9–25.

39 Tom Boellstorff, *Coming of Age in Second Life: An Anthropologist Explores the Virtually Human* (Princeton: Princeton University Press, 2015), 81.

40 Lorenzo Brutti, "Aesthetics versus Knowledge: An Ambiguous Mixture of Genres in Visual Anthropology," *Reviews in Anthropology* 37, no. 4 (2008): 291.

41 Lila Abu-Lughod, "Writing Against Culture," in *Recapturing Anthropology: Working in the Present*, ed. Richard G. Fox (Santa Fe: School of American Research Press, 2016), 474.

42 Hastrup, "Getting it Right," 469.

43 Wanda Vrasti, "Dr Strangelove, or How I Learned to Stop Worrying about Methodology and Love Writing," *Millennium* 39, no. 1 (2010): 79.

44 Acquiring the tools of poetic craft—while aiming for cultural knowledge—provides the ethnographer with multiple genres with which to see the world truthfully, and, as Emily Dickinson advises, to "tell it slant." In this way, ethnographic poets will become aware of writers who approach culturally significant themes in their work with revelation and surprise: Maynard & Cahnmann-Taylor, "Anthropology at the Edge of Words," 13.

45 Jessica Symons and Rodolfo Maggio, "'Based on a True Story': Ethnography's Impact as a Narrative Form," *Journal of Comparative Research in Anthropology and Sociology* 5, no. 2 (2014): 2.

46 Renato Rosaldo, *The Day of Shelly's Death: The Poetry and Ethnography of Grief* (Durham: Duke University Press, 2014), 105.

47 Peter Pels, "After Objectivity: An Historical Approach to the Intersubjective in Ethnography," *HAU: Journal of Ethnographic Theory* 4, no. 1 (2014): 222.

48 Elizabeth Chin, "Using Fiction to Explore Social Facts: The Laboratory of Speculative Ethnology," in *The Routledge Companion to Digital Ethnography*, eds. Larissa Hjorth et al. (London: Routledge, 2017), 478–489.

49 Beth Baker-Cristales, "Poiesis of Possibility: The ethnographic Sensibilities of Ursula K. Le Guin," *Anthropology and Humanism* 37, no. 1 (2012): 15–26.

50 Rosaldo, *The Day of Shelly's Death*.

51 See also the work of L. Shelley Rawlins, "Poetic Existential. A Lyrical Autoethnography of Self, Others, and World." *Art/Research International: A Transdisciplinary Journal* 3, no. 1 (2018): 155–177.

52 Leonora C. Angeles, "Ethnographic Poetry and Social Research: Problematizing the Poetics/Poethics of Empathy in Transnational Cross-Cultural Collaborations," *GeoHumanities* 3, no. 2 (2017): 351–370.

53 Lynn Butler-Kisber, "Inquiry through Poetry: The Genesis of Self-Study," in *Just Who Do We Think We Are? Methodologies for Autobiography and Self-study in Teaching*, ed. Claudia Mitchell, Sandra Weber, Kathleen O'Reilly-Scanlon (London: Routledge, 2013), 95–110.

54 See Michael B. Snyder, Women in Literature: Reading through the Lens of Gender (London: Greenwood Press, 2003).

55 See Renate Papke, *Poems at the Edge of Differences: Mothering in New English Poetry by Women* (Göttingen: Göttingen University Press, 2018).

56 See Annie Finch, *The Body of Poetry: Essays on Women, Form, and the Poetic Self* (Ann Arbor: The University of Michigan Press, 2005).

57 See Casey Golomski, "Breathing Room: Poetic Form as Resistance to Convention in the Ethnography of Suffering," *Somatosphere: Science, Medicine, and Anthropology*

(8 March 2019), http://somatosphere.net/2019/breathing-room-poetic-form-as-resistance-to-convention-in-the-ethnography-of-suffering.html/

58 Marc Augé, *The War of Dreams: Studies in Ethno Fiction* (London: Pluto Press, 1999), 102.

59 Sarah Pink, "Digital Visual Anthropology: Potentials and Challenges," in *Made to Be Seen: Perspectives on the History of Visual Anthropology*, ed. Marcus Banks and Jay Ruby (Chicago: University of Chicago Press, 2011), 209–233.

60 Giovanni da Col, "Turns and Returns," *Hau: Journal of Ethnographic Theory* 4, no. 1 (2014): i–v.

61 Bob W. White and Kiven Strohm, "Preface. Ethnographic Knowledge and the *Aporias* of Intersubjectivity," *HAU: Journal of Ethnographic Theory* 4, no. 1 (2014): 190–191.

62 Rapport, "Representing Anthropology," 34.

63 Madden cited in Jessica Smartt Gullion, *Writing Ethnography* (Rotterdam: Sense Publishers, 2016), 12.

Index

For Product Safety Concerns and Information please contact our EU
representative GPSR@taylorandfrancis.com
Taylor & Francis Verlag GmbH, Kaufingerstraße 24, 80331 München, Germany

www.ingramcontent.com/pod-product-compliance
Lightning Source LLC
Chambersburg PA
CBHW061629220326
41598CB00026BA/3941

9 7 8 1 0 3 2 0 3 7 0 6 6